Recent Developments in Power and Energy Engineering

Recent Developments in Power and Energy Engineering

Edited by **Helena Walker**

Willford Press
New York

Published by Willford Press,
118-35 Queens Blvd., Suite 400,
Forest Hills, NY 11375, USA
www.willfordpress.com

Recent Developments in Power and Energy Engineering
Edited by Helena Walker

International Standard Book Number: 978-1-68285-215-6 (Hardback)

Printed in China.

Contents

	Preface	**VII**
Chapter 1	**Fuzzy Controller for a Voltage-Regulated Solar-Powered MPPT System for Hybrid Power System Applications** Jaw-Kuen Shiau, Yu-Chen Wei and Min-Yi Lee	**1**
Chapter 2	**Optimized Adaptive Perturb and Observe Maximum Power Point Tracking Control for Photovoltaic Generation** Luigi Piegari, Renato Rizzo, Ivan Spina and Pietro Tricoli	**22**
Chapter 3	**China's Low-Carbon Scenario Analysis of CO_2 Mitigation Measures towards 2050 Using a Hybrid AIM/CGE Model** Wei Li, Hao Li and Shuang Sun	**41**
Chapter 4	**A Novel Hybrid-Flux Magnetic Gear and Its Performance Analysis Using the 3-D Finite Element Method** Yiduan Chen and Weinong Fu	**68**
Chapter 5	**Magnetic Field Equivalent Current Analysis-Based Radial Force Control for Bearingless Permanent Magnet Synchronous Motors** Huangqiu Zhu and Hui Li	**83**
Chapter 6	**Residual Mix Calculation at the Heart of Reliable Electricity Disclosure in Europe—A Case Study on the Effect of the RE-DISS Project** Markus Klimscheffskij, Thierry Van Craenenbroeck, Marko Lehtovaara, Diane Lescot, Angela Tschernutter, Claudia Raimundo, Dominik Seebach and Christof Timpe	**106**
Chapter 7	**A New Protection System for Islanding Detection in LV Distribution Systems** Anna Rita Di Fazio, Mario Russo and Sara Valeri	**136**
Chapter 8	**Enhanced Load Power Sharing Accuracy in Droop-Controlled DC Microgrids with Both Mesh and Radial Configurations** Yiqi Liu, Jianze Wang, Ningning Li, Yu Fu and Yanchao Ji	**155**
Chapter 9	**Improved Performance of Electrical Transmission Tower Structure Using Connected Foundation in Soft Ground** Doohyun Kyung, Youngho Choi, Sangseom Jeong and Junhwan Lee	**170**
Chapter 10	**Optimization and Analysis of a Hybrid Energy Storage System in a Small-Scale Standalone Microgrid for Remote Area Power Supply (RAPS)** Fengbing Li, Kaigui Xie and Jiangping Yang	**190**

Chapter 11 **Designing Structure-Dependent MPC-Based AGC Schemes Considering Network Topology** 215
Young-Sik Jang, JoonHyung Park and Yong Tae Yoon

Chapter 12 **Dynamic Power Management for Portable Hybrid Power-Supply Systems Utilizing Approximate Dynamic Programming** 233
Jooyoung Park, Gyo-Bum Chung, Jungdong Lim and Dongsu Yang

Permissions

List of Contributors

Preface

This book includes some of the vital pieces of work being conducted across the world, in various areas related to power and energy engineering. While understanding the long-term perspectives of the topics such as energy conversion, generation, transmission, distribution and utilization, etc. it makes an effort in highlighting their impact as a modern tool for the progress of energy engineering. Scientists and students actively engaged in this field will find this book full of crucial and unexplored concepts. For all readers who are interested in power and energy engineering, the case studies included in this book will serve as an excellent guide to develop a comprehensive understanding of this crucial subject.

This book is a comprehensive compilation of works of different researchers from varied parts of the world. It includes valuable experiences of the researchers with the sole objective of providing the readers (learners) with a proper knowledge of the concerned field. This book will be beneficial in evoking inspiration and enhancing the knowledge of the interested readers.

In the end, I would like to extend my heartiest thanks to the authors who worked with great determination on their chapters. I also appreciate the publisher's support in the course of the book. I would also like to deeply acknowledge my family who stood by me as a source of inspiration during the project.

Editor

Fuzzy Controller for a Voltage-Regulated Solar-Powered MPPT System for Hybrid Power System Applications

Jaw-Kuen Shiau *, Yu-Chen Wei and Min-Yi Lee

Department of Aerospace Engineering, Tamkang University, Tamsui, New Taipei City 25137, Taiwan; E-Mails: 601430662@s01.tku.edu.tw (Y.-C.W.); 601430688@s01.tku.edu.tw (M.-Y.L.)

* Author to whom correspondence should be addressed; E-Mail: shiauj@mail.tku.edu.tw

Academic Editor: Paul Stewart

Abstract: This paper presents the design of a fuzzy-logic-based voltage-regulated solar power maximum power point tracking (MPPT) system for applications involving hybrid power systems. The system contains a solar power system and battery as the primary and secondary power sources, respectively. The solar system alone supplies power to the electric motor and maintains the output voltage at a predetermined level when it has sufficient power. When the solar power is insufficient, the solar system is operated at its maximum power point (MPP) and the battery is engaged to compensate for the insufficiency. First, a variant of the incremental conductance MPP condition was established. Under the MPP condition, the voltage-regulated MPPT system was formulated as a feedback control system, where the MPP condition and voltage regulation requirements were used as the system inputs. Next, a fuzzy controller was developed to perform the voltage-regulated MPPT function for the hybrid power system. A simulation model based on Matrix laboratory (MATLAB)/SIMULINK (a block diagram environment for multi-domain simulation and model-based design) and a piecewise linear electric circuit simulation (PLECS) tool for controlling the dc motor velocity was developed to verify the voltage-regulated solar power MPPT system.

Keywords: maximum power pointing tracking; fuzzy controller; circuit simulation

---

1. Introduction

Depending on specific mission objectives and performance requirements, unmanned aerial vehicles (UAVs) can be equipped with cameras, sensors, communications equipment, and other payloads. UAVs are widely used in civil, commercial, military, and science applications. The development of UAVs is expected to continue to receive attention from researchers in the future [1,2]. A key area of UAV design is the endurance capabilities of such vehicles. Solar power could potentially redefine the endurance capabilities of UAVs because sunlight is an inexhaustible energy source. Since the successful flight of the first solar-powered aircraft, Sunrise [3], in 1974, designing solar-powered aircraft has been a topic of interest for the general public and UAV communities [4–9]. Solar-powered airplanes are used in developing scientific technology and for commercial exploitation purposes. The effectiveness of using solar energy depends on the efficiency of how solar power management systems handle the acquisition and distribution of solar energy.

The solar power delivered from the photovoltaic (PV) cells depends on environmental conditions (solar irradiation changes, incident angle of sunlight, and cell temperature) and load conditions. Maximum power point tracking (MPPT) schemes are typically employed in such systems to ensure that the maximum power is obtained from the PV system. Common MPPT techniques include perturbing and observing [10], incremental conductance [11], and fuzzy logic-based MPPT [12–19]. For continuous operation and safety considerations, hybrid power systems are typically designed to power electric motors for a long endurance UAV. Hybrid power systems incorporate several different but complementary power sources. PV system is the basic power source of the hybrid system. Analysis and design of power management system for hybrid power systems received great attention recently [20–25]. A comprehensive power management function including the MPPT to manage and control the power flow among the hybrid system is critical for maximizing the efficiency of energy usage. Results using combination of neural network and fuzzy logic algorithms for energy management for hybrid power systems were reported in [22–24]. In this study, solar and battery hybrid power systems are considered. Typical solar power management systems contain solar power MPPT and power conversion functions [26,27]. The MPPT function obtains the maximum power from the PV system, and the power conversion function converts the voltage from the MPPT stage to a required voltage level to support motor operation. Both functions are typically accomplished using a certain type of power converter, resulting in additional power consumption.

Motivated by the design of a small solar powered UAV, this study designed a novel voltage-regulated solar power MPPT system. The system contains a solar power system and battery as the primary and secondary power sources, respectively. The main purpose was to combine the MPPT and power converter functions as an integrated module and use only one power converter for the solar and battery hybrid-power system. The solar system alone supplied power for the electric motor and maintains the output voltage at a predetermined level when it has sufficient power for motor operation. When the power was insufficient, the solar system was operated at its maximum power point (MPP) and the battery was engaged to compensate for the insufficiency.

Three major accomplishments are presented in this paper. First, the voltage-regulated MPPT system was formulated as a feedback control system, where the MPP condition and voltage regulation are used as the control system inputs. To define the feedback control system, a variant of the incremental conductance MPP condition was first established. Instead of computing the sum of conductance and incremental conductance, the sum of the arctangent of the conductance and the arctangent of the incremental conductance [16] was investigated. At MPP, the sum was 180°. In addition, the range of the sum was confined to 90° to 270°; thus, the voltage-regulated MPPT system was formulated as a feedback control system with two reference inputs (*i.e.*, 180° as the reference for the MPPT function and a reference voltage for the desired output voltage). Second, a buck-boost converter based fuzzy controller was successfully designed for the voltage-regulated MPPT system. The error of the MPP condition and the error of the voltage regulation were used as input variables for the fuzzy controller. MPPT and voltage regulation were achieved through continuously adjusting the duty ratio of the power switch of a buck-boost converter. Finally, this research successfully developed a simulation model for the system. In the proposed algorithm, the MPPT and voltage regulation functions are integrated into a single standard feedback control system that requires only one power converter. With this standard feedback formulation, not only the efficiency of the solar energy is increased, but design and implementation of the real time control system is also simpler.

A simulation model based on MATLAB/SIMULINK and a piecewise linear electric circuit simulation (PLECS) tool for controlling the velocity of the dc motor was developed to verify the voltage-regulated solar power MPPT system. Three cases were selected for the computer simulation, namely, varying the regulation voltages, solar irradiation, and motor speed. The system was also used to power two motors that were operated at different speeds. The results demonstrated the success of the proposed fuzzy controller design for the voltage-regulated MPPT system.

2. Hybrid Power System Using the Photovoltaic (PV) System as the Primary Source

The problem considered in this study arose from developing an experimental solar-powered UAV. For endurance and safety considerations, the hybrid power system shown in Figure 1 was selected for the UAV. As indicated in Figure 1, the PV power source and the battery are connected using a diode-OR circuit. The output voltage of the PV system was regulated at a level slightly higher than the battery voltage. When the PV system supplied sufficient power for system operation, the output of the PV system was adequately regulated at the designated voltage without drawing power from the battery. This is due to the fact that the PV voltage is higher than the battery voltage, so the diode on the battery side is reverse-biased (non-conducting). Thus, no power is drawn from the battery. By contrast, when the power was insufficient (particularly during take-off or highly dynamic maneuvers), the battery supplied power to the motor. When the battery powered the system, the load for the PV system was reduced, which lead to an increase in the output voltage from the PV system. The output voltage from the PV system and battery were balanced when they power the motor simultaneously. Power from the PV panels depended on atmospheric conditions and the incident angle of sunlight, and a buck-boost converter was selected to regulate the voltage to accommodate voltage changes resulting from variations in these conditions. An MPPT system was also required for maximizing the use of the solar energy. The design and simulations of the voltage-regulated MPPT system are detailed in the following sections.

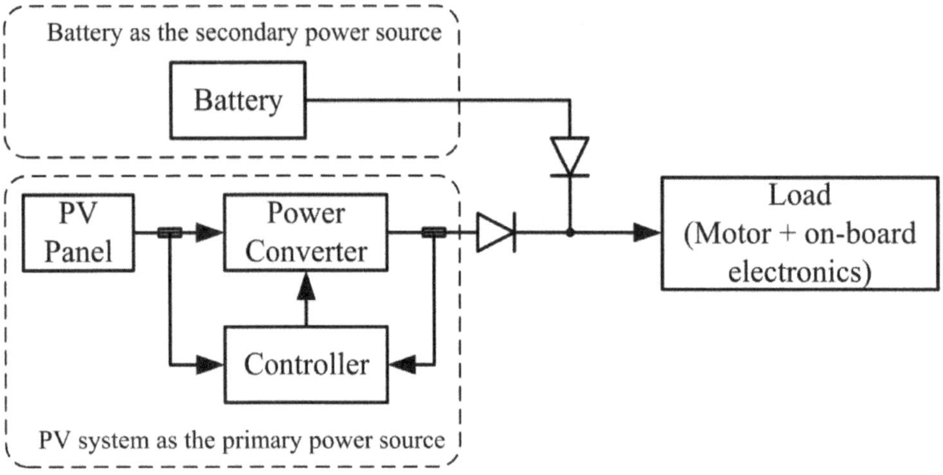

Figure 1. Block diagram of the hybrid power system.

3. PV Characteristics and MPPT Conditions

A single-diode equivalent circuit (Figure 2) was selected as the PV model in this study. The circuit model comprised a current source I_{ph}, diode, shunt resistor R_P, and series of resistor R_S. The relationship between the current I_{PV} and voltage V_{PV} from the PV panel is expressed in the following equations:

$$I_{PV} = I_{ph} - I_D - \frac{V_{PV} + R_S I_{PV}}{R_P} \tag{1}$$

$$I_D = I_0 \left[exp\left(\frac{q(V_{PV} + R_S I_{PV})}{nAKT} \right) - 1 \right] \tag{2}$$

The parameters are defined as follows:

I_{ph}: Current generated by incident light;
I_D: Diode current;
I_0: Reverse saturation current;
q: Electron charge (1.602×10^{-19} C);
K: Boltzmann constant (1.38×10^{-23} J/K);
T: Cell's operating temperature in Kelvins (K);
A: Diode ideality constant;
n: Number of diodes in series to form the single-diode model.

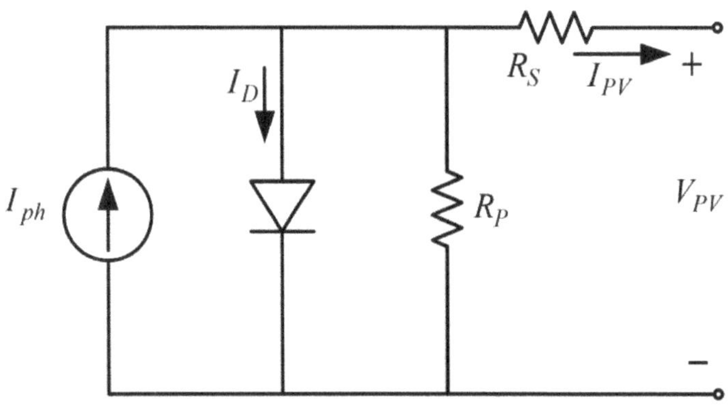

Figure 2. Single-diode equivalent circuit model.

Based on the parameters in [28] for the Solarex MSX60 PV module, Figure 3 shows the characteristics of the PV model at 25 °C where irradiation level $G = 1$. These characteristics were used as the power source for the buck-boost converter and establish the MPPT condition and design of a voltage-regulated MPPT algorithm by using fuzzy logic.

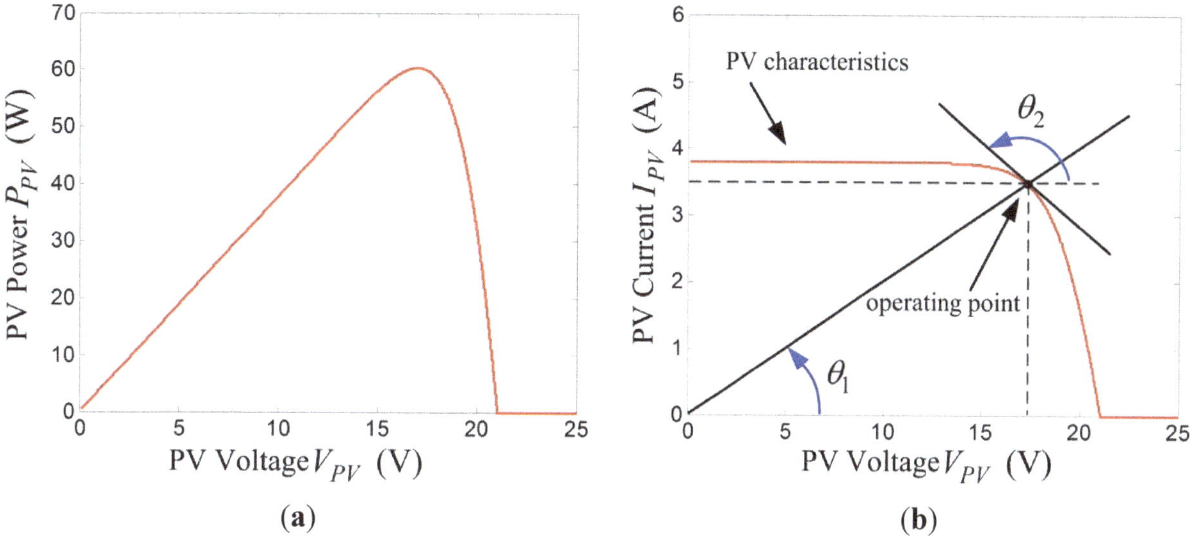

Figure 3. PV characteristics. (**a**) PV power *versus* voltage; and (**b**) PV current *versus* voltage.

Figure 3a shows that the MPP occurs when the derivative of PV power with respect to PV voltage is zero (*i.e.*, at the MPP):

$$\frac{dP_{PV}}{dV_{PV}} = 0 \Rightarrow I_{PV} + V_{PV}\frac{dI_{PV}}{dV_{PV}} = 0 \Rightarrow \frac{I_{PV}}{V_{PV}} + \frac{dI_{PV}}{dV_{PV}} = 0 \tag{3}$$

The condition of the MPP in Equation (3) has been widely used for designing MPPT algorithms. In this study, we aimed at establishing an alternative representation for the condition of MPP. Given:

$$\theta_1 = \tan^{-1}\left(\frac{I_{PV}}{V_{PV}}\right), \qquad \theta_2 = \tan^{-1}\left(\frac{dI_{PV}}{dV_{PV}}\right) \tag{4}$$

as indicated in Figure 3b, fundamental trigonometric functions were used to obtain:

$$\tan(\theta_1 + \theta_2) = \frac{\tan\theta_1 + \tan\theta_2}{1 - \tan\theta_1 \tan\theta_2} = \frac{\frac{I_{PV}}{V_{PV}} + \frac{dI_{PV}}{dV_{PV}}}{1 - \tan\theta_1 \tan\theta_2} \tag{5}$$

From Equations (3) and (5), the condition $\tan(\theta_1 + \theta_2) = 0$ is satisfied at MPP. In addition, from the current–voltage characteristics (Figure 3b), we have the following constraints for θ_1 and θ_2:

$$0 \le \theta_1 \le 90^o, 90^o \le \theta_2 \le 180^o \tag{6}$$

Hence, from the constraints shown in Equation (6), an alternative condition of the PV system to operate at MPP is $\theta_1 + \theta_2 = 180^o$, which is expressed as follows:

$$\tan^{-1}\left(\frac{I_{PV}}{V_{PV}}\right) + \tan^{-1}\left(\frac{dI_{PV}}{dV_{PV}}\right) = 180^o \tag{7}$$

The range of the $\theta_1 + \theta_2$ was confined to the interval (90°, 270°) rather than (−∞, ∞) for the range of the MPP evaluation condition $(I_{PV}/V_{PV} + dI_{PV}/dV_{PV})$ as given in Equation (3). The MPP condition in

Equation (7) was used in conjunction with the voltage regulation requirement for the voltage-regulated MPPT design in this study.

Compared to the other artificial intelligent MPPT algorithms [22–24], the MPP condition of the proposed algorithm is well defined and the range of the MPP evaluation condition is confined to a finite interval (90°, 270°) that simplifies the design for determining the range of the input membership functions. Besides, the process of this algorithm in locating the operating point was more direct and would not require the use of variations of the input variables or search algorithms to predict operating point locations, allowing for a fast tracking speed.

4. Fuzzy Controller for Voltage-Regulated MPPT System

Figure 4 depicts the voltage-regulated MPPT system considered in this study. The system contains a PV system and battery as the primary and secondary power sources, respectively. The goal was to obtain as much power as possible from the PV system while maintaining the output voltage at a designated voltage level. The PV system alone powered the motor when it had sufficient power. Otherwise, the secondary power source was engaged to power the motor. To ensure that the PV system supplied maximal power to the motor, we implemented both MPP tracking and voltage regulation simultaneously. To optimize the performance, a compromise between the MPP operation and voltage regulation was required. The MPP condition in Equation (7) was used for the MPPT design. The MPP condition enabled formulation of the voltage-regulated MPPT system in the form of feedback control (Figure 4). The inputs of the feedback control system were $\tan^{-1}(I_{PV}/V_{PV}) + \tan^{-1}(dI_{PV}/dV_{PV})$ and output voltage V_{out} (Figure 4). The reference inputs of the feedback control system are 180° (the MPP condition) and the desired output voltage. A Zeta type buck-boost converter in [29] (Figure 5) was used to achieve the MPPT and voltage regulation functions. The controller continuously adjusted the duty ratio command for the buck-boost converter to maintain the voltage regulation and MPPT function. The converter parameters were selected such that the converter was operated in continuous conducting mode. The terminal voltage of the PV system, denoted by V_{PV}, was highly sensitive to variation in the duty ration command D for power-switching of the buck-boost converter. Increasing the duty ratio command resulted in a decrease in the PV voltage and vice versa. Thus, maximum power was obtained by adjusting the duty ratio command. The output voltage of the converter depended on the converter input voltage. Neglecting the internal resistances of the components, the relationship of the input and output voltages in steady state is given as:

$$V_{out} = \frac{D}{1-D} V_{PV} \tag{8}$$

The input–output relation in Equation (8) was used for the design of the fuzzy controller for the voltage-regulated MPPT system.

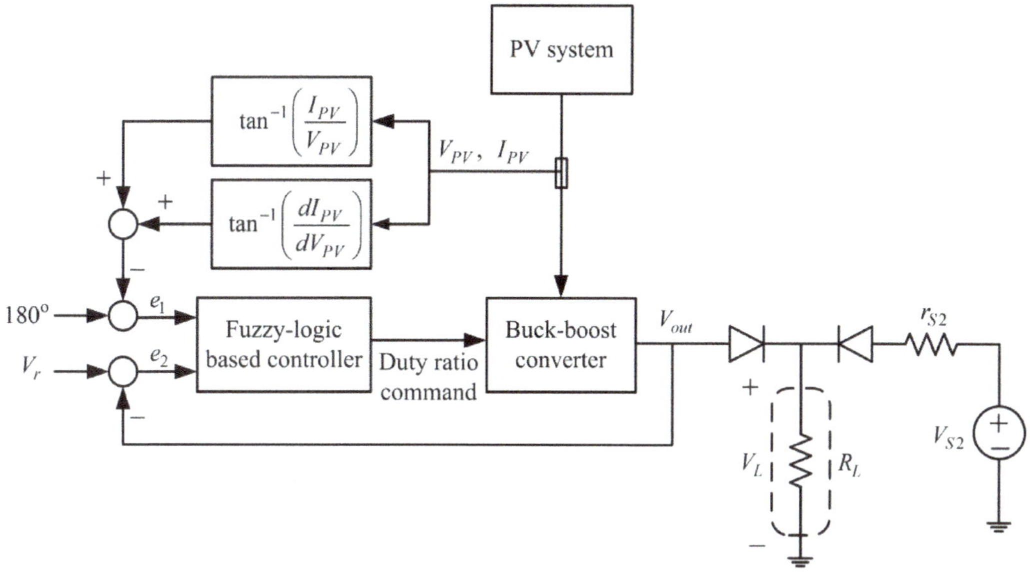

Figure 4. Block diagram of the voltage-regulated MPPT system.

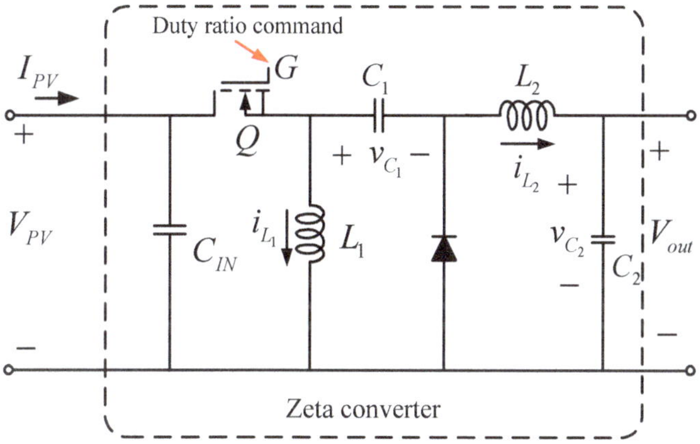

Figure 5. Zeta-type buck-boost converter.

As indicated in Figure 4, the voltage-regulated MPPT requirement was achieved using a fuzzy controller, the computation flowchart of which is shown in Figure 6. The variables required for the fuzzy controller are voltage V_{PV} and current I_{PV} (from the PV system) and output voltage V_{out} (from the buck-boost converter). The fuzzy input variables e_1 and e_2 are defined as follows:

$$e_1 = 180^o - \left[\tan^{-1}\left(\frac{dI_{PV}}{dV_{PV}}\right) + \tan^{-1}\left(\frac{I_{PV}}{V_{PV}}\right)\right], \tag{9}$$

$$e_2 = V_r - V_{out} \tag{10}$$

The output variable of the fuzzy controller is the increment Δd of the duty ratio command for the power switch of the buck-boost converter. The linguistic variables e_2 and Δd were described using the following seven-term fuzzy set: Positive big (PB), positive medium (PM), positive small (PS), zero (ZE), negative small (NS), negative medium (NM) and negative big (NB). The variable e_1 was described using the following five-term fuzzy set: PB, PS, ZE, NS, and NB. The fuzzy rules defined for the fuzzy computation (Figure 7) were divided into three regions (Regions 1–3). The rules in Region 1 were mainly defined to regulate the voltage. In this region, the output voltage was higher than the desired voltage, implying that the PV system alone can supply sufficient power to the motor. Thus, the MPPT function was not critical for this condition. The rules in Region 2 were for the condition of voltage regulation being achieved. The rules for Region 3 were mainly for the MPPT operation. In Region 3, the output voltage V_{out} was lower than the reference voltage V_r. Thus, the secondary power source, the battery, started supplying current to the motor. Under this condition, the output voltage was maintained at the battery's terminal voltage. Hence, V_{out} was almost fixed. Under a constant V_{out}, the output voltage from the PV system, denoted by V_{PV}, was adjusted by controlling the duty ratio of the power switch of the buck-boost converter.

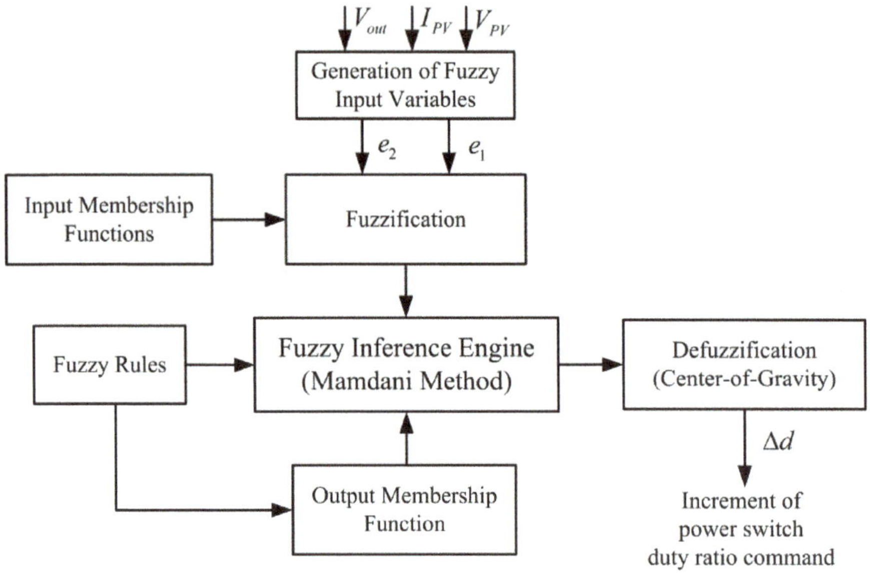

Figure 6. Computation flowchart of the fuzzy controller.

Fuzzy Rules		$e_1 : 180^\circ - \left[\tan^{-1}\left(\frac{I_{PV}}{V_{PV}}\right) + \tan^{-1}\left(\frac{dI_{PV}}{dV_{PV}}\right)\right]$				
		NB	**NS**	**ZE**	**PS**	**PB**
$e_2 : V_r - v_{out}$	**NB**	NB	NB	NB	NB	NB
	NM	NM	NS	NS	NS	NM
	NS	NS	NS	NS	NS	NS
	ZE	ZE	ZE	ZE	ZE	ZE
	PS	NS	NS	ZE	PS	PS
	PM	NM	NS	ZE	PS	PM
	PB	NM	NS	ZE	PS	PM

Region 1: voltage regulation region
Region 2: voltage regulation achieved
Region 3: MPPT region

Figure 7. Fuzzy rules for the voltage-regulated MPPT system. The green and yellow arrows are the direction of state transition.

The relationship in a steady state condition is expressed as:

$$V_{PV} = \frac{1 - D}{D} V_{out} \tag{11}$$

Thus, the MPP can be reached through appropriate control of the duty ratio. Figure 8a shows the PV power-voltage relation along with the relation of Equation (11). When V_{out} was fixed (the green-line in Figure 8a), increasing the duty ratio caused a decrease in V_{PV}, and *vice versa*. MPP was determined by examining the MPP condition established in Equation (7). Figure 8b shows the relationship between the PV current–voltage and MPP condition. Therefore, we continuously controlled the PV voltage by adjusting the duty ratio and examining the MPP condition to ensure that the PV system supplied maximum power to the system as necessary.

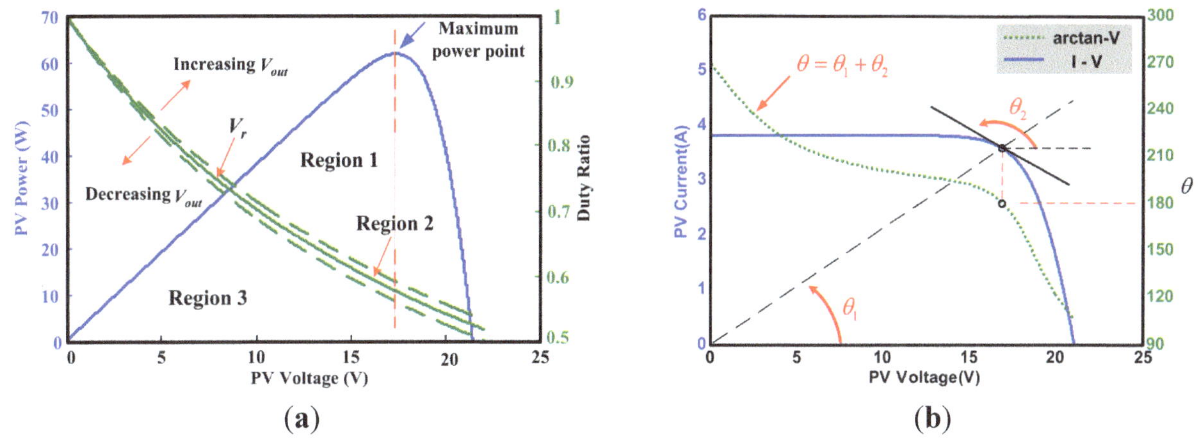

Figure 8. PV characteristics and MPP condition. (**a**) PV power-voltage curve (blue) and duty ratio *versus* PV voltage for fixed a fixed output voltage (green); and (**b**) PV current-voltage curve (blue) and MPP condition using angle information. The black dashed line is the I–V characteristic of the resistive load.

Figure 9 shows the membership functions for the input and output variables. The most commonly used fuzzy inference systems are Mamdani and Sugeno procedures [30]. The main difference is the way the crisp output is computed. The Sugeno method uses weighted average to generate the crisp output. It is computationally efficient and works well with optimization and adaptive techniques. On the other hand, Mamdani method uses defuzzification procedure to obtain the crisp output. Mamdani procedure is widely accepted for capturing expert knowledge. It allows us to describe the expertise in more intuitive, more human-like manner [31]. Thus, Mamdani fuzzy inference is selected for this study. The common forms of the composition operation are maximum of minimum composition and maximum of product composition. Max-min composition is computationally simpler. This method of composition effectively expresses the approximate and interpolative reasoning used by humans when they employ linguistic propositions for deductive reasoning [30]. Most commonly used techniques for defuzzifying fuzzy output function are center of gravity and weighted average method. The weighted average method is more computationally efficient. However, it is usually restricted to symmetrical output membership functions. The center of gravity method is the most prevalent and physically appealing of all the defuzzification methods [30]. Therefore, the maximum of minimum composition technique was used for the inference, and the center of gravity was selected for the defuzzification process in this study.

The fuzzy controller presented here was employed to control the dc motor velocity by using the described solar and battery hybrid-power system. The details of the simulation and results are presented in the following section.

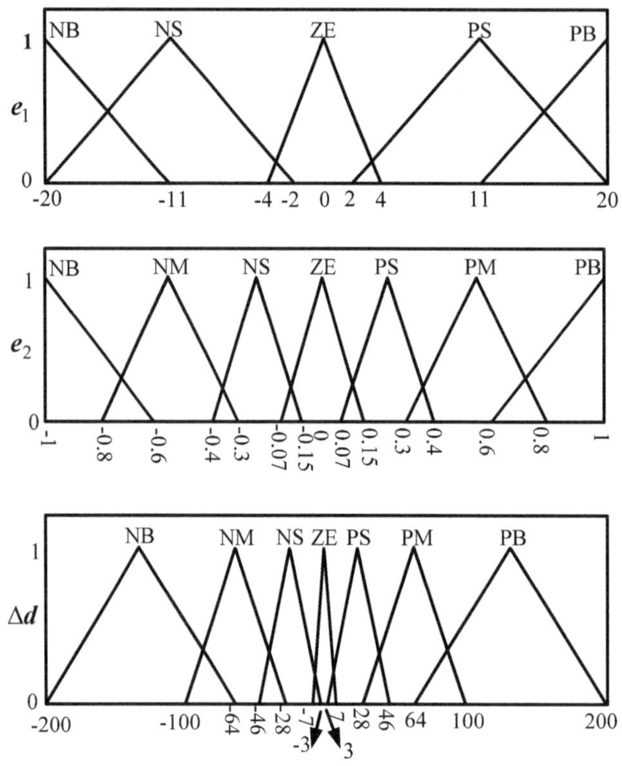

Figure 9. Membership function for the voltage-regulated MPPT system.

5. Simulation Model

A dc motor velocity control scheme powered by solar and battery hybrid system was formulated based on the described fuzzy controller. Figure 10 shows a block diagram of the hybrid powered motor control system. The system contains a PV system and battery as the primary and secondary power sources, respectively, and a dc motor velocity control system. The two power sources were diode ORed to supply power to the motor control. The goal was to have the PV system supply maximum power while using the battery as a back-up power source. The PV system included a PV panel, buck-boost converter, and fuzzy logic controller to regulate the voltage and perform the MPPT functions. The secondary power source contains an ideal voltage source V_{S2} with a series output resistance r_{S2} (Figure 10).

Using the PV model described in Section 3 and the proposed fuzzy controller, a simulation model for the hybrid solar-powered motor control system was developed (Figure 11). The top-level block diagram of the simulation model (Figure 11a) mainly contains the PV model, fuzzy controller, PLECS circuit model for the buck-boost converter, secondary power source, and motor control system. The PLECS circuit simulation model is delineated in Figure 11b. Figures 12 and 13 respectively show the details of the circuit model for the buck-boost converter and motor control system. The motor control system comprises a simulation model of the dc motor, PI controller, and buck power converter for power amplification.

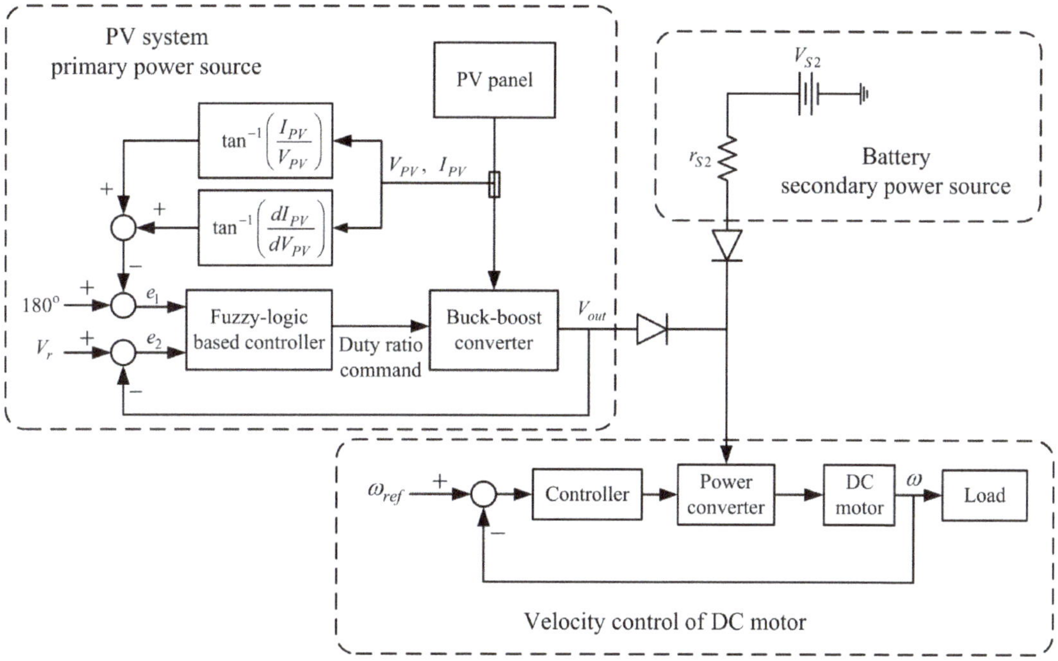

Figure 10. Block diagram of the hybrid powered motor control system.

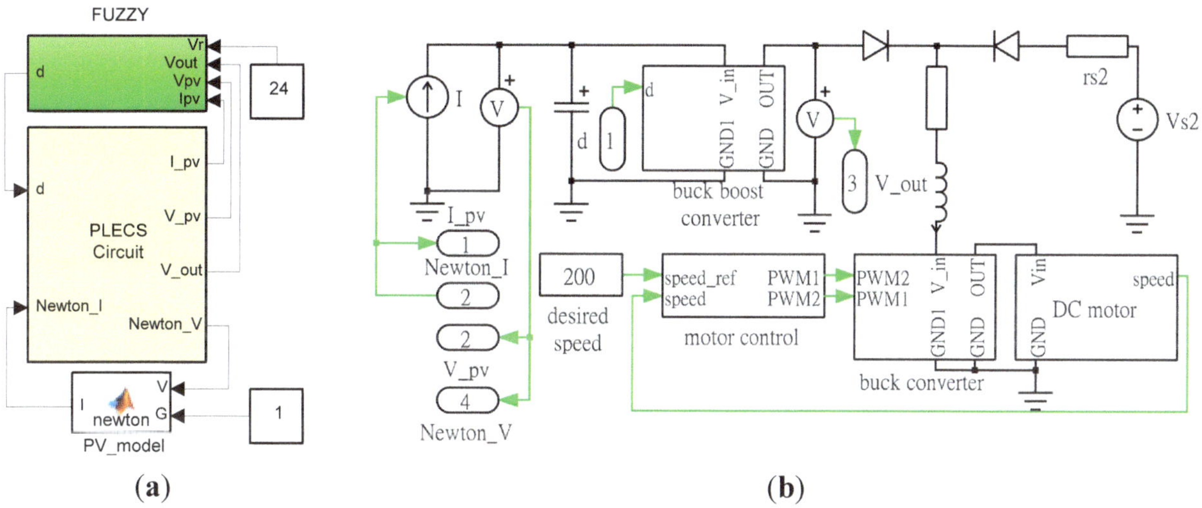

Figure 11. Simulation model of the hybrid powered motor control system. (**a**) Top level functions of the simulation model; and (**b**) Piecewise linear electric circuit simulation (PLECS) circuit simulation model.

The governing equations of the DC motor are:

$$\begin{cases} e_a = R_a i_a + L_a \dfrac{di_a}{dt} + e_b \\ T_m = k_i i_a \\ e_b = k_b \dfrac{d\theta_m}{dt} = k_b \omega_m \\ T_m - B\dfrac{d\theta_m}{dt} = J\dfrac{d^2\theta_m}{dt^2} \end{cases} \tag{12}$$

where e_a is the applied voltage; i_a is the armature current; R_a is the armature resistance; L_a is the armature inductance; e_b is the back electromotive force (emf); T_m is the motor torque; k_i is the torque constant; k_b is the back emf constant; θ_m is the rotor displacement; ω_m is rotor angular velocity; J is moment of inertia of the motor rotor with attached mechanical load; B is the viscous-friction coefficient of motor rotor with attached mechanical load.

In this simulation, the PV model was operated in the voltage-to-current mode. In other words, the PV voltage V_{PV} was fed back to the PV model to compute the PV current I_{PV}, which was then supplied to the buck-boost converter. The induced voltage at the input terminal of the buck-boost converter was returned to the PV model to compute the PV current for subsequent iterations. To achieve the design goal, the fuzzy controller obtained the PV voltage and current from the PV model and determined the duty ratio command for controlling the power switch of the buck-boost converter. Figure 14 shows a simulation block diagram of the fuzzy controller. The proposed fuzzy controller described in Section 4 was implemented in this simulation. Figure 15 details the generation of the fuzzy input variables.

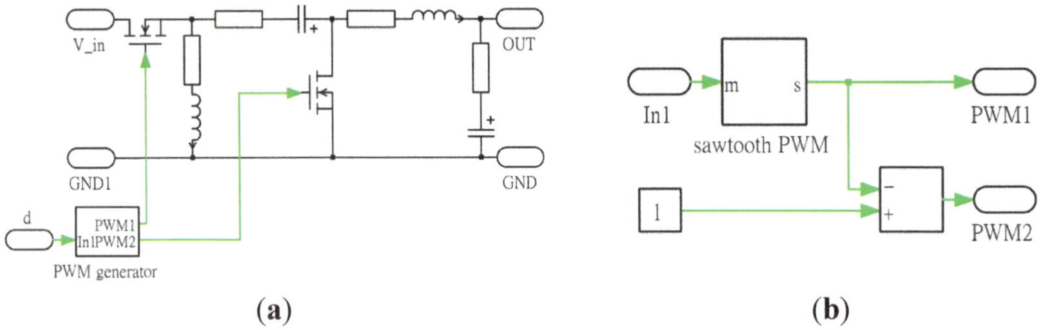

Figure 12. Circuit model for a buck-boost converter. (**a**) Buck-boost converter; (**b**) PWM generator. The green arrows are the control signals.

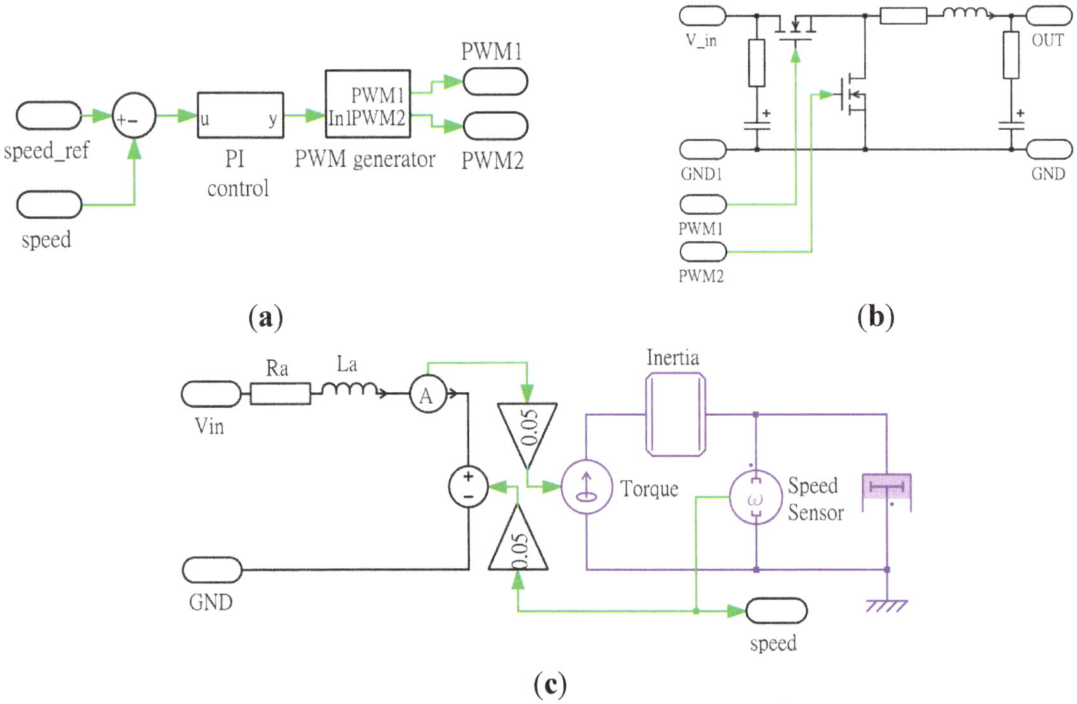

Figure 13. Circuit simulation model for the motor control system. (**a**) Motor control; (**b**) Buck converter; and (**c**) DC motor. The green arrows are the control signals.

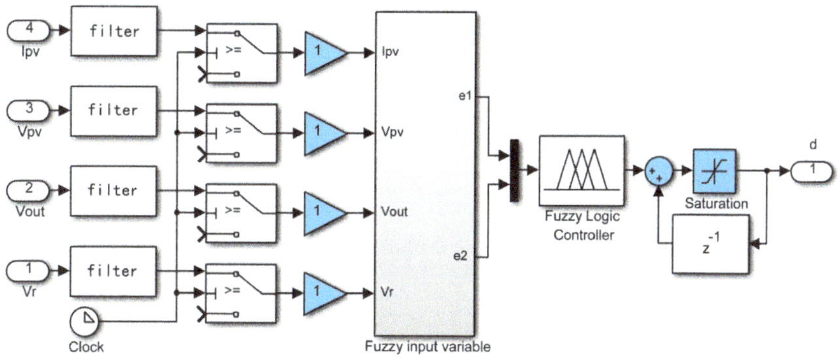

Figure 14. Simulation model of the fuzzy controller.

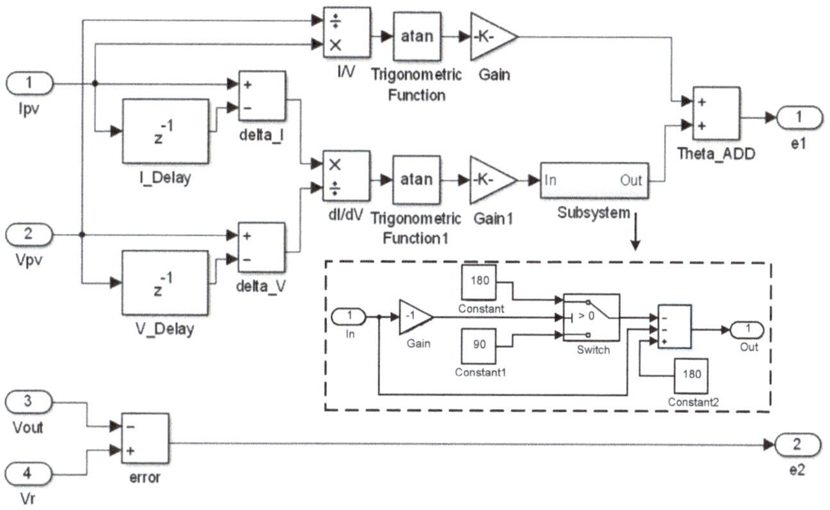

Figure 15. Preprocessing of the fuzzy input variable.

6. Computer Simulation and Results

Based on the proposed fuzzy controller and the simulation model presented in Sections 4 and 5, three cases were selected to demonstrate the design results. In Case 1, we set the regulated voltage to $V_r = 24\ V$ and the battery voltage to 23.76 V (assuming six Li-ion battery cells in a series with a terminal voltage of 3.96 V for each battery). The solar irradiation conditions and motor velocity were subjected to changes according to the sequence (Figure 16). The total duration of the simulation was 5 s. The test conditions were designated as Conditions 1–5, with each condition lasting for 1 s. In the first 3 s (*i.e.*, Conditions 1–3), the solar irradiation was fixed to $G = 1$, the motor velocity was controlled to start from $\omega = 250$ rad/s for 1 s, adjusted to $\omega = 200$ rad/s for a further 1 s, and then adjusted back to $\omega = 250$ rad/s. During the final 3 s (*i.e.*, Conditions 3–5), the motor velocity was fixed at 250 rad/s and the solar irradiation level was adjusted from $G = 1$ to $G = 0.6$, and then back to $G = 1$. In Case 2, the regulated voltage was set to $V_r = 16\ V$ and the battery voltage was set to 15.84 V (four cells in series). The test conditions for the solar irradiation and motor velocity were the same as for Case 1. In Case 3, two motors powered by the hybrid power system were operated at different speeds ($\omega = 250$ rad/s and $\omega = 200$ rad/s). The test duration for this case was 3 s. The solar irradiation level was adjusted from $G = 1$ to $G = 0.6$, and then back to $G = 1$. For this simulation, the parameters used for the buck-boost converter were $L_1 = L_2 = 470\ \mu H$, input and output capacitors $C_{IN} = C_2 = 1000\ \mu F$, and coupling capacitor $C_1 = 200$ μF. The parameters of the buck converter for the motor control were $C = 1000$ μF and $L = 470$ μH. Table 1 shows the parameters of the dc motor. The parameters of the PI controller for DC motor velocity control are $K_P = 0.1$ and $K_I = 100$. For the simulations, the update rate of the fuzzy controller was set to 100 Hz.

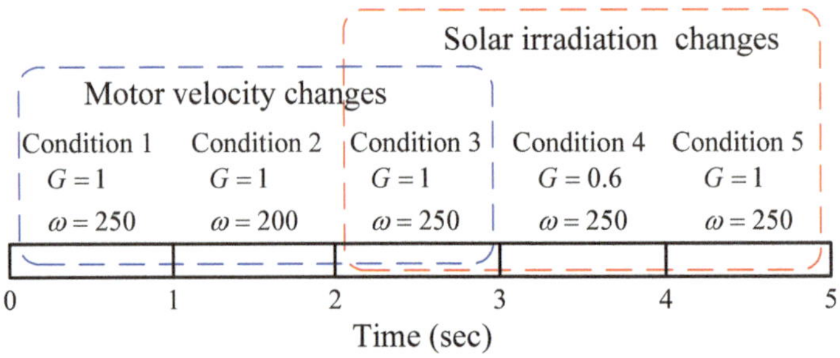

Figure 16. Solar radiation and motor velocity conditions for simulation.

Table 1. Motor parameters.

Parameters	Value	Unit
Armature resistance (R_a)	0.3	Ω
Armature inductance (L_a)	0.006	H
Electromotive force constant	0.05	V/rad/s
Motor torque constant	0.05	Nm/A
Moment of inertia of the rotor	4×10^{-5}	Kgm^2
Motor viscous friction constant	0.001	Nms

Regarding the simulation results, Figure 17 shows the voltage, current, and power output from the PV system, Figure 18 shows the voltage, current, and power from the output of the buck-boost converter, Figure 19 shows the battery outputs, and Figure 20 shows the motor power and velocity. For Condition 1, the solar irradiation was $G = 1$ and the desired motor speed was 250 rad/s, and the maximum power of the PV system was 61.77 W. However, because the power required for the motor to maintain 250 rad/s was 71.91 W, the PV system had insufficient power for the motor. Thus, the battery must be engaged to compensate for this insufficiency.

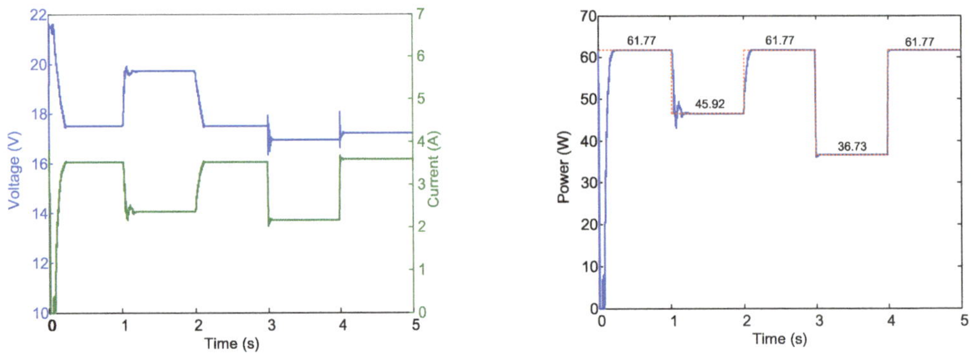

(a) (b)

Figure 17. PV system outputs for Case 1. (**a**) PV voltage and PV current; and (**b**) PV power output. Dotted line (red) is the ideal PV output power.

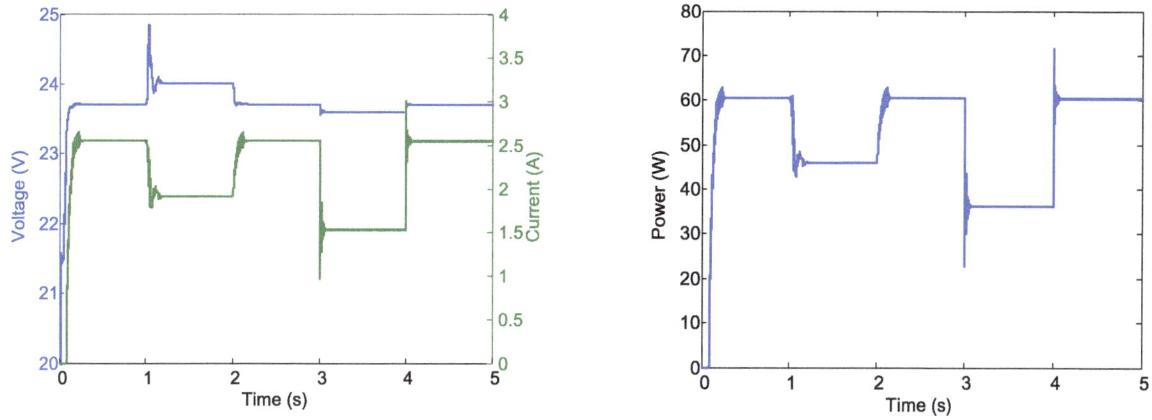

(a) (b)

Figure 18. Buck-boost converter outputs for Case 1. (**a**) Buck-boost converter output voltage and current; and (**b**) Buck-boost converter output power.

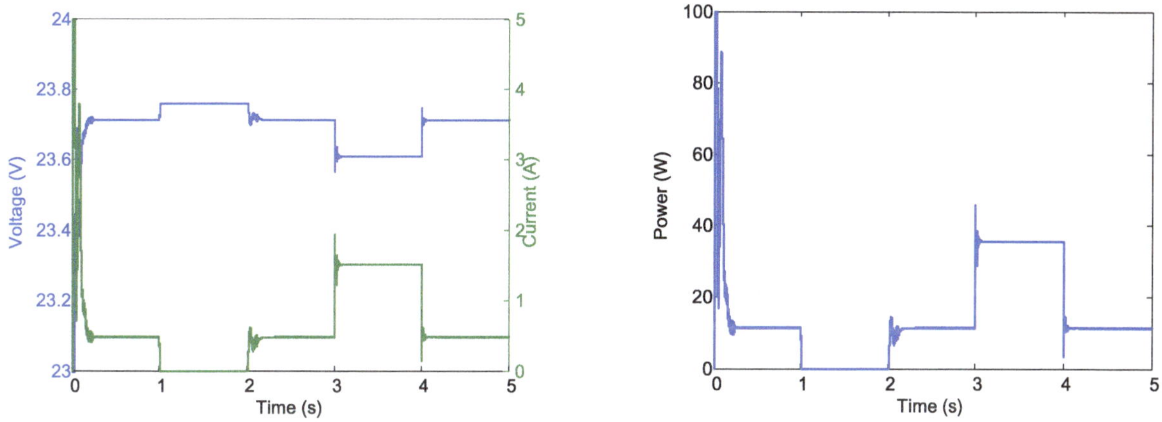

(a) (b)

Figure 19. Battery output for Case 1. (**a**) Battery output voltage and current; and (**b**) Battery output power.

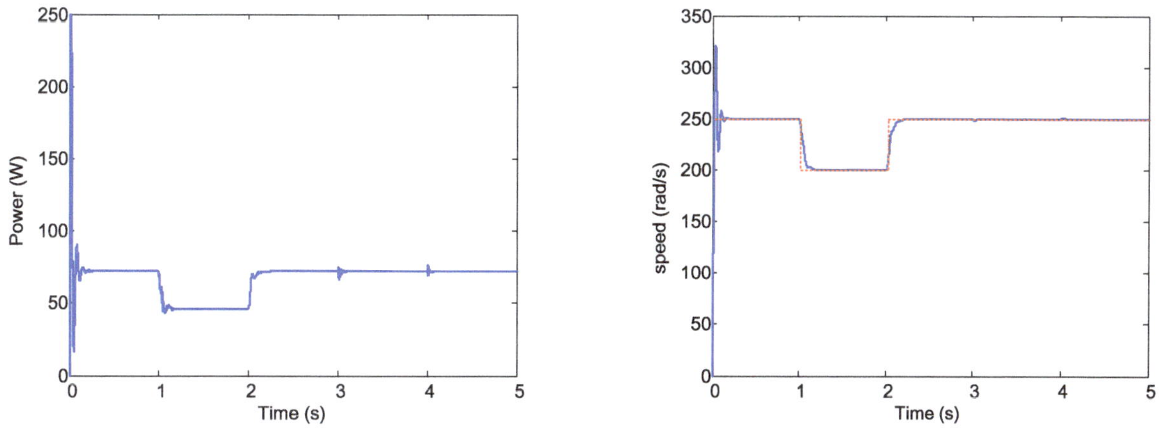

(a) (b)

Figure 20. Motor input power and speed for Case 1. (**a**) Motor input power; and (**b**) Motor speed. Dotted line (red) is the motor command.

As indicated in Figure 17b, after the power-up transient phase, the PV panel outputs 61.75 W steadily, as anticipated. The power output from the buck-boost converter was 60.5 W (Figure 18b). A 2.02% power loss in the buck-boost converter was caused by the internal resistance modeled for the inductors and capacitors. Because the PV system alone cannot supply sufficient power, it would be unable to maintain the converter output voltage of 24 V. When the battery was added to the system, the output voltage of the buck-boost converter was in agreement with the output voltage of the battery, which was slightly less than the battery source voltage (23.76 V) because of the voltage drop of the modeled internal resistance ($r_{S2} = 0.1\ \Omega$ in this simulation). Figure 19b shows that following the motor's power-up transient phase, the battery supplied a steady 11.41 W to the motor to compensate for the aforementioned insufficiency. The voltage from the battery and buck-boost converter achieved equilibrium at 23.71 V. Figure 20b shows that the motor speed was regulated at $\omega = 250\ \mathrm{rad/s}$ after the short transient phase. For Condition 2, the required motor speed was reduced to $\omega = 200\ \mathrm{rad/s}$, requiring only 45.91 W for proper operation. This indicated that the PV system alone supplied sufficient power for the motor, and thus did not require the MPP condition; thus, no power was drawn from the battery to operate the motor. Figure 19 confirms this situation. The output voltage of the power converter was regulated at 24 V (Figure 18a). In the transient instant at $t = 1$ (*i.e.*, the moment where the motor speed was adjusted from $\omega = 250\ \mathrm{rad/s}$ to $\omega = 200\ \mathrm{rad/s}$), a voltage peak appeared at the buck-boost converter output (Figure 18a). The occurred because the power required for the motor was reduced, resulting in the PV operating point moving away from its MPP and increasing the PV output voltage. The buck-boost converter output voltage increased accordingly because the converter was operated in boost mode for this condition. The voltage was soon regulated to the desired voltage of 24 V approximately 0.14 s after the control actions were initiated.

For test Condition 3, the motor speed was switched back to $\omega = 250\ \mathrm{rad/s}$. The simulation revealed that a steady state (identical to that in Condition 1) was achieved within 0.11 s. For Condition 4, the solar irradiation was adjusted from $G = 1$ to $G = 0.6$. At the instant when the condition changed, the PV system falls off from the MPP, and competition between the PV system and battery occurred (the battery was activated in both conditions). Reducing the power from the PV system caused the battery to deliver more power (and current) to the motor, which caused an increase in the voltage drop on the resistance r_{s2} on the battery side. The controller then adjusted the duty ratio for the converter so that maximum power was delivered to the motor and reach voltage balance on converter and battery outputs. Similar actions occurred for test condition changes from Condition 4 to 5. The results in Figure 20 show that the motor speed was regulated following the short power-up transient phase. The steady state results of the Case 1 simulation (Table 2) clearly indicate the success of the design. Figure 21 shows the steady-state operating points of the PV system, including the steady-state PV voltage and power, duty ratio commands, and output-voltage regulation zone.

For the Case 2 simulation, the output reference voltage was set to $V_r = 16\ V$ to ensure that the converter operated in the buck mode under the steady-state condition. The battery voltage was set to 15.84 V. The solar irradiation and motor speed conditions were identical to those in the Case 1 simulation. The simulation results were similar to those obtained for Case 1.

Table 2. Case 1 simulation results.

Test Condition	PV Outputs			Converter Outputs			Battery Outputs			Motor	
	V (V)	*I* (A)	*P* (W)	*V* (V)	*I* (A)	*P* (W)	*V* (V)	*I* (A)	*P* (W)	*P* (W)	ω (rad/s)
Condition 1: $0 \le t < 1, G = 1, \omega_r = 250$	17.49	3.53	61.75	23.71	2.55	60.50	23.71	0.48	11.41	71.91	250
Condition 2: $1 \le t < 2, G = 1, \omega_r = 200$	19.69	2.36	46.53	24.02	1.91	45.91	23.76	0	0	45.91	200
Condition 3: $2 \le t < 3, G = 1, \omega_r = 250$	17.49	3.53	61.76	23.71	2.55	60.50	23.71	0.48	11.41	71.91	250
Condition 4: $3 \le t < 4, G = 0.6, \omega_r = 250$	16.94	2.16	36.66	23.61	1.53	36.19	23.61	1.51	35.73	71.91	250
Condition 5: $4 \le t < 5, G = 1, \omega_r = 250$	17.22	3.59	61.72	23.71	2.55	60.45	23.71	0.48	11.47	71.91	250

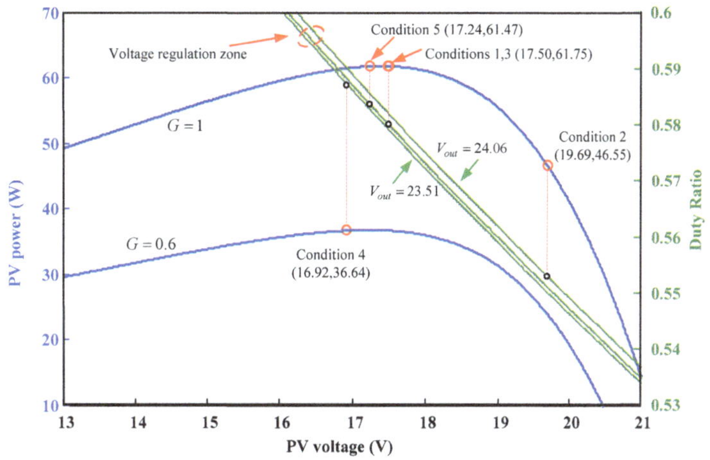

Figure 21. Steady-state operating points of the PV system (Case 1 simulation).

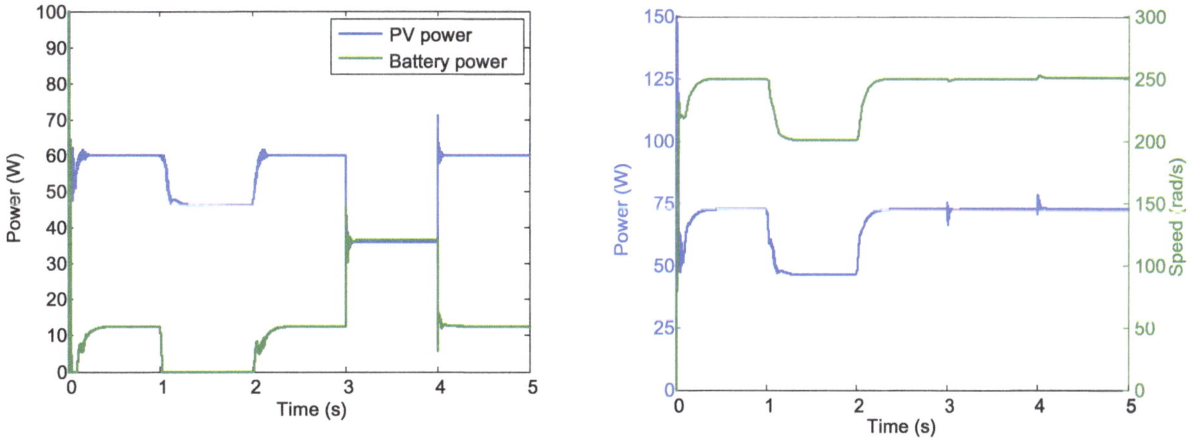

(**a**) (**b**)

Figure 22. Case 2 simulation results. (**a**) PV system and Battery power outputs; and (**b**) Motor power (blue) and speed (green).

Figure 22 shows the results for PV system and battery power outputs (Figure 22a) and motor power and speed (Figure 22b). The steady-state results of the Case 2 simulation (Table 3) clearly indicate the success of the design.

Table 3. Simulation results for Case 2.

Test Condition	PV Outputs			Converter Outputs			Battery Outputs			Motor	
	V (V)	*I* (A)	*P* (W)	*V* (V)	*I* (A)	*P* (W)	*V* (V)	*I* (A)	*P* (W)	*P* (W)	ω (rad/s)
Condition 1: $0 \le t < 1, G = 1, \omega_r = 250$	17.54	3.52	61.73	15.76	3.80	59.95	15.76	0.80	12.53	72.49	250
Condition 2: $1 \le t < 2, G = 1, \omega_r = 200$	19.66	2.39	47.07	16.07	2.87	46.14	15.84	0	0	46.14	200
Condition 3: $2 \le t < 3, G = 1, \omega_r = 250$	17.41	3.55	61.77	15.76	3.81	59.97	15.76	0.79	12.52	72.49	250
Condition 4: $3 \le t < 4, G = 0.6, \omega_r = 250$	16.91	2.17	36.64	15.61	2.31	35.97	15.61	2.34	36.54	72.49	250
Condition 5: $4 \le t < 5, G = 1, \omega_r = 250$	17.25	3.58	61.74	15.76	3.80	59.93	15.76	0.80	12.56	72.49	250

For the Case 3 simulation, the hybrid power system was used to power two motors that were operated at different speeds (*i.e.*, $\omega = 250$ rad/s and $\omega = 200$ rad/s). The duration of this test was 3 s. The solar irradiation level was adjusted from $G = 1$ to $G = 0.6$, and then back to $G = 1$. The simulation results were similar to those obtained for Cases 1 and 2. Figure 23 shows the results for the PV system, battery power output (Figure 22a), and motor speed (Figure 22b). The steady-state results of the Case 3 simulation (Table 4) clearly indicate the success of the design.

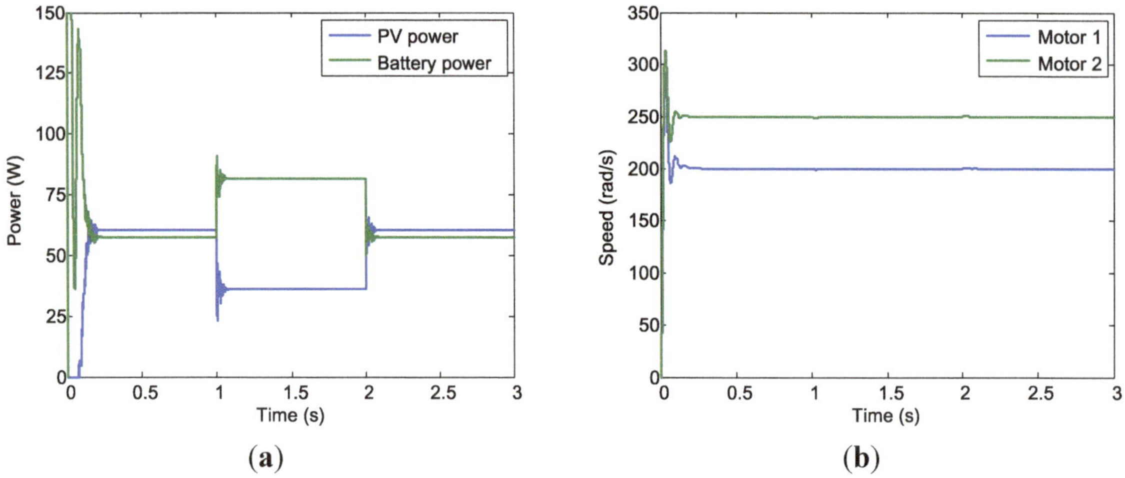

Figure 23. Case 3 simulation results. (**a**) PV and battery power; and (**b**) Motor speed.

Table 4. Case 3 simulation results.

Test Condition	PV Outputs			Converter Outputs			Battery Outputs			Motor		
	V (V)	*I* (A)	*P* (W)	*V* (V)	*I* (A)	*P* (W)	*V* (V)	*I* (A)	*P* (W)	*P* (W)	ω_1 (rad/s)	ω_2 (rad/s)
Condition 1: $0 \le t < 1, G = 1$	17.49	3.53	61.76	23.52	2.57	60.50	23.52	2.44	57.35	143.84	200	250
Condition 2: $1 \le t < 2, G = 0.6$	16.93	2.17	36.65	23.41	1.55	36.18	23.41	3.49	81.67	143.84	200	250
Condition 3: $2 \le t < 3, G = 1$	17.23	3.58	61.73	23.52	2.57	60.45	23.52	2.44	57.40	143.84	200	250

In summary, we successfully developed a voltage-regulated MPPT system for a solar and battery hybrid-power system. The proposed control system was verified through three simulation cases that included buck and boost operations. Changes in solar irradiation and motor speed were also successfully demonstrated in the simulations.

The proposed system is formulated in a standard feedback control system. The measurement inputs to the control system are PV voltage, PV current, and regulated output voltage. The MPPT and voltage regulation functions have been integrated into a single voltage-regulated MPPT algorithm and verified through circuit simulation using PLECS tool. The simple and standard control structure and integrated computation algorithm allow us to implement the proposed system using microcontroller as the core with only voltage and current inputs. The proposed system will be used for the power management for our subsequent development of solar-powered UAV.

7. Conclusions

The design of a fuzzy-logic based voltage-regulated solar power MPPT system for hybrid-power system application was presented in this paper. The system contains a solar power system and battery as the primary and secondary power sources, respectively. The system was used to supply power to a dc motor. The solar system alone supplied power to the motor and maintained the output voltage at a predetermined level when sufficient power was available. When the power was insufficient, the solar system operated at its MPP and the battery was engaged to compensate for the insufficiency. A variant of the incremental conductance MPP condition was used for the MPPT design. Instead of computing the sum of conductance and incremental conductance, the sum of the angles of the arctangent of the conductance and the arctangent of the incremental conductance was investigated. At MPP, the sum was 180°. Moreover, the range of the sum was confined to 90° to 270°. The voltage-regulated solar power MPPT function was formulated in the form of feedback control. A fuzzy controller was then developed to perform the voltage-regulated MPPT function for the hybrid power system.

A simulation model based on MATLAB/SIMULINK and a PLECS tool for controlling the velocity of the dc motor velocity was developed to verify the voltage-regulated solar power MPPT system. Three cases were selected for the computer simulation, including various regulation voltages, solar irradiation changes, and motor speed variations. The system was used to power two motors operated at different speeds. The results demonstrated the success of the proposed fuzzy controller design.

Acknowledgments

This research was supported by the National Science Council, Taiwan, under Grant NSC 102-2221-E-032-017.

Author Contributions

Jaw-Kuen Shiau designed the main parts of the study. Jaw-Kuen Shiau was also responsible for writing the paper. Yu-Chen Wei mainly worked on the design of the fuzzy controller. Min-Yi Lee helped in the PLECS circuit model development and simulation.

Conflicts of Interest

The authors declare no conflict of interest.

References

1. Cambone, S.A.; Krieg, K.J.; Pace, P.; Wells, L., II. *Unmanned Aircraft Systems Roadmap 2005–2030*; Office of the Secretary of Defense: Washington, DC, USA, 2005.

2. Cai, G.; Dias, J.; Seneviratne, L. A survey of small-scale unmanned aerial vehicles: Recent advances and future development trends. *Unmanned Syst.* **2014**, *2*, 1–25.
3. Boucher, R.A. Sunrise, the world's first solar-powered airplane. *J. Aircr.* **1985**, *22*, 840–846.
4. MacCready, P.B.; Lissaman, P.B.S.; Morgan, W.R.; Burke, J.D. Sun-powered aircraft design. *J. Aircr.* **1983**, *20*, 487–493.
5. Brandt, S.A.; Gilliam, F.T. Design analysis methodology for solar-powered Aircraft. *J. Aircr.* **1995**, *32*, 703–709.
6. NASA Armstrong Fact Sheet: Solar-Power Research. Available online: http://www.nasa.gov/centers/armstrong/news/FactSheets/FS-054-DFRC.html (accessed on 17 April 2015).
7. Noth, A. Design of Solar Powered Airplane for Continuous Flight. Ph.D. Thesis, Swiss Federal Institute of Technology Zurich, Zürich, Switzerland, 2008.
8. Shiau, J.K.; Ma, D.M.; Shie, J.R.; Chiu, C.W. Optimal sizing and cruise speed determination for a solar powered airplane. *J. Aircr.* **2010**, *47*, 622–629.
9. Zhu, X.; Guo, Z.; Hou, Z. Solar-powered airplanes: A historical perspective and future challenges. *Prog. Aerosp. Sci.* **2014**, *71*, 36–53.
10. Femia, N.; Petrone, G.; Spagnuolo, G.; Vitellio, M. Optimization of perturb and observe maximum power point tracking method. *IEEE Trans. Power Electron.* **2005**, *20*, 963–973.
11. Hussen, K.H.; Muta, I.; Hoshino, T.; Osakada, M. Maximum photovoltaic power tracking: An algorithm for rapidly changing atmospheric conditions. *IEE Gener. Transm. Distrib.* **1995**, *142*, 59–64.
12. Alajmi, B.N.; Ahmed, K.H.; Finney, S.J.; Williams, B.W. Fuzzy-logic-control approach of a modified hill-climbing method for maximum power point in microgrid standalone photovoltaic system. *IEEE Trans. Power Electron.* **2011**, *26*, 1022–1030.
13. Takun, P.; Kaitwanidvilai, S.; Jettanasen, C. Maximum power point tracking using fuzzy logic control for photovoltaic systems. In Proceedings of the International MutiConference of Engineers and Computer Scientists, Hong Kong, China, 16–18 March 2011; Volume II, IMECS 2011.
14. Khateb, A.E.; Rahim, N.A.; Selvaraj, J.; Uddin, M.N. Fuzzy logic controller based SEPIC converter of maximum power point tracking. In Proceedings of the IEEE Industry Application Society Annual Meeting (IAS) 2012, Las Vegas, NV, USA, 7–11 October 2012; pp. 1–9.
15. Liu, C.-L.; Chen, J.-H.; Liu, Y.-H.; Yang, Z.-Z. An asymmetrical fuzzy-logic-control-based MPPT algorithm for photovoltaic systems. *Energies* **2014**, *7*, 2177–2193.
16. Sakly, A.; Ben Smida, M. Adequate fuzzy inference method for MPPT fuzzy control of Photovoltaic systems. In Proceedings of the 2012 International Conference on Future Electrical Power and Energy Systems, Lecture Notes in Information Technology, Information Engineering Research Institute, Newark, DE, USA, February 2012; Volume 9, pp. 457–468.
17. Altas, I.H.; Sharaf, A.M. A novel maximum power fuzzy logic controller for photovoltaic solar energy systems. *Renew. Energy* **2008**, *33*, 388–399.
18. Gounden, N.A.; Peter, S.A.; Nallandula, H.; Krithiga, S. Fuzzy logic controller with MPPT using line-commutated photovoltaic systems. *Renew. Energy* **2009**, *34*, 909–915.
19. Bos, M.J.; Abhijith, S.; Aswin, V.; Basil, R.; Dhanesh, R. Fuzzy logic controlled PV powered buck converter with MPPT. *Int. J. Adv. Res. Electr. Electron. Instrum. Eng.* **2014**, *3*, 9370–9377.

20. Ou, T.C. A novel unsymmetrical faults analysis for microgrid distribution systems. *Int. J. Electr. Power Energy Syst.* **2012**, *43*, 1017–1024.
21. Ou, T.C. Ground fault current analysis with a direct building algorithm for microgrid distribution. *Int. J. Electr. Power Energy Syst.* **2013**, *53*, 867–875.
22. Hong, C.M.; Ou, T.C.; Lu, K.H. Development of intelligent MPPT (Maximum Power Point Tracking) control for a grid-connected hybrid power generation system. *Energy* **2013**, *50*, 270–279.
23. Ou, T.C.; Hong, C.M. Dynamic operation and control of microgrid hybrid power systems. *Energy* **2014**, *66*, 314–323.
24. Natsheh, E.M.; Albarbar, A. Hybrid power systems energy controller based on neural network and fuzzy logic. *Smart Grid Renew. Energy* **2013**, *4*, 187–197.
25. Swati Negi, A.; Mathew, L. Hybrid enewable energy system: A review. *Int. J. Electron. Electr. Eng.* **2014**, *7*, 535–542.
26. Shiau, J.K.; Ma, D.M.; Yang, P.Y.; Wang, G.F.; Gong, J.H. Design of a solar power management system for an experimental UAV. *IEEE Trans. Aerosp. Electron. Syst.* **2009**, *45*, 1350–1360.
27. Gao, X.Z.; Hou, Z.X.; Guo, A.; Liu, J.X.; Chen, X.Q. Energy management strategy for solar-powered high-altitude long-endurance aircraft. *Energy Convers. Manag.* **2013**, *70*, 20–30.
28. Walker, G. Evaluating MPPT converter topologies using a MATLAB PV model. *J. Electr. Electron. Eng.* **2001**, *21*, 49–55.
29. Falin, J. Designing DC/DC converters based on ZETA topology. Texas Instruments Incorporated. 2Q 2010. *Analog Appl. J.* Available online: http://www.ti.com/lit/an/slyt372/slyt372.pdf (accessed on 17 April 2015).
30. Ross, T.J. *Fuzzy Logic with Engineering Applications*, 3rd ed.; John Wiley & Sons, Ltd.: West Sussex, UK, 2010.
31. Kaur, A.; Kaur, A. Comparison of mamdani-type and sugeno-type fuzzy inference systems for air conditioning system. *Int. J. Soft Comput. Eng.* **2012**, *2*, 323–325.

2

Optimized Adaptive Perturb and Observe Maximum Power Point Tracking Control for Photovoltaic Generation

Luigi Piegari [1,†], Renato Rizzo [2,†], Ivan Spina [2,*] and Pietro Tricoli [3,†]

1 Department of Electronics, Information and Bioengineering, Politecnico di Milano, Piazza Leonardo da Vinci 32, 20133 Milan, Italy; E-Mail: luigi.piegari@polimi.it
2 Department of Electrical Engineering and Information Technologies, University of Naples Federico II, Via Claudio 21, 80125 Naples, Italy; E-Mail: renato.rizzo@unina.it
3 School of Electronic, Electrical and Computer Engineering, University of Birmingham, Gisbert Kapp Building, Birmingham B15 2TT, UK; E-Mail: p.tricoli@bham.ac.uk

† These authors contributed equally to this work.

* Author to whom correspondence should be addressed; E-Mail: ivanspina@gmail.com

Academic Editor: Enrico Sciubba

Abstract: The power extracted from PV arrays is usually maximized using maximum power point tracking algorithms. One of the most widely used techniques is the perturb & observe algorithm, which periodically perturbs the operating point of the PV array, sometime with an adaptive perturbation step, and compares the PV power before and after the perturbation. This paper analyses the most suitable perturbation step to optimize maximum power point tracking performance and suggests a design criterion to select the parameters of the controller. Using this proposed adaptive step, the MPPT perturb & observe algorithm achieves an excellent dynamic response by adapting the perturbation step to the actual operating conditions of the PV array. The proposed algorithm has been validated and tested in a laboratory using a dual input inductor push-pull converter. This particular converter topology is an efficient interface to boost the low voltage of PV arrays and effectively control the power flow when input or output voltages are variable. The experimental results have proved the superiority of the proposed algorithm in comparison of traditional perturb & observe and incremental conductance techniques.

Keywords: Maximum Power Point Tracking (MPPT); Perturb & Observe (P&O); photovoltaic generation; push-pull converter

---

1. Introduction

Photovoltaic (PV) power generation has received significant attention from scientists over the last few years to help reduce the environmental pollution inherently associated with traditional electric generators. The economic convenience of PV generation is directly connected to the cost of the cells and the amount of energy that the arrays are capable of supplying over their life. The former is influenced mainly by the price of the feedstock for the PV industry and the improvement of fabrication technologies. The latter is connected to the efficiency of the power conversion system, which is normally necessary when PV arrays are used for power generation. At the present state of the art, the control algorithms used for the maximization of the power extracted from PV arrays are widely known as maximum power point tracking (MPPT) algorithms. These power conditioners guarantee that PV arrays operate close to their point of maximum efficiency under any weather conditions. This is particularly important for variable irradiation levels, because the efficiency of the PV modules is very low when the operating point is far from the maximum power point.

The MPPT techniques proposed so far in the technical literature can be classified mainly as Perturb & Observe (P&O), Incremental Conductance (IC), and Temperature Gradient (TG) techniques [1]. The P&O algorithm is very popular because of its easy implementation [2–5]. The operating voltage of the converter is changed slightly at every control step and the consequent variation of the power output is measured. If the perturbation has caused a power increase, the voltage is further changed in the same direction; if not, a backwards perturbation is applied. A typical disadvantage of this technique is that the operating point always oscillates around the maximum, losing some part of the energy actually available [6]. Smaller perturbation steps can mitigate the problem, although the algorithm becomes slow at finding the new maximum when irradiation conditions change suddenly.

Several improvements on the P&O algorithm have been proposed in order to reduce the oscillations around the maximum power point (MPP) at steady-state, but all of them slow down the response speed of the algorithm when atmospheric conditions change and reduce the efficiency during cloudy days [7]. An adaptive MPPT algorithm for a field-programmable gate array has been proposed in [8]. This adjustable algorithm is based on an improved P&O method that does not need external sensory units. Despite its cost-effectiveness and high-speed processing, the system performance has been demonstrated to be outstanding.

In the original MPPT P&O controller, the adjustment of the operating point is achieved by changing the reference voltage of the controller. However, this adjustment can be made through the duty ratio of the converter such as described in [8]. The experimental results reported in the recent technical literature show that the adaptive algorithm has better performance than conventional P&O techniques in terms of dynamics and efficiency. In particular, an efficiency of the algorithm above 96% has been reported in [9,10]. In [10] several adaptive perturbation functions are considered and the final choice is made by randomly setting the coefficients for each function and comparing the results obtained from numerous tests. Since the function associated with best results shows stability problems in the right side of the maximum power point, the algorithm has been modified by associating zero values for the unstable operation range. In [11] the authors investigated and compared the characteristics and performance of several P&O techniques, showing that some are not truly adaptive, though they present improved performance in comparison to their fixed perturbation step counterparts [12–14]. Other techniques propose truly adaptive modifications, but suffer from a high computational load caused by aggressive calculation of derivatives [15,16] and need initial user-dependent constants for the perturb adaptation [17,18]. Novel techniques show improved performance using nonlinear equations [19], fuzzy logic [20], and complicated optimization algorithms [21–23].

This paper proposes a design criterion to choose the adaptive perturbation step, in order to assure the most suitable perturbation amplitude for a P&O MPPT algorithm in different operating conditions, while keeping a good dynamic response. In addition to the work developed in [10], this paper proposes a modified P&O algorithm with a properly chosen perturbation function. Experimental tests on the proposed algorithm are carried out on a dual input inductor push-pull converter (DIIPPC) [24] with the aim of comparing the results with both classic P&O and incremental conductance (IC) algorithms. DIIPPCs achieve a high boost ratio and provide isolation between the PV arrays and the load, optimising the power transfer for every irradiation and temperature level. Moreover, these converters present lower output voltage ripple, lower voltage across the main switches, and a lower volt-ampere rating of the transformer than current-fed push-pull converter [24].

This paper is organised as follows: Section 2 presents the DIIPPC used for PV generation and its basic operating characteristics; Section 3 describes the adaptive P&O MPPT algorithm; Section 4 reports the parameter identification of PV panel used for the experiments; Section 5 provide details about the MPPT implementation; Section 6 discusses the experimental validation of the proposed technique and the testing methodologies; Section 7 introduces a discussion on the MPP estimation error.

2. Dual Inductor Push-Pull Converter

The basic configuration of a DIIPPC [24] is illustrated in Figure 1. DIIPPCs operate similarly to converters with a single inductor.

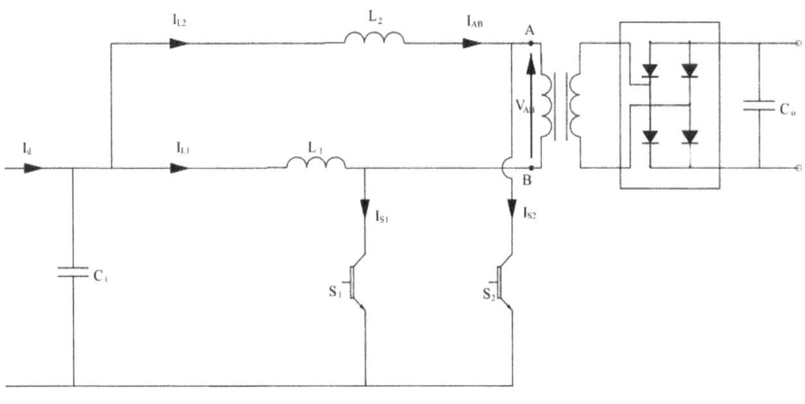

Figure 1. Circuit diagram of the dual input inductor push-pull converter.

The switches of the input stage are both turned on to charge the input inductors and are alternately turned off to transfer the energy to the primary winding of transformer. The two components are driven with the same control signal delayed by one half of the period. Therefore, a duty cycle greater than 0.5 is sufficient to guarantee the operating principle just described.

Figure 2 shows the control signals for the switches S1 and S2, the current and the voltage across the transformer, the source current, input inductance currents and the current across one of the main switches for a typical operating condition. In order to make the converter suitable to be used for interfacing PV panels, according with [25,26] an input capacitor is connected. In the hypothesis that this capacitance is high enough, during one switching period the input voltage can be considered constant.

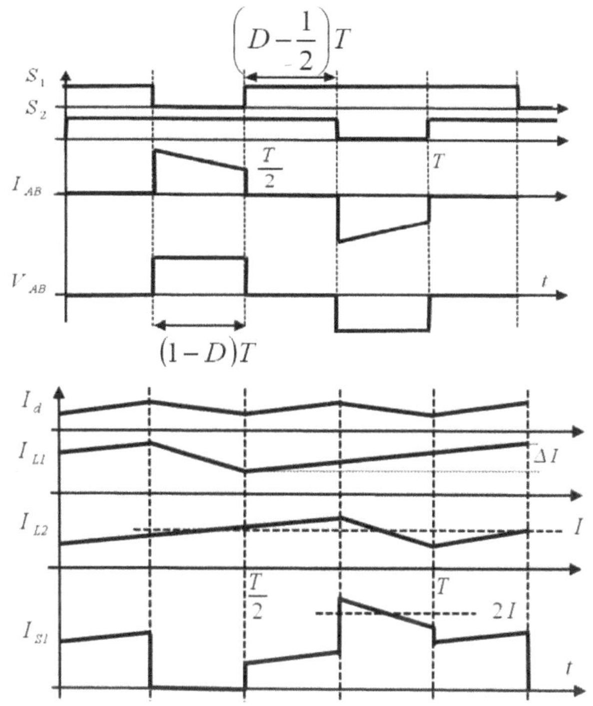

Figure 2. Control signals and primary current and voltage across the transformer (upper diagram); source current, inductor currents and current across the first switch (lower diagram).

The output voltage of this converter is increasing with the duty cycle of the switches, because the input stage acts as a boost converter. The output voltage of the converter can be expressed as:

$$V_0 = \frac{N_2}{N_1} \frac{V_i}{1-D} \quad (1)$$

where V_i is the input voltage, D is the duty cycle (defined as the ratio between the on-state time and the switching period, T_{on}/T), and N_1 and N_2 are the numbers of turns of the primary and secondary windings, respectively.

The rapid change of the inductor current, due to the turn-off of the switch, causes an overvoltage across the switch itself. For this reason, a snubber circuit may be added to reduce the overvoltage. Since the efficiency of the converter is not the main focus of the paper, a traditional dissipative snubber has been used.

3. Adaptive Perturbation Function for P&O Algorithm

The MPPT algorithm proposed by the authors in [10] is a variation of the classic P&O method and has been set up and implemented for the control of the DIIPPC. The main limitation of the P&O algorithm is the compromise between a fast dynamic response and a reduced oscillation at steady-state. Large perturbation steps imply a fast dynamic performance, whereas small perturbation steps ensure little oscillations at steady-state. In particular conditions, during cloudy days, the solar irradiation can vary many times and quickly. This situation can cause fault or reduction of efficiency in MPP trackers. In order to address this problem some studies has been proposed in the technical literature [27,28]. In these conditions, it is necessary to achieve a good dynamic response to avoid losing a great quantity of energy. The efficiency of the P&O technique depends on the perturbation step. It is particularly difficult to find the optimal perturbation step, because this depends on the panel characteristics, working conditions and the converter's topology. To obtain a faster dynamic with small oscillations without knowing accurately the technical details of the PV system, the proposed adaptive P&O method operates with variable perturbation step adapted to the actual operating conditions. In particular, when the operating point is far from the MPP, large perturbation steps are chosen, whereas small ones are used in proximity to the maximum. Therefore, the algorithm needs to know the MPP position. However, a precise knowledge is not required because the P&O technique always works around the real maximum. In the proposed method, the perturbation step is defined by the following equation:

$$\Delta V(V) = f\left(V_{\mathrm{M,est}} - V\right) + \Delta V_{\mathrm{min}} \tag{2}$$

where ΔV is the perturbation step; $V_{\mathrm{M,est}}$ is the estimated voltage at MPP; V is the measured voltage; ΔV_{min} is the minimum value of the perturbation.

In order to guarantee a high dynamic performance and limited steady-state oscillations, the function f must be a positive function and must assume a high value when the working point is far from the maximum; moreover, it must be nearly zero in proximity to the maximum. The condition of working in proximity to the MPP is assured by setting ΔV_{min} at an appropriate value. If the working point is near to the value of $V_{\mathrm{M,est}}$, $\Delta V \approx \Delta V_{\mathrm{min}}$; when $V \equiv V_{\mathrm{M,est}}$, $f(0) = 0 \rightarrow \Delta V(0) = \Delta V_{\mathrm{min}}$.

In the previous work [10], $f(V_{\mathrm{M,est}} - V)$ has been chosen to be a quadratic function in the left side of the MPP, and zero in the right side of MPP for stability problems. This procedure for choosing the function given by Equation (2) has not been fully justified and no correlation between $f(V_{\mathrm{M,est}} - V)$ and ΔV_{min} has been considered. In this paper a criterion to properly choose the adaptive perturbation function is discussed taking into account the possible correlation between the two addends of Equation (2) and the estimation error in evaluating the MPP.

If the MPP voltage estimation were perfect ($V_{\mathrm{M,est}} \equiv V_{\mathrm{M}}$, where V_{M} is the actual position of the MPP), and the converter did not introduce delays, the best expression for f and ΔV_{min} would be:

$$\begin{aligned} f\left(V_{\mathrm{M,est}} - V\right) &= \left|V_{\mathrm{M,est}} - V\right| \\ \Delta V_{\mathrm{min}} &= 0 \end{aligned} \tag{3}$$

In this ideal condition, the MPP would be reached after one iteration. However, the MPP voltage estimation introduces an error. Therefore, in order to make the algorithm capable of reaching the actual

MPP, it is necessary to have $\Delta V_{\min} \neq 0$. In addition, when the measured voltage V is close to V_M, the function f should yield a value lower than $\Delta V_{\min}$ when $V \equiv V_M$:

$$f\left(V_{\mathrm{M,est}} - V_{\mathrm{M}}\right) < \Delta V_{\min} \tag{4}$$

The condition given by Equation (4) guarantees that f helps to increase the rapidity of the MPPT algorithm far from the maximum power point, but does not worsen the oscillations when the PV array is operating around V_M. Indicating with k the maximum estimation error in p.u. it is:

$$\frac{\left|V_{\mathrm{M,est}} - V_{\mathrm{M}}\right|}{V_{\mathrm{M}}} \leq k \tag{5}$$

The algorithm has to be capable of correctly matching the MPP also in case of wrong estimation of V_M. For this reason, a minimum perturbation step has to be introduced, as reported in Equation (2). Being k the maximum allowed estimation error, the algorithm could go to work in a point far from the real MPP of, at worst, kV_M. In order to achieve the correct matching of the MPP it has to be:

$$\Delta V_{\min} > kV_{\mathrm{M}} \tag{6}$$

In this way, the P&O action can compensate the estimation error in two steps.

Equation (6) proves that the linear expression (3) cannot be used in practice because estimation errors would be too high to ensure small oscillations at steady state. For this reason, polynomial expressions with higher order or exponential expressions are preferred in the practical implementation of the algorithm. Choosing, for example, a quadratic function:

$$f\left(V_{\mathrm{M,est}} - V\right) = a\left(V_{\mathrm{M,est}} - V\right)^2 \tag{7}$$

It is possible to set a high value for f in order to have a quick answer when the voltage is far from the MPP. In particular, when V is zero (the furthest position), the MPP can be reached in two steps if the function f is selected as:

$$f\left(V_{\mathrm{M,est}}\right) = 0.5V_{\mathrm{M,est}} \tag{8}$$

This last equation, together with Equation (7), gives the value of parameter a:

$$a = \frac{0.5}{V_{\mathrm{M,est}}} \tag{9}$$

Substituting the expression of f (given by Equation (7)) into Equation (4), and using Equation (9) for parameter a, it is possible to obtain the following condition on $\Delta V_{\min}$:

$$\Delta V_{\min} > \frac{0.5}{V_{\mathrm{M,est}}}\left(V_{\mathrm{M,est}} - V_{\mathrm{M}}\right)^2 = A \tag{10}$$

where the quantity A systemizes the right term of the inequality. Moreover, from Equation (5) the following condition can be obtained:

$$A = \frac{0.5}{V_{\mathrm{M,est}}}\left(V_{\mathrm{M,est}} - V_{\mathrm{M}}\right)^2 \leq \frac{0.5}{V_{\mathrm{M,est}}}V_{\mathrm{M}}^2k^2 = B \tag{11}$$

where the quantity B synthesizes the right term of the inequality. In order to satisfy condition (10) it is possible to choose $\Delta V_{\min}$ higher than B, *i.e.*, if $\Delta V_{\min} > B \Rightarrow \Delta V_{\min} > A$ (since $A \leq B$):

$$\Delta V_{\min} > B \quad \Rightarrow \quad \begin{cases} \Delta V_{\min} > A \\ \text{with:} A \leq B \end{cases} \tag{12}$$

Thus, $\Delta V_{\min} > B$ is a sufficient (but not necessary) condition to satisfy Equation (10). Solving Equation (5) it is possible to deduce the following conditions on $V_{\mathrm{M,est}}$, and, consequently on the quantity B:

$$\frac{\left|V_{\mathrm{M,est}} - V_{\mathrm{M}}\right|}{V_{\mathrm{M}}} \leq k \;\Rightarrow\; \begin{cases} V_{\mathrm{M,est}} \in \left]V_{\mathrm{M}}, (1+k)V_{\mathrm{M}}\right] \\ \text{or} \\ V_{\mathrm{M,est}} \in \left[(1-k)V_{\mathrm{M}}, V_{\mathrm{M}}\right[\end{cases} \;\Rightarrow\; V_{\mathrm{M,est}} \geq (1-k)V_{\mathrm{M}} \;\Leftrightarrow\; B \leq \frac{0.5k^2}{(1-k)} V_{\mathrm{M}} = C \tag{13}$$

where the quantity C synthesizes the right term of the last inequality. In order to satisfy $\Delta V_{\min} > B$ it is possible to choose $\Delta V_{\min}$ higher than C, *i.e.*, if $\Delta V_{\min} > C \Rightarrow \Delta V_{\min} > B$ (since $B \leq C$):

$$\Delta V_{\min} > C \quad \Rightarrow \quad \begin{cases} \Delta V_{\min} > B \\ \text{with:} B \leq C \end{cases} \tag{14}$$

which, substituting the expression of the term C, gives:

$$\Delta V_{\min} > 0.5 \frac{k^2}{(1-k)} V_{\mathrm{M}} \tag{15}$$

Equation (15) returns a minimum variable step of 0.56% for a 10% estimation error, which ensures reasonably small oscillations at steady state.

4. PV Panel Identification

The PV array is composed of 36 cells of monocrystalline silicon wafers, connected in series and measuring a total active area of 0.54 m^2. The technical data of the PV array are reported in Table 1. The PV array model is mathematically described by the following equations [10]:

$$\begin{aligned} V &= \frac{AKT}{q} \ln\left(\frac{I - I_{\mathrm{ph}} - I_{\mathrm{sat}}}{I_{\mathrm{sat}}}\right) - r_s I \\ I_{\mathrm{ph}} &= \left[I_{\mathrm{scr}} + k_i (T - T_r)\right] s \\ I_{\mathrm{sat}} &= I_{\mathrm{rr}} \left(\frac{T}{T_r}\right)^3 \exp\left[\frac{q E_g}{K}\left(\frac{1}{T_r} - \frac{1}{T}\right)\right] \end{aligned} \tag{16}$$

where I_{scr} is the short circuit current, I_{rr} is the reverse current at the reference temperature, q is the electronic charge, A is the junction ideality factor, K is the Boltzmann constant, T is the absolute temperature of the array, T_{r} is a constant equal to 273.14 K, r_{s} is the series resistance, s is the p.u. irradiation, E_{g} is the band gap energy of the semiconductor, and k_i is the short-circuit current temperature coefficient.

The curves of the PV array have been evaluated experimentally for several temperatures, ranging from 23 °C to 40 °C with a step of 1 °C. However, only some of these curves are presented in Figure 3 for a better visualisation. The tests were made with an artificial light to control the level of irradiance and keep the irradiance smooth during the test. The open circuit voltage of the panel, V_{oc}, presents a decreasing rate of 2.3 mV/°C, whereas the short-circuit current, I_{scr}, presents an increasing rate of 0.2%/°C. This is reflected by a power peak reduction, which is equal to 0.6%/°C–0.7%/°C. The experimental data have been fitted using a least mean square numeric procedure in order to obtain the parameters of the model. The results of this procedure are shown in Table 2. The curves obtained by the model are also plotted in Figure 3, showing very good agreement with the experimental data.

Table 1. Characteristics of PV array.

Type	Aluminum-Tedlar125M36
Open circuit voltage, V_{oc}	21.85 V
Short circuit current, I_{sc}	4.52 A
Maximum power, P_M	69.03 W
Maximum power voltage, V_M	17.74 V
Maximum power current, I_M	3.89 A

Table 2. Main data of the PV panel used.

A [–]	I_{rr} [nA]	r_s [Ω]	I_{scr} [A]	k_i [A/K]
57	−15.8	0.25	3.5	0.001

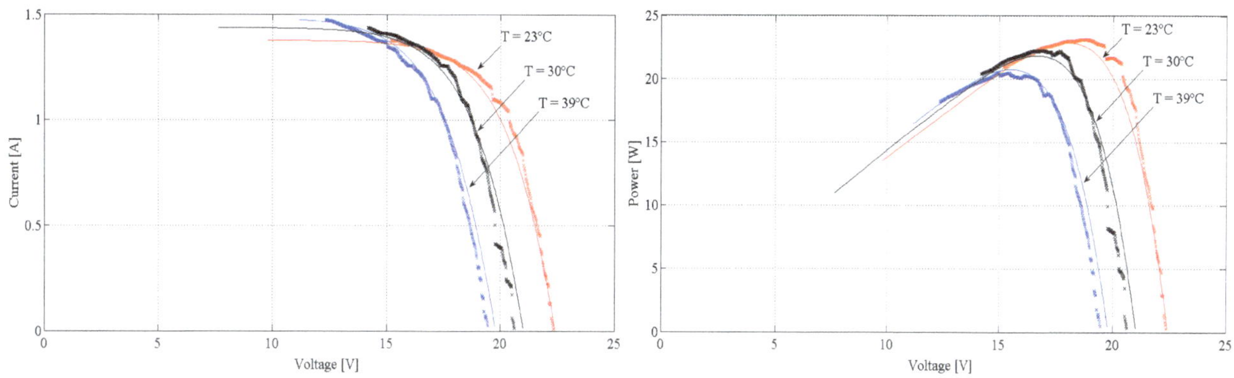

Figure 3. PV current and power curves for different array temperatures and their fitting.

5. Algorithm Implementation

The MPPT algorithm presented in this paper requires the estimation of the maximum V_M. Once the panel has been identified according to the procedure of the previous section, the theoretical position of V_M can be determined by the mathematical model of the panel. In practical applications the characteristic of the panel changes during its life for ageing and dust. For this reason, a periodical characterization has to be performed and the experimental characteristic has to be registered. A good time interval for characterization of the panel can be one month. During one month the efficiency degradation of a PV panel has been estimated to be 7% [29]. Anyway, with the proposed algorithm, an error in the estimation of the theoretical maximum is compensated by the P&O function. As analyzed in Section 7 a 10% error in the estimation of V_M does not practically affect the performances of the proposed algorithm. It is worth noting that, the chosen converter, allows the complete catching of panel characteristic because it can make the panel work from short circuit (duty cycle equal to 1) to open circuit (duty cycle equal to 0).

In order to estimate the MPP voltage and current, it is necessary to measure the irradiation and the temperature. Although it is quite easy and inexpensive to measure the temperature, the measurement of the irradiation greatly increases the cost of the MPPT control system. For this reason, the power-voltage characteristic has been preferred to the power-current characteristic. Indeed, the voltage at the MPP is weakly variable with the irradiation, whereas the current at the MPP is weakly variable with the temperature. In the following figures, the irradiation is normalised to 1 kW/m^2. Figure 4 shows the MPP voltage as a function of the irradiation for different temperatures.

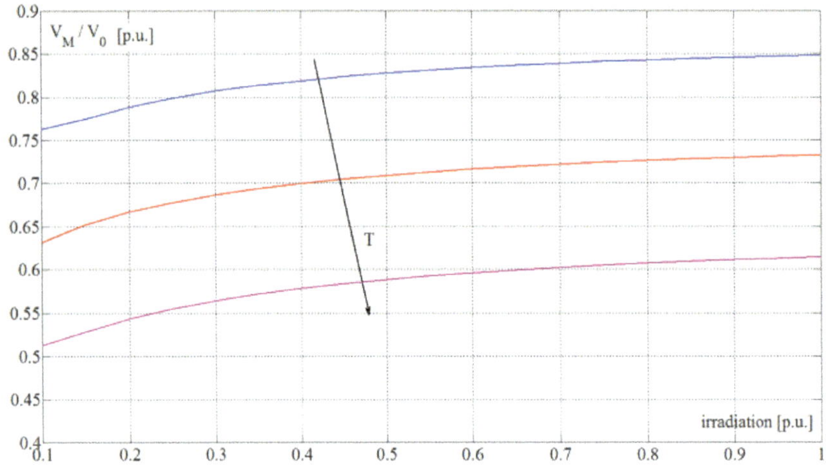

Figure 4. MPP voltage *vs.* irradiation for different temperatures.

The power-voltage characteristic has been chosen for the algorithm implementation, in order to use only one temperature sensor for the individuation of the maximum position. Using the parameters of the panel, reported in Table 2, the error committed in the MPP voltage estimation is reported in Figure 5.

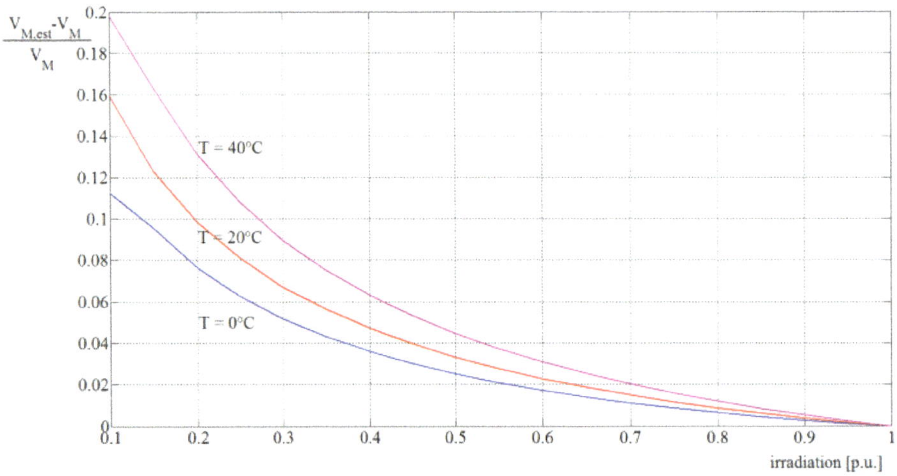

Figure 5. MPP voltage estimation error in p.u. *vs.* irradiation for different temperatures.

Figure 5 shows that when the irradiation is higher than 0.3 p.u. the MPP voltage estimation error is lower than 10%. Without a P&O algorithm, a MPP voltage estimation error implies a power loss with respect to the maximum power that can be generated. Figure 6 shows this power loss in per unit of the nominal array power as a function of the p.u. irradiation for different temperatures.

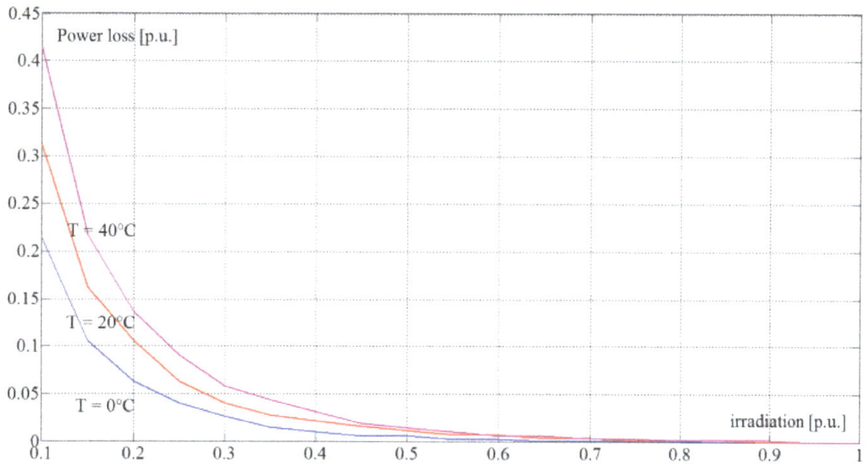

Figure 6. Power loss in p.u. *vs.* irradiation for different temperatures.

If the irradiation is higher than 0.3 p.u., the power losses are lower than 10% over a wide temperature variation range. Thus, the proposed algorithm has efficiency higher than 90% during quick transients. This error is reduced by the P&O action and is very low at a steady-state for a low value of ΔV_{min}.

6. Test Bench and Experimental Results

The MPPT technique proposed in this paper has been experimentally tested on a laboratory prototype of a PV system, shown in Figure 7. Figures 8 and 9 show also a close-up view of the DIIPPC used for experiments. The test bench permits the simulation of different operating conditions of solar irradiations and temperatures.

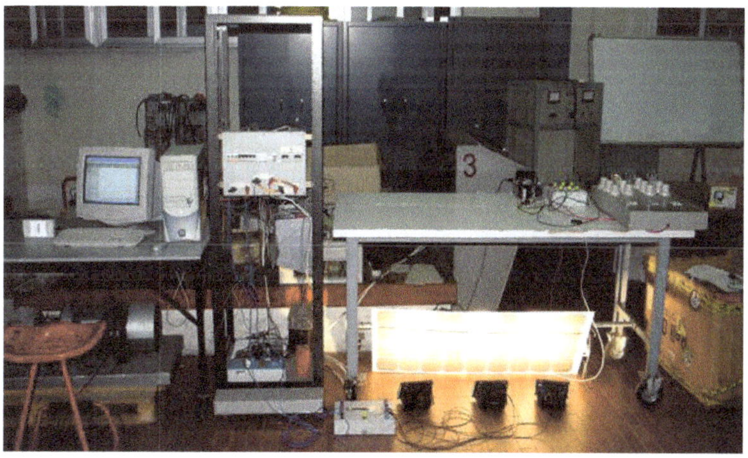

Figure 7. Photograph of the test bench used for the experimentation.

The ratings of the DIIPPC are reported in Table 3. The ripple of the current is below 1 A under rated conditions. Snubber circuits, with capacitance and resistance of 47 μF and 220 Ω, respectively, are used to protect the power Mosfets 170N10P, manufactured by IXYS (Milpitas, CA, USA). These switches have a nominal voltage of 100 V and a nominal current of 170 A. The capacitance value has been selected to store an amount of energy double than that stored in one input inductor under rated conditions. The resistance was selected on the basis of the energy discharged by the snubber capacitor at the rated voltage.

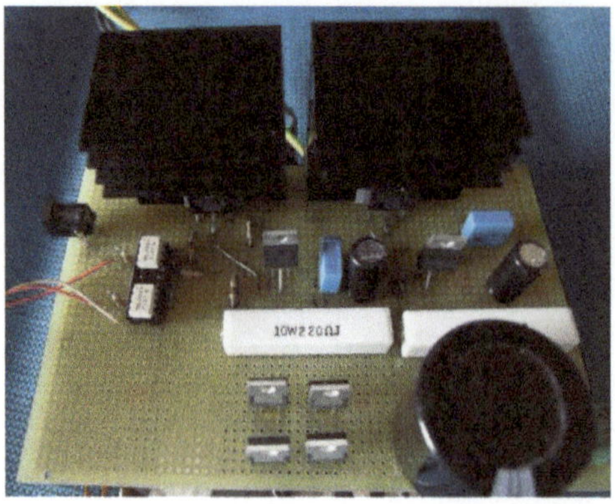

Figure 8. Close-up front view of the DIIPPC.

Table 3. Characteristics of the DC-DC converter.

Rated Power	Rated Input Voltage	Rated Output Voltage	Switching Frequency	Input Current Ripple	Drain to Source Mosfet Voltage	Inductances Value	Transformer Ratio
300 W	10 V	100 V	100 kHz	<3%	100 V	80 μH	4:12

The control of the converter is based on the PIC 18F4550 microcontroller by Microchip (Chandler, AZ, USA), which features a PWM module for the gate drivers of the switches. The microcontroller inputs are the array temperature measured by a thermocouple, the PV voltage and the PV current. The thermocouple is placed on the back of the PV array, whereas the current is measured using a shunt resistor.

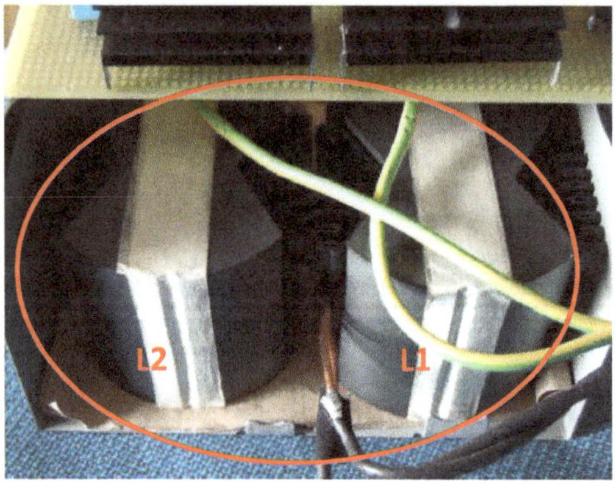

Figure 9. Close-up rear view of the DIIPPC.

The solar irradiation has been simulated by means of three halogen lamps, each one having a power of 500 W. Different solar conditions are simulated regulating the power irradiated by the lamps. The use of halogen lamps is not ideal for testing PV systems. Anyway, the use of artificial light is the only method for ensuring the same variable conditions for testing different MPPT algorithms. Moreover, being the same the irradiation change, also if it is not the ideal light for optimizing the panels, the results, in terms of comparison between different MPPTs can be considered correct. That's why, also in [30] the authors use halogen lamps for simulating controllable irradiation conditions.

In order to evidence the differences between the different algorithms, some step changes in the solar irradiation have been simulated. Also if this condition is not so frequent, in nature, the step change in solar irradiation represents a good test for comparing the dynamic response of different MPPT algorithms [27].

A 100 Ω resistor has been used as a load. The actual load seen by PV array is dependent on the duty-cycle of the converter according to the following relationship:

$$R_{\text{actual}} = R_{\text{load}} \frac{(1-D)^2 N_1^2}{N_2^2} \tag{17}$$

A block diagram of the system control is represented in Figure 10. The MPPT algorithm is implemented by changing the duty-cycle of the DIIPPC via a proportional-integral regulator. The constants of the regulator have been selected in order to obtain a good compromise between good dynamic of MPPT and reduced steady-state oscillations around the MPP. After several tests, have been chosen the values: $k_p = -0.03$ and $k_i = 1.5$. The two constants are negative because, for the electrical characteristic of the PV source, an increasing duty cycle of the converter implies a reduction of the output voltage of the PV system.

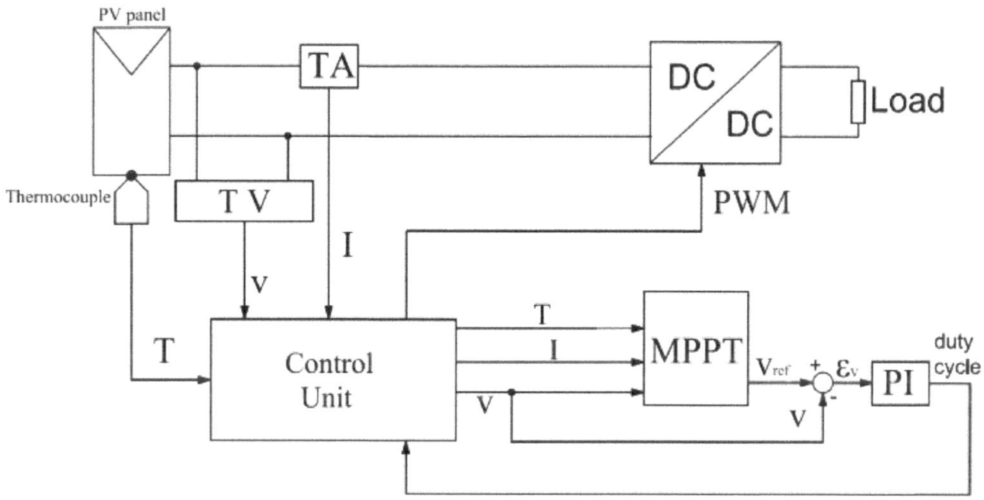

Figure 10. Block diagram of the control system.

The modified P&O algorithm proposed in this paper has been experimentally compared to a traditional P&O. In order to enhance the performances of the traditional P&O, some preliminary tests were carried out in order to find the best perturbation step, *i.e.* the step capable of extracting the maximum average power from the PV panel. To this aim, a sample irradiation pattern was created, consisting of 30 s of full irradiation, 30 s of 40% irradiation, and another 30 s of full irradiation. At a temperature of 23 °C, the best step for the traditional P&O is 0.3 V, whereas at 32 °C it is 0.7 V.

The performances of the modified P&O algorithm are strongly dependent on the function *f*, which defines the adaptive step of the algorithm. Experimental tests showed that the best results are obtained when *f* is a quadratic polynomial function, with a minimum perturbation step of 0.05 V at the maximum power point voltage under rated conditions. This result is in accordance with the consideration reported in Section 3.

The results of the comparison are presented in Figure 11 for a temperature of 23 °C. The left side of Figure 11 clearly shows that the proposed modified P&O algorithm is better than the standard P&O at steady-state, because the oscillations are smaller and the average power level is higher. During transients (starting at $t = 30$ s and $t = 60$ s) the proposed method shows the same reactivity of the standard P&O, as shown in the right side of Figure 11.

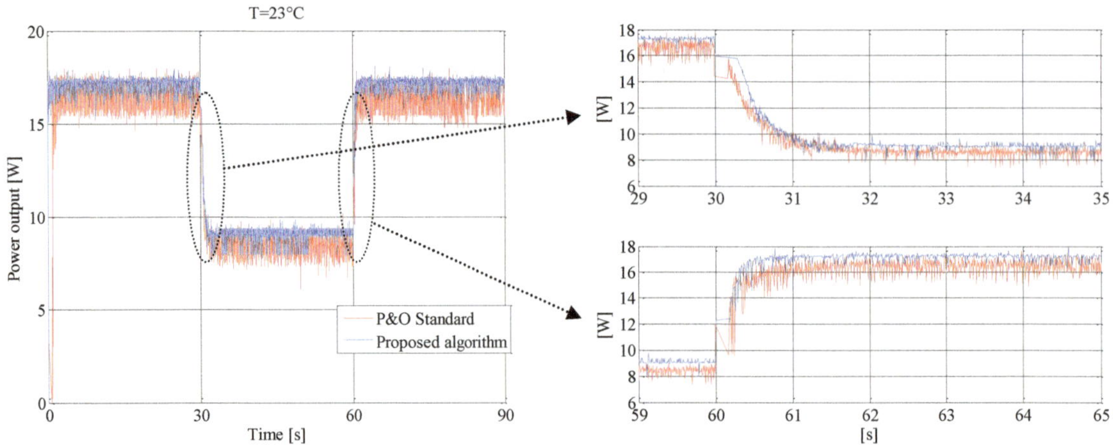

Figure 11. Comparison between the proposed algorithm and a standard P&O at 23 °C.

This means that the enhanced oscillations around the maximum of the proposed method is not paid with a slower dynamic capability, as it would happen for a traditional P&O algorithm. Indeed the perturbation step is higher when the irradiation changes suddenly and is lower when the operating point is close to the maximum power point. Moreover, the good performance at steady-state is also guaranteed by the smooth variation of the perturbation step, which is obtained by the quadratic shape of the function *f*. Figure 12 shows a similar behaviour for a temperature of 32 °C.

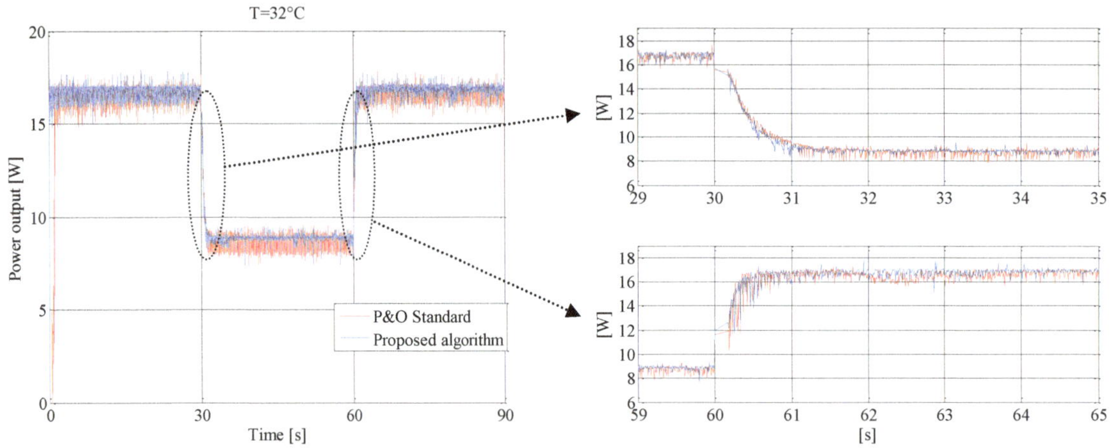

Figure 12. Comparison between the proposed algorithm and a standard P&O at 32 °C.

It is important to notice that at 32 °C the perturbation voltage step for the traditional P&O has been changed from 0.3 to 0.7 in order to assure the maximum average power and a good dynamic; this adjustment was not necessary for the proposed modified P&O technique, which shows the same reactivity for different temperatures. The proposed MPPT algorithm has also been compared to the IC algorithm. As for the case of the traditional P&O, the IC algorithm was tuned in order to define the best perturbation step. The comparison between the proposed algorithm and IC algorithm is reported in Figure 13 for a temperature of 23 °C.

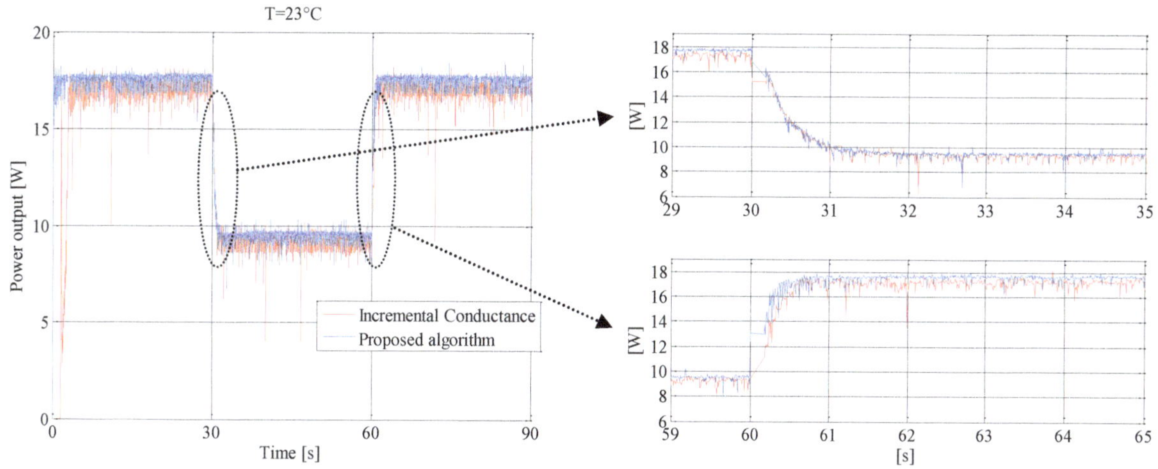

Figure 13. Comparison between the proposed algorithm and IC method at 23 °C.

The diagrams show that situation is similar to that of Figure 11, where the proposed algorithm shows smaller oscillations and higher average power level together with comparable dynamics during transients. Figure 14 shows that a similar behaviour is observed when the temperature is 32 °C. Finally Table 4 summarizes the average power experimentally obtained from the panel using the three algorithms under the solar irradiations of Figures 10–14.

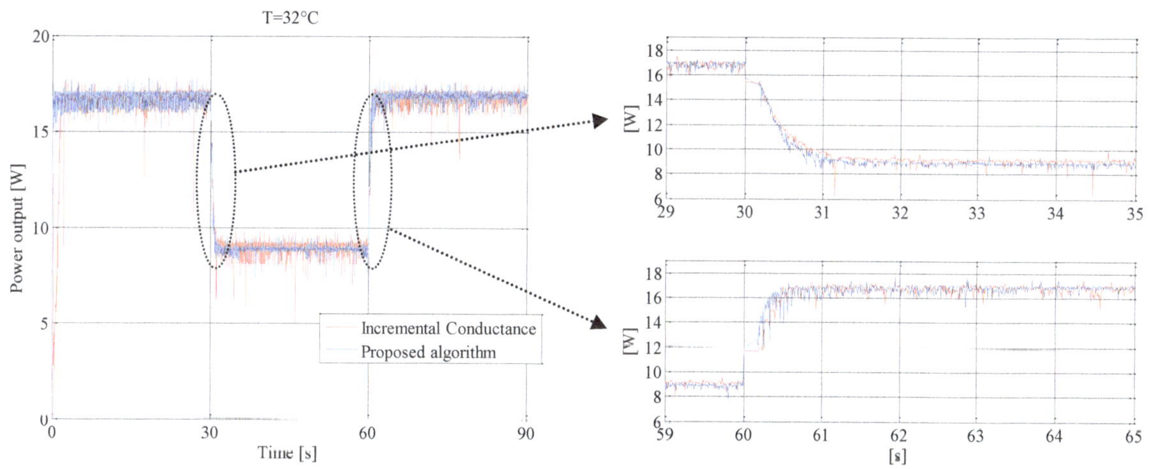

Figure 14. Comparison between the proposed algorithm and IC method at 32 °C.

Table 4. Average power obtained from different algorithms.

Temperature	Standard P&O	Incremental Conductance	Modified P&O
23 °C	14.1 W	14.1 W	15.0 W
32 °C	14.1 W	14.1 W	14.5 W

As it can be seen from the values of Table 4, the proposed algorithm reaches higher average power in comparison to traditional methods; the increment is 3.6% depending on the operating temperature.

An extensive study was carried out in order to evaluate the conversion efficiencies of the three algorithms considered in this paper. For each operating temperature, the solar irradiation was estimated by the short circuit current, which is directly proportional to the irradiation. For the nominal irradiation of 1000 W/m^2 and temperature of 25 °C, the short circuit current is about 300 A/m^2. For different temperatures, the short circuit current has been modified by the factor 0.2%/°C, as explained in Section 4. Therefore, the power reaching the panel is the product of the instantaneous irradiation times the area of the array A_c. If the power output of the panel is indicated as p(t) and the solar irradiation as $I(t)$, the efficiency of the panel, η, can be evaluated as follows:

$$\eta = \frac{\int_0^T p(t)\,dt}{A_c \int_0^T I(t)\,dt} \tag{18}$$

Table 5 gives the results of the comparison of the three algorithms for several operating temperatures in terms of panel efficiencies. From the examination of Tables 4 and 5, it is clear that the modified P&O algorithm proposed in this paper has efficiency higher than that of the algorithms traditionally described in the technical literature. The shape of the smoothing function, which modifies the perturbation step, plays a key role in improving the transitions after a change in the solar irradiation and reduces the steady-state oscillations around the MPP.

Table 5. Panel efficiencies obtained from different algorithms.

Temperature	23 °C	26 °C	28 °C	30 °C	32 °C	35 °C	36 °C	40 °C	44 °C
Standard P&O	8.9%	9.0%	9.2%	9.0%	8.9%	9.0%	9.0%	8.8%	8.4%
Incremental Conductance	9.1%	9.2%	8.8%	9.1%	8.9%	9.0%	9.1%	8.5%	8.5%
Modified P&O	9.5%	9.6%	9.3%	9.3%	9.2%	9.2%	9.2%	8.9%	8.6%

7. Additional Discussion on the MPP Estimator Error

As additional investigations a numerical analysis is carried out supposing that the actual MMP is estimated with a ±10% of error; that produces a ΔV_{min} according to Equation (15). In these simulations the irradiation starts from value of 1 p.u., at time of 0.5 sec. decrease to 0.45 p.u. and gets back to 1 p.u. at time of 1.0 sec. For the proposed modified P&O algorithm, Figure 15 shows the output power in the cases $V_{M,est} = 0.9\ V_M$ and $V_{M,est} = 1.1\ V_M$, together with the perfect estimation case $V_{M,est} = V_M$. Since the performance of a traditional P&O method is strongly influenced by the voltage perturbation step, several simulations were carried out. In Figure 16 the result of a traditional P&O algorithm with the settings $\Delta V = \Delta V_{min}$, $\Delta V = 2\ \Delta V_{min}$ and $\Delta V = 4\ \Delta V_{min}$, is reported in terms of output power. The simulations shows that, for the irradiation pattern selected, the traditional P&O method is unable to reach the MPP when $\Delta V = \Delta V_{min}$. Therefore the minimum step has to be at least $2{\cdot}\Delta V_{min}$, with consequent reduction of the steady-state performance of the MPP algorithm. Thus, a setting of $2{\cdot}\Delta V_{min}$ has been chosen to make comparison with the proposed method; for the proposed method the setting $V_{M,est} = 0.9\ V_M$ is chosen, corresponding to the worst case (minimum mean output power).

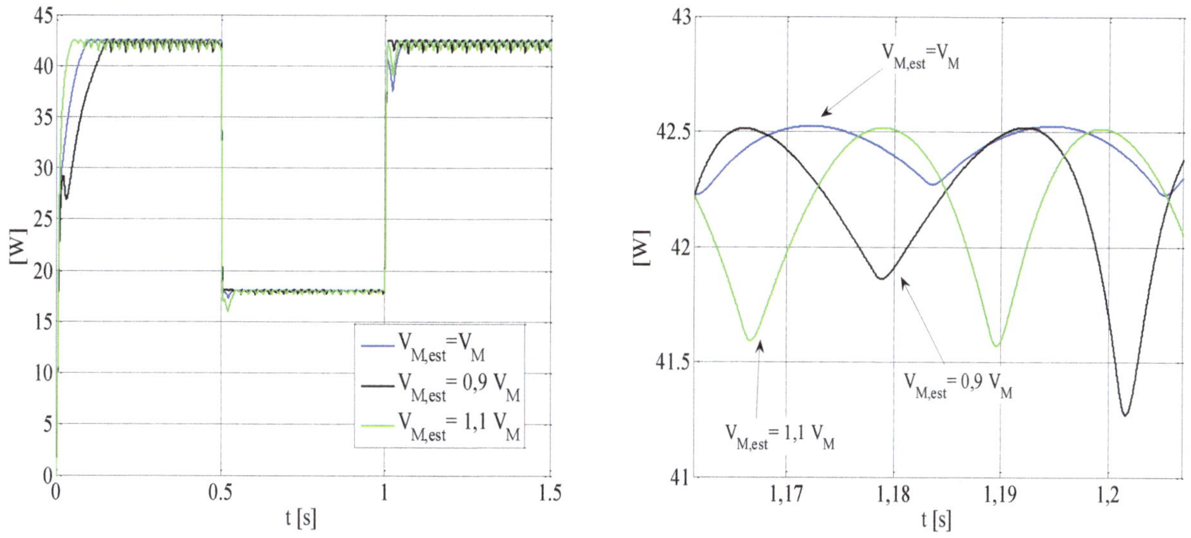

(a) (b)

Figure 15. Output power when using the proposed method for three different estimation errors; complete simulation (**a**) and steady-state analysis (**b**).

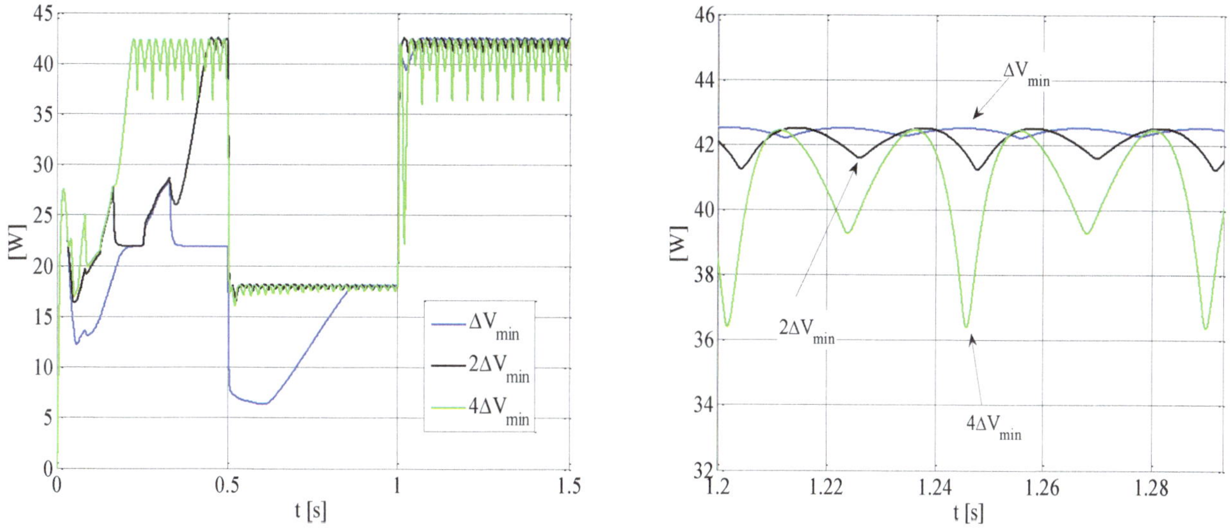

(a) (b)

Figure 16. Output power when using the standard P&O method for three different perturbation steps; complete simulation (**a**) and steady-state analysis (**b**).

In Figure 17 the best result of traditional P&O algorithm ($\Delta V = 2\ \Delta V_{\text{min}}$) is compared to the worst case of the proposed algorithm ($V_{\text{M,est}} = 0.9\ V_{\text{M}}$).

From Figure 17 it can be deduced that the proposed algorithm is faster than the traditional P&O method during transient and ensures smaller oscillation in steady-state conditions. As a consequence, the proposed method produces a higher mean power value, even when an estimation MPP error of 10% is introduced.

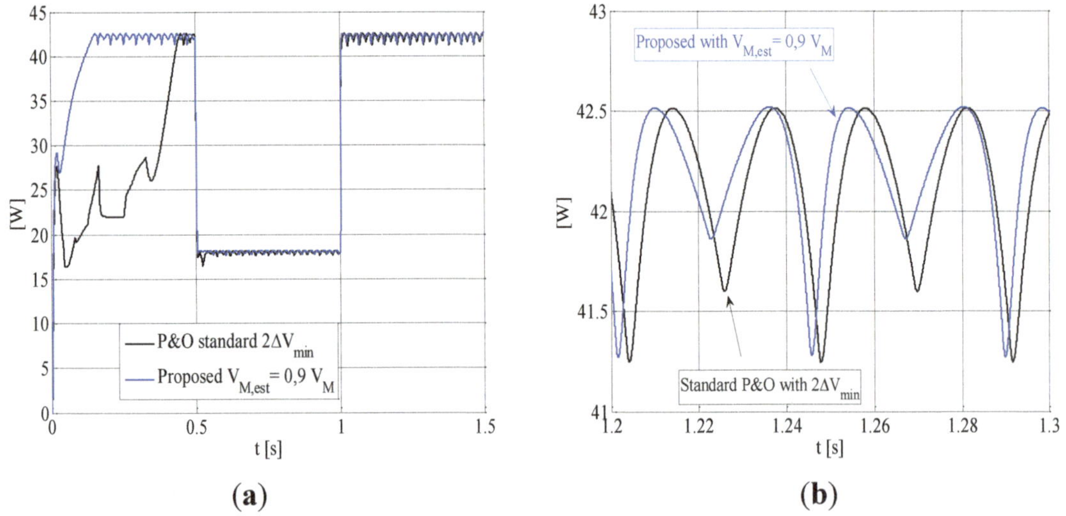

Figure 17. Comparison of output power between traditional P&O and proposed method with 10% estimation error; complete simulation (**a**) and steady-state analysis (**b**).

8. Conclusions

A variation of the classic maximum power point tracking perturb & observe method for photovoltaic power systems has been proposed, implemented and tested. The main limitation of perturb & observe algorithms is the compromise between fast dynamic response and steady-state stability. To improve the dynamic performance and ensure good stability, a criterion for choosing the optimal perturbation step and its coefficients has been discussed. The chosen function enables more flexibility of algorithm to the actual photovoltaic system conditions.

The modified perturb & observe algorithm has been implemented on a dual input inductor push-pull converter and experimentally compared to traditional P&O and IC techniques. For both these algorithms, the perturbation step was adjusted with tuning tests in order to extract the maximum average power from the PV panel and, consequently, penalizing the proposed system in the comparison.

The experimental results have demonstrated clearly that the proposed modified perturb & observe algorithm is better than the standard P&O and IC at steady-state and shows comparable reactivity during transients, even if no adjustment is operated on the perturbation voltage step for different operating conditions. This is due to the intrinsic adaptive characteristic of the algorithm: the perturbation step is higher when the irradiation changes suddenly and is lower when the operating point is close to the maximum power point. From the analysis of the results, it is evident that the proposed perturb & observe algorithm has higher efficiencies than the algorithms already proposed in the technical literature.

Author Contributions

The authors participated at the paper through a wide and balanced cooperation. They all worked at the mathematical model, the simulation set-up, the experiments and writing the manuscript. They have given equal contribution regarding all aspects of the paper.

Conflicts of Interest

The authors declare no conflict of interest.

References

1. Esram, T.; Chapman, P.L. Comparison of photovoltaic array maximum power point tracking techniques. *IEEE Trans. Energy Convers.* **2007**, *22*, 439–449.
2. Liu, X.; Lopes, L.A.C. An improved perturbation and observation maximum power point tracking algorithm for PV arrays. In Proceedings of the Power Electronics Specialists Conference PESC, Aachen, Germany, 20–25 June 2004; pp. 2005–2010.
3. Kim, I.S.; Kim, M.B.; Youn, M.J. New maximum power point tracker using sliding-mode observer for estimation of solar array current in the grid-connected photovoltaic system. *IEEE Trans. Ind. Electron.* **2006**, *53*, 1027–1035.
4. Cabal, C.; Alonso, C.; Cid-Pastor, A.; Estibals, B.; Seguier, L.; Leyva, R.; Schweitz, G.; Alzieu, J. Adaptive digital MPPT control for photovoltaic applications. In Proceedings of the ISIE, Vigo, Spain, 4–7 June 2007; pp. 2414–2419.
5. Sera, D.; Kerekes, T.; Teodorescu, R.; Blaabjerg, F. Improved MPPT method for rapidly changing environmental conditions. In Proceedings of the IEEE, Montreal, QC, Canada, 9–13 July 2006; pp. 1420–1425.
6. Femia, N.; Petrone, G.; Spagnuolo, G.; Vitelli, M. Optimization of perturb and observe maximum power point tracking method. *IEEE Trans. Power Electron.* **2005**, *20*, 963–973.
7. Wu, T.; Chang, C.; Chen, Y. A fuzzy-logic-controlled single stage converter for PV-powered lighting system applications. *IEEE Trans. Ind. Electron.* **2000**, *47*, 287–296.
8. Khaehintung, N.; Wiangtong, T.; Sirsuk, P. FPGA Implementation of MPPT using variable step-size P&O algorithm for PV applications. In Proceedings of the ISCIT, Bangkok, Thailand, 20 September–18 October 2006; pp. 212–215.
9. Femia, N.; Granozio, D.; Petrone, G.; Spagnuolo, G.; Vitelli, M. Predictive and adaptive MPPT perturb and observe method. *IEEE Trans. Aerosp. Electron. Syst.* **2007**, *43*, 934–950.
10. Piegari, L.; Rizzo, R. Adaptive perturb and observe algorithm for photovoltaic maximum power point tracking. *IET Renew. Power Gener.* **2010**, *4*, 317–328.
11. Abdelsalam, A.K.; Massoud, A.M.; Ahmed, S.; Enjeti, P.N. High-performance adaptive perturb and observe MPPT technique for photovoltaic-based microgrids. *IEEE Trans. Power Electron.* **2011**, *26*, 1010–1021.
12. Al-Amoudi, A.; Zhang, L. Optimal control of a grid-connected PV system for maximum power point tracking and unity power factor. In Proceedings of the 7th International Conference Power Electronics and Variable Speed Drives, London, UK, 21–23 September 1998; pp. 80–85.
13. Zhang, L.; Al-Amoudi, A.; Bai, Y. Real-time maximum power point tracking for grid-connected photovoltaic systems. In Proceedings of the 8th International Conference Power Electronics and Variable Speed Drives, London, UK, 18–19 September 2000; pp. 124–129.
14. Patel, H.; Agarwal, V. MPPT scheme for a PV-fed single-phase single stage grid-connected inverter operating in CCM with only one current sensor. *IEEE Trans. Energy Convers.* **2009**, *24*, 256–263.
15. Chiang, M.-L.; Hua, C.-C.; Lin, J.-R. Direct power control for distributed PV power system. In Proceedings of the Power Conversation Conference, Osaka, Japan, 2–5 April 2002; pp. 311–315.
16. Pandey, A.; Dasgupta, N.; Mukerjee, A.K. High-performance algorithms for drift avoidance and fast tracking in solar MPPT system. *IEEE Trans. Energy Convers.* **2008**, *23*, 681–689.

17. Xiao, W.; Dunford, W.G. A modified adaptive hill climbing MPPT method for photovoltaic power systems. In Proceedings of the IEEE 35th Annual PESC, Aachen, Germany, 20–25 June 2004; pp. 1957–1963.
18. Wolfs, P.J.; Tang, L. A single cell maximum power point tracking converter without a current sensor for high performance vehicle solar arrays. In Proceedings of the IEEE 36th PESC, Recife, Brazil, 16 June 2005; pp. 165–171.
19. Jain, S.; Agarwal, V. A new algorithm for rapid tracking of approximate maximum power point in photovoltaic systems. *IEEE Power Electron. Lett.* **2004**, *2*, 16–19.
20. D'Souza, N.S.; Lopes, L.A.C.; Liu, X. An intelligent maximum power point tracker using peak current control. In Proceedings of the IEEE 36th PESC, Recife, Brazil, 16 June 2005; pp. 172–177.
21. Manna, B.; Saha, N.; Palit, D. Optimization of solar power using servo mechanism by Perturb & Observe method. In Proceedings of the 2014 Annual IEEE India Conference (INDICON), Pune, India, 11–13 December 2014; pp. 1–6.
22. Femia, N.; Petrone, G.; Spagnuolo, G.; Vitelli, M. A technique for improving P&O MPPT performances of double-stage grid-connected photovoltaic systems. *IEEE Trans. Ind. Electron.* **2009**, *56*, 4473–4482.
23. Desai, H.P.; Patel, H.K. Maximum power point algorithm in PV generation: An overview. In Proceedings of the 7th International Conference Power Electron Drive System, Bangkok, Thailand, 27–30 November 2007; pp. 624–630.
24. De Aragao Filho, W.C.; Barbi, P.I. A comparison between two current-fed push-pull DC-DC converters-analysis, design and experimentation. In Proceedings of the 18th INTELEC '96, Boston, MA, USA, 6–10 October 1996; pp. 313–320.
25. Nousiainen, L.; Puukko, J.; Mäki, A.; Messo, T.; Huusari, J.; Jokipii, J.; Viinamäki, J.; Lobera, D.T.; Valkealahti, S.; Suntio, T. Photovoltaic generator as an input source for power electronic converters. *IEEE Trans. Power Electron.* **2013**, *28*, 3028–3038.
26. Suntio, T.; Leppäaho, J.; Huusari, J.; Nousiainen, L. Issues on solar-generator interfacing with current-fed MPP-tracking converters. *IEEE Trans. Power Electron.* **2010**, *25*, 2409–2419.
27. Sera, D.; Teodorescu, R.; Hantschel, J.; Knoll, M. Optimized maximum power point tracker for fast changing environmental conditions. In Proceedings of the IEEE International Symposium on Industrial Electronics, Cambridge, UK, 30 June 2008–2 July 2008; pp. 2401–2407.
28. Khanaki, R.; Radzi, M.A.M.; Marhaban, M.H. Comparison of ANN and P&O MPPT methods for PV Applications under Changing Solar Irradiation. In Proceedings of the 2013 IEEE Conference on Clean Energy and Technology (CEAT), Lankgkawi, Malesia, 18–20 November 2013; pp. 287–292.
29. Zorrilla-Casanova, J.; Piliougine, M.; Carretero, J.; Bernaola, P.; Carpena, P.; Mora-López, L.; Sidrach-de-Cardona, M. Analysis of dust losses in photovoltaic modules. In Proceedings of the World Renewable Energy Congress, Linköping, Sweden, 8–13 May 2011; pp. 2985–2992.
30. Galan, D.M.; Starzak, L.; Torzewicz, T.; Piotrowicz, M.; Maranda, W. Laboratory setup for investigation of MPPT algorithms of photovoltaic modules under non-uniform insolation. In Proceedings of the 20th International Conference Mixed Design of Integrated Circuits and Systems (MIXDES), Gdynia, Polonia, 20–22 June 2013; pp. 603–606.

China's Low-Carbon Scenario Analysis of CO_2 Mitigation Measures towards 2050 Using a Hybrid AIM/CGE Model

Wei Li [1,2,*], Hao Li [1] and Shuang Sun [1]

[1] Department of Economics and Management, North China Electric Power University, No. 619 Yonghua Street, Baoding 071003, Hebei, China; E-Mails: ncepulh@126.com (H.L.); ncepuss@126.com (S.S.)

[2] Philosophy and Social Science Research Base of Hebei Province, North China Electric Power University, Baoding 071003, Hebei, China

* Author to whom correspondence should be addressed; E-Mail: ncepulw@126.com

Academic Editor: Vincenzo Dovì

Abstract: China's emissions continue to rise rapidly in line with its mounting energy consumption, which puts considerable pressure on China to meet its emission reduction commitments. This paper assesses the impacts of CO_2 mitigation measures in China during the period from 2010 to 2050 by using a computable general equilibrium method, called AIM/CGE. Results show that renewable energy makes a critical difference in abating emissions during the period from 2010 to 2020. The scenarios with emission trading would drive more emission reductions, whereby the emission-cutting commitment for 2020 would be achieved and emission reductions in 2050 would be more than 57.90%. Meanwhile, the share of non-fossil energy increases significantly and would be more than doubled in 2050 compared with the BAU scenario. A carbon tax would result in a significant decline in emissions in the short term, but would have an adverse effect on economic growth and energy structure improvements. It is also observed that the integrated measures would not only substantially decrease the total emissions, but also improve the energy structure.

Keywords: emission mitigation; AIM/CGE model; renewable energy; emission trading; carbon taxes

---

1. Introduction

There is a growing awareness that global climate change caused by the increase in greenhouse gas emissions into the biosphere is one of the main challenges in the 21st century. In order to meet the challenge of global climate change and to achieve energy-environment-economy development, the building of a low carbon society is a priority in the national development strategies of many countries. The concept of a low carbon society concerns all aspects of the society, including economy, culture and life [1]. The European Union (EU) has officially declared that it would cut emissions by at least 80% by 2050 [2]. Meanwhile, the U.S. and Japan have also made it clear that substantial investment in clean energy development is expected to prompt the reduction of CO_2 emissions by 80% in 2050 compared with those in 2005 [3]. Developing countries represented by the BRICS nations (Brazil, Russia, India, China and South Africa) have also expressed a strong willingness to control emissions, while putting forward their own specific plans or programs for emission mitigation. According to China's decarbonisation roadmap, the emissions per unit of gross domestic product (GDP) are required to be reduced by 40% to 45% compared with those in 2005, and renewable energy consumption should account for 15% of the total by 2020 [4]. However, China's emissions are more likely to continue rising rapidly in line with its ongoing industrialization and urbanization, which will result in more energy consumption [5], hence, China will face tremendous challenges in achieving its mitigation targets provided effective measures are not taken. In response, the Chinese central government has made unremitting efforts towards exploring mitigation policies and other measures.

In the context of pursuing climate policy targets for 2020, a number of studies are underway to deeply probe different emission abatement policies and assess the corresponding compliance costs. Economists exceptionally value market-based climate management tools, to strive to minimize the economic costs in realizing the given emission reduction goals, and to provide dynamic incentives for long-term development of cheaper abatement technologies. The legislative package is designed to curb carbon emissions and to maintain low-cost abatement incentives in which carbon renewable, taxes and emission trading may all play a prominent role.

There is evidence that a broader set of energy and climate policies must be implemented to lower the abatement costs as much as possible. Research on imposing duties on carbon emissions has been done in the past few decades, mainly to identify the right tax rate that would help reduce carbon emissions with the lowest adverse impact on economic operations, and consequently alleviate climate change [6–8]. Emission trading is the most important regulatory instrument that makes emission allowances obligatory for CO_2 emissions from electricity generation as well as energy-intensive industries [9–11]. There is extensive literature on emission reduction targets, possible mitigation policies and costs in developed countries [12,13], but few studies are focused on the impacts of energy and climate policies in the context of emission mitigation in China. Undoubtedly, it is very necessary to quantitatively assess the potential results of the deployment of mitigation policies to reach the objective of impacting both the national economy and the energy mix to a large extent. Impact assessments of emission reduction commitments and non-fossil energy target plans for 2020, have been performed in [5,14].

Currently, renewable energy, considered as the main way to achieve energy savings and emission cuts, will see a substantial expansion over the next several decades. Subsequently, under a reasonable policy guidance for key economic sectors, technological changes will have great economic implications through improving energy efficiency whilst reducing the reliance on fossil fuels and cutting emission intensity [15]. Different from the carbon market in developed countries, an absolute cap on emissions needs to be set in China, which is eventually to be implemented by the different sectors and enterprises. Both carbon taxes and ET encourage enterprises to mitigate emissions in the production process by means of market forces. It is expected that timely introduction of a carbon tax from 2015 to 2020 which will achieve its impact through price measures will curb mounting emissions, while emission trading, a cap mechanism based on market competition, provides cost-effective and flexible environmental compliance for energy systems.

Increases of carbon emissions and their possible negative effects are major concerns in the imposition of any policy strategy [16]. Considering that climate policies including renewable energy plans, carbon taxes and emission trading are particularly helpful to combat climate change, it is very necessary to investigate the implications of alternative climate policies on economic and social development, so as to consequently provide policy recommendations for the state.

In recent years, as a popular policy simulation tool, computable general equilibrium (CGE) models have been widely employed in the analysis of climate policy, in particular, the assessment of long-term effects on economic consequences, sector productivity and energy systems [14,17,18]. This study aimed to provide an in-depth analysis of these mitigation measures, especially the impacts of renewable energy, ET and carbon taxes, on economic development, sectoral output, energy systems, *etc.* by establishing a new and revised hybrid static Asia-Pacific Integrated Model/computable general equilibrium (AIM/CGE) model.

The analysis is organized as follows: Section 2 briefly describes the AIM/CGE model and design scenarios. Section 3 presents the acquired simulation results, including the impacts of mitigation commitment on GDP, sectoral output, energy supply, and energy consumption by sectors and by types, energy intensity, carbon intensity and carbon prices and the impacts of different time levels to cancel the electricity price subsidies, increasing carbon taxes and limited emission trading. Finally, the results are summarized in brief, which also shows the useful policy implications of the study.

2. Methodology

2.1. AIM/CGE Model

The AIM/CGE model was developed by the National Institute for Environmental Studies (NIES), Japan. The AIM/CGE model is a recursive dynamic model, used to investigate the influence of climate policy at both the global [19,20] and country levels [21,22].

In this study, based on the "Standard CGE model", an improved AIM/CGE model is established to evaluate the impact of China's climate commitments, incorporating alternative energies and climate policies. The structure of China's AIM/CGE is approximately the same as that of the standard version [23]. It consists of four parts: production block, income block, expenditure block and market block. The present version of AIM/CGE is shown in Figure 1.

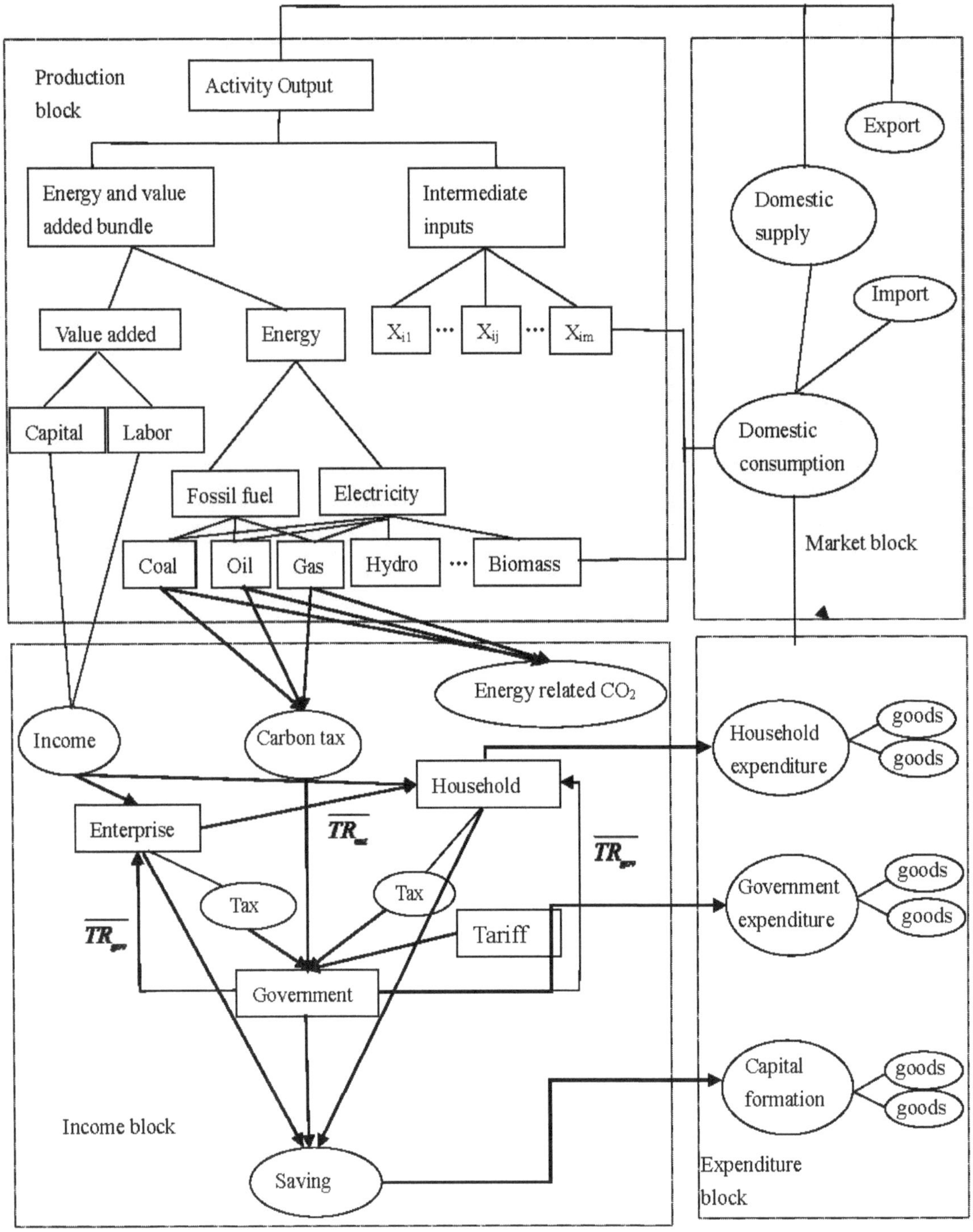

Figure 1. Overview of China's AIM/CGE model structure (for one region).

2.1.1. Production Block

In the standard AIM/CGE model, the default classification has 19 non-energy sectors and 19 energy sectors. In this model, in consideration of the study purpose and convenience, the sector classifications are disaggregated to 24 sectors, including 17 non-energy sectors and seven energy sectors. There are three basic types of inputs, including intermediate commodities, energy commodities and primary factors of capital and labor, regardless of land. It is assumed that the labor and capital are fully employed, perfectly competitive and freely mobile. The aggregation of intermediate commodities is in Leontief form while the composite of energy is a constant elasticity of substitution (CES) aggregation of fossil fuel (the bundle of six kinds of primary energy) and electricity. For the specific value of elasticity of substitution for inputs and outputs, see [5]. The following mathematical terms describes the production module at the lowest level:

$$F_i = A_i^f [\sum_{k=1}^{8} \alpha_{k,i} EC_{k,i}^{\frac{\sigma_{f,i}-1}{\sigma_{f,i}}}]^{\frac{\sigma_{f,i}}{\sigma_{f,i}-1}} \tag{1}$$

where eight kinds of energies, including coal, coke, crude oil, gasoline, kerosene, diesel, fuel oil, and natural gas, with the sum of share coefficients being 1. Moreover, produced emissions owing to the consumption of fossil energy would be calculated as follows:

$$EM_i = \sum_{k=1}^{8} c_k EC_{i,k} \tag{2}$$

At the second level, the CES function represents the substitution possibilities between the demand for aggregation of fossil fuels and electricity as well as the value added composite of capital and labor:

$$E_i = A_i^e [\alpha_{F,i} F_i^{\frac{\sigma_{E,i}-1}{\sigma_{E,i}}} + (1-\alpha_{F,i}) EL_i^{\frac{\sigma_{E,i}-1}{\sigma_{E,i}}}]^{\frac{\sigma_{E,i}}{\sigma_{E,i}-1}} \tag{3}$$

$$KL_i = A_i^{kl} [\alpha_{L,i} K_i^{\frac{\sigma_{KL,i}-1}{\sigma_{KL,i}}} + (1-\alpha_{L,i}) L_i^{\frac{\sigma_{KL,i}-1}{\sigma_{KL,i}}}]^{\frac{\sigma_{KL,i}}{\sigma_{KL,i}-1}} \tag{4}$$

Subsequently, the value-add and energy aggregates a value added energy bundle with the CES composite, subject to a constant elasticity of substitution σ_{KLE}:

$$KLE_i = A_i^{kle} [\alpha_{KLE,i} KL_i^{\frac{\sigma_{KLE,i}-1}{\sigma_{KLE,i}}} + (1-\alpha_{KLE,i}) E_i^{\frac{\sigma_{KLE,i}-1}{\sigma_{KLE,i}}}]^{\frac{\sigma_{KLE,i}}{\sigma_{KLE,i}-1}} \tag{5}$$

However, the composite of intermediate inputs trades off with the production function of Leontief forms with a fixed proportion of input factors, which is different from others:

$$M_i = \sum_{j-1}^{m} \alpha_{i,j} X_{i,j} \tag{6}$$

At the last level, the output comes from a CES nesting indicating the demand of sectors for value-added energy bundle and intermediate products:

$$Q_i = A_i^q [\alpha_{KLEM,i} KLE_i^{\frac{\sigma_{KLEM,i}-1}{\sigma_{KLEM,i}}} + (1-\alpha_{KLEM,i}) M_i^{\frac{\sigma_{KLEM,i}-1}{\sigma_{KLEM,i}}}]^{\frac{\sigma_{KLEM,i}}{\sigma_{KLEM,i}-1}} \tag{7}$$

2.1.2. Income Block

The income module consists of three sectors, namely, households, government and enterprises. The representative household acquires income from the rental of basic factors and the lump-sum transfer from the government and enterprises. The enterprises or corporations mainly derive their income from capital income and lump-sum transfers from the government. The government collects its earning mainly from various taxes, including carbon taxes and uses the income to purchase commodities for its consumption and for transfers to other institutions. This block mainly describes the income allocation and distribution. The equations below give an excellent description of the earnings that institutions acquire:

$$Y_h = \sum_{i=1}^{24} W_i \cdot L_i + r_i \cdot HK + \beta \cdot \overline{TR_{gov}} + \overline{TR_{ent}} \tag{8}$$

$$Y_{ent} = \sum_{i=1}^{24} r_i \cdot K_i + \gamma \cdot \overline{TR_{gov}} \tag{9}$$

$$Y_{gov} = \sum_{i}^{24} (\tau_i \cdot P_i \cdot Q_i) + \tau_r \cdot Y_h + \tau_{ent} \cdot Y_{ent} + TARIFF + TAX_{carbon} \tag{10}$$

It must be pointed out that institutions, including households and enterprises, should pay the carbon tax based on their CO_2 emissions, seen in Figure 1. The CO_2 emissions by each sector and household are estimated according to their fossil energy consumption. In the basic model version, the revenues are assumed to be received by the households. However, in this model, the revenues collected through carbon taxes are distributed to households and enterprises in the form of temporary tax cuts and transfer payments. Meanwhile, the returns are consistent with the calculation of confirmed taxable income rate, respectively. This study ignores the impact of carbon tax on households and focuses on the impact on sectors.

In this study, we follow the assumption that direct taxes and transfers to other domestic institutions are defined as fixed shares in the basic model version. In addition, the assumption of the balance of government financial revenues and expenditure is applied in this model. The taxation tariffs are also an important tax revenue and a control means for the central government. Products brought into one country are often subject to import taxes, or customs duties. Accordingly, imports are controlled by the adjustment of customs duty rates.

2.1.3. Expenditure Block

The expenditure block describes household expenditures, government spending and capital formation, whose final consumption is goods. The income of the representative household is spent on commodities or investments. It is natural that households intend to maximize commodity utility, which is subject to the constraints of income and prices. In terms of investment, household expenditures are assumed to grow linearly at the same pace as the GDP growth rate. Government spending and capital formation is assumed as a constant coefficient function. Both purchases and transfer of the government are introduced as an exogenous variable, so that the consuming quantity is fixed.

2.1.4. Market Block

The market block represents the process of commodity trading. Domestic commodities are disaggregated into domestic and export markets. Based on the Armington assumption, the aggregation of domestically produced and imported commodities with an imperfect substitute supplies the domestic market, derived on the assumption that domestic demanders minimize cost. The derived demands for imported commodities are met by international supplies that are infinitely elastic at given world prices. The import prices paid by domestic demanders also include import tariffs (at fixed *ad valorem* rates) and the cost of a fixed quantity of transactions services per import unit, covering the cost of moving the commodity from the border to the demander. Similarly, the derived demand for domestic output is met by domestic suppliers. The key difference with the standard version is that certified emission reductions (CERs) are regarded as a commodity in this model. We are working on the assumption that CERs are a commodity produced by enterprises. The amount of CERs owned by enterprises is determined by the emissions gap between the limits set by the government and the real value. The CERs can be traded in the carbon market. The carbon market demands are for a composite commodity made up of imports and domestic output. Developers of emission reduction projects act as a buyer transacting the entity of emissions reduction at a lower cost.

The AIM/CGE model treats the volume of all commodities as monetary units. Domestic energy sectors produce energy commodities and ensure supplies if not imported. The energy sectors in this study are classified into seven kinds. In the basic version, the share of imported and domestic consumption is determined by the ratio of the current price to the previous year's price. In this study, we solve the CES aggregation function to find the optimal value. The Equations (11)–(14) show how the optimal allocation of all commodities including energy between the imports and domestic supply:

$$QC_l = A_l^{am}[\alpha_{m,l} QM_l^{\sigma_{m,l}} + (1-\alpha_{m,l}) QAM_l^{\sigma_{m,l}}]^{\frac{1}{\sigma_{m,l}}} \tag{11}$$

$$\frac{PM_l}{PAM_l} = \frac{\alpha_{m,l}}{(1-\alpha_{m,l})}\left(\frac{QAM_l}{QM_l}\right)^{1-\sigma_{m,l}} \tag{12}$$

$$PC_l \cdot QC_l = PM_l \cdot QM_l + PAM_l \cdot QAM_l \tag{13}$$

$$PM_l = (1+\tau_{m,l}) \cdot \overline{pwm_l} \cdot \overline{EXR} \tag{14}$$

2.2. Dataset

The dataset used in this model includes the 2005 input-output table [24], energy balance table [25], and CO_2 emission factors of fossil fuels [26]. The year of 2005 is thus used as a base year. In this study, we use a single country social accounting matrix (SAM) which covers the typical sectors. The SAM in 2005 in this model is from the prolonged input-output table. In China's input output table, petroleum and natural gas are treated as one sector, the disaggregation of which in this study is done by using the consumption information of petroleum and natural gas among sectors in the energy balance table. The SAM describes physical volume transaction data as physical unit volumes. The volume of energy consumption is calculated based on the monetary volume transaction data and the energy price in the base year.

In the CGE model, energy consumption is commonly treated as a monetary term. However, in this paper, energy consumption is discussed in physical terms. Data from the energy balance table in 2010 is used to turn monetary terms into physical counterparts. SAM is formulated based on the energy balance table, whose balance is adjusted by a cross-entropy method [27]. Relevant coefficients include carbon emission factors and elasticity of substitution. It is assumed that the annual GDP growth rate is 7.5% from 2011 to 2015 in accordance with the policy guidance [28]. For the total GDP and growth rate after 2015, we refer to [29]. The current population growth rate in China is slow. If the total fertility rate is held at 1.8, the population in China will reach a peak of 1500 million persons around 2033, but then drop to 1383 million in 2050 [30]. Consequently, the annual growth rate of population can be roughly calculated (see Table 1).

Table 1. GDP and population growth.

Year	GDP		Population	
2005 (base year)	Billion USD	2903.75	Million persons	1307.56
2006–2010		9.57%		0.80%
2011–2015		7.50%		0.61%
2016–2020		6.15%		0.61%
2021–2025		5.82%		0.14%
2026–2030	Growth rate	5.50%	Growth rate	0.12%
2031–2035		5.20%		−0.08%
2036–2040		4.90%		−0.13%
2041–2045		4.61%		−0.33%
2046–2050		3.05%		−0.52%

The labor grows at exactly the same rate as population change. The capital stock is dynamically determined by the previous year's stock, capital formation and depreciation. The relevant data are based on [31]. The efficiency coefficients in the CES energy and value-added function, known as the total factor productivity (TFP), stand for the technology status. The efficiency coefficients change autonomously at 1.5% per year in agriculture, construction and the service sector. However, the efficiency coefficients in the industrial sector are determined by labor and capital stocks and will be calculated every year [14]. The direct subsidies to the production of renewable electricity are used to assess the impact of the renewable plan. The feed-in tariff and subsidy rate for the renewable during the study period are displayed in Table 2 [15].

Table 2. The feed-in tariff and subsidy rate for the renewable during the study period.

Renewable electricity	Renewable electricity feed-in tariff (Yuan/kwh)	Fossil fuel electricity sale price (Yuan/kwh)	Subsidy rate (%)
Wind	0.55	0.40	38
Solar	0.95	-	138
Biomass	0.75	-	88

Looking ahead, the non-fossil energy share is predicted to be at 11.4% in 2015 and 15% in 2020, the specific target of which is shown in Figure 2 [32]. The total CO_2 trading volume is likely to exceed 24 million tons this year, which will surge to about 227 million tons in 2015, nearly nine times higher than the current level. Furthermore, the implementation of subsidies for renewable electricity will occur between 2010 and 2020, and possibly be phased out in 2020, referring to the corresponding data [15].

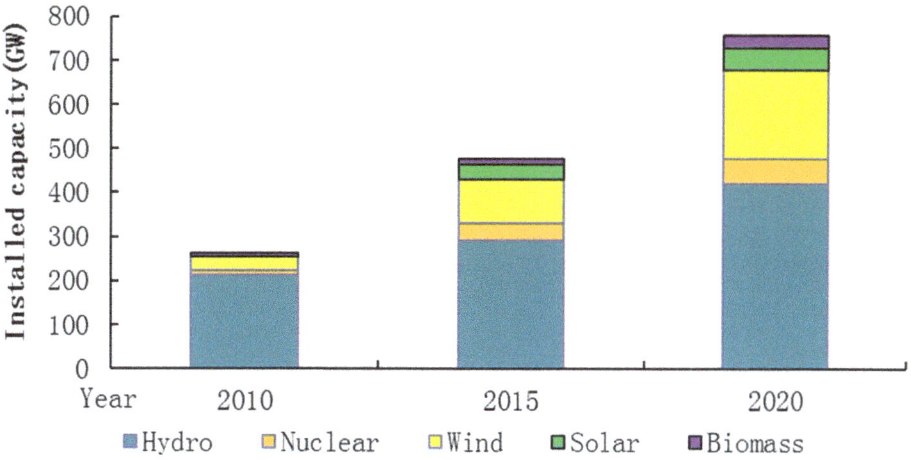

Figure 2. Published situation and targets for installed capacity of renewable energy towards 2015 and 2020 in China.

2.3. CO_2 Mitigation Scenarios

In this paper, a total of 11 scenarios are analyzed to investigate the impacts of three key policy measures. In the business as usual (BAU) scenario, it is assumed that the three CO_2 mitigation countermeasures or policies have not been introduced during the study phase. The GDP growth and population change follow the assumptions of Table 1. In addition, some variables including labor productivity changes, energy efficiency and capital productivity improvement are introduced to indicate technical progress and the corresponding values refer to [5]. The fossil fuel consumption from 2010 to 2050 is adopted from [33]. The other ten scenarios are designed to assess the impacts of policy combinations in fulfilling CO_2 mitigation commitments in contrast to the BAU scenario by using the AIM/CGE model. China has made an ambitious commitment to cut down its carbon dioxide emissions of GDP per unit from 40% to 45% by 2020 compared with that in 2005. Accordingly, all scenarios are divided into CM40 and CM50, which are roughly consistent with the national mitigation targets, *i.e.*, 40% and 50% reduction of the total CO_2 emissions. In this paper, all sectors should cut their emissions by 40% or 50%, respectively, compared with those in 2005. The CO_2 emissions come from the fossil energy consumption when enterprises produce their products. All sectors should take emission reduction into account and then decide the appropriate activity level. The CO_2 emissions will restrict the activity level as an exogenous variable. Since laws and regulations promoting the development of renewable energy were formulated in 2006, all scenarios have included the renewable energy, excluding the BAU scenario. The National Development and Reform Commission (2014) announced that a carbon market would probably be launched during 2016–2020 [34]. Considering the stage and uncertainty of this policy implementation, it is estimated that emission trading will be established throughout the country in 2016. A carbon tax is to be collected by 2020. When the national emission trading market is started in 2016, CERs transactions will be conducted freely on a global scale. In this study, carbon tax rate is designed as a constant, and always remains at 5 \$/t. The "E", "T", and "C" in the end of scenario names mean renewable energy, emission trading and carbon tax, respectively. For example, the emission trading policy will be taken into account in the CM40ET, CM40ECT, CM50ET and CM50ECT scenarios. Detailed information about all scenarios is shown in Table 3.

Table 3. CO_2 emission reduction measures in all scenarios.

Scenario	Reduction level (%)			Renewable energy	Emission trading	Carbon tax
	0%	40%	50%			
BAU	+	–	–	–	–	–
CM40	–	+	–	–	–	–
CM50	–	–	+	–	–	–
CM40E	–	+	–	+	–	–
CM50E	–	–	+	+	–	–
CM40ET	–	+	–	+	+	–
CM40EC	–	+	–	+	–	+
CM50ET	–	–	+	+	+	–
CM50EC	–	–	+	+	–	+
CM40ETC	–	+	–	+	+	+
CM50ETC	–	–	+	+	+	+

Note: "+" indicates the given scenario containing the appropriate policy and "–" indicates that corresponding measure is not considered.

3. Simulation Results

Our model allows us to assess the impacts of emission mitigation measures on macroeconomics, sector output, energy system and carbon prices. The results of each scenario are presented under the background of three alternative mitigation policies to support the exploration of China's future mitigation strategy.

3.1. Carbon Intensity

3.1.1. Carbon Intensity in This Model

China's sky-rocketing energy consumption causes high emissions. The current situation shows that emission mitigation is relatively difficult. Nevertheless, China's government has made an ambiguous commitment to lower emissions per unit of GDP from 40% to 45% compared with those in 2005. Over the past five years, the total CO_2 emissions have increased from 5170 million tons in 2005 to 8332 million tons in 2010, a 61.16% rise. In terms of emissions for GDP per unit from 2005 to 2010, they only dropped from 1.36 to 1.29 tons per 1000 US dollars (t/1000 USD, a 5.14% drop) meaning that there are enormous pressures in achieving reduction targets.

Results of CO_2 emission intensity during 2020–2050 are illustrated in Figure 3. It is shown that CO_2 emission intensity changes significantly after the deployment of renewable energy, a drop of 33.33% in CM40E and 36.43% in CM50E by 2020, respectively, compared with 1.36 t/1000 USD in 2005. On the basis of renewable energy, emission trading will reduce emission intensity by 43.31% in CM40ET and 46.53% in CM50ET by 2020, which obviously shows a high probability of achieving China's CO_2 emission mitigation target commitments made in Copenhagen.

The CO_2 emissions in the BAU scenario remain on a downward trend, but the value in 2050 is apparently higher than that in other scenarios. The carbon intensity is less than 0.70 t/1000 USD in the scenarios with renewable energy and emission trading, which means that the targets proposed in

Copenhagen are achievable. It is apparent that the carbon tax overwhelmingly contributes to lower carbon emissions when implemented. However, the effect of emission trading measures on emission intensity reduction obviously surpasses that of the carbon tax. Besides, the impacts of carbon tax on reducing emission intensity work in the short term, for example, the annual average rate of decrease of emission intensity is 1.1%–1.4% in CM40EC, CM40ETC, CM50EC, and CM50ETC during 2020–2030, while it is only 0.2%–0.5% during 2030–2050. In the short run, energy intensive enterprises have to reduce their production to fulfill the emission reductions in the face of the absolute increase in production costs because fixed costs remain constant and the introduction of new technology or switching to other production may be unrealistic. Therefore, the emission reduction and carbon intensity drop are obvious. However, in the long run, introduced low carbon technology and turning over to other sector makes the carbon tax less efficient. Moreover, energy efficient enterprises lack any motivation to reduce emissions because they pay lower carbon taxes. Hence, the carbon intensity would not sink rapidly in the long term.

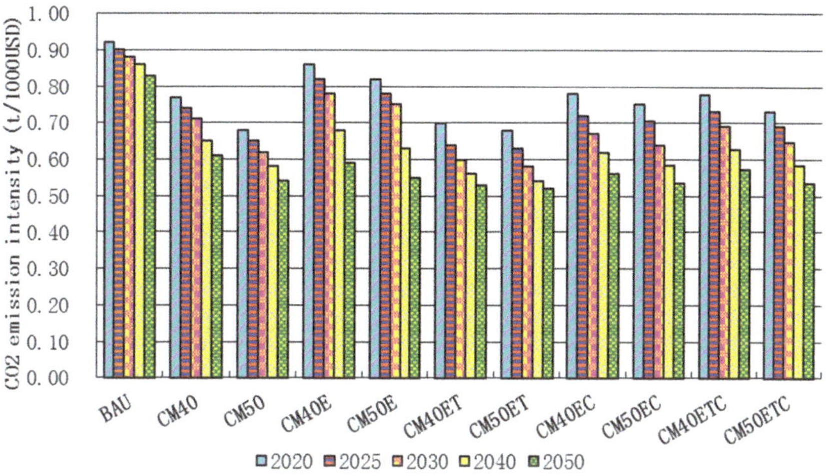

Figure 3. CO_2 emission intensity from 2020 to 2050.

In contrast, the emission intensity in scenarios without a carbon tax falls steadily during the research period. The estimated value of emission intensity falls to 0.54 t/1000 USD in CM40ET and 0.52 t/1000 USD in CM50ET, respectively, by 2050. The CM50ECT scenario sees roughly the same effect as CM50ET.

3.1.2. Renewable Energy Subsidies and Carbon Intensity

In the early stages, the development of renewable energy is faced with immense difficulties due to the high costs. Given that renewable energy plays a key role in the development of any national energy strategy, most advanced countries and regions have established appropriate capital subsidy mechanisms. As the largest developing country, China has adopted effective measures to support the development of renewable energy, mainly including renewable energy feed-in tariffs. The continuing financial support for promoting the development of renewable energy will unquestionably increase the government's financial burden. The tariff subsidies are estimated to be 16.7 billion dollars in 2015 and 25.3 billion dollars in 2020. Tariff subsidies are predictably eliminated based on the renewable energy development trend. Nevertheless, a sober assessment is needed to establish the best specific time to cancel subsidies.

In this study, diverse temporal patterns are used to assess the impacts when the government removes its tariff subsidies at different times. It is assumed that subsidies are removed in 2015, 2020 and 2025, respectively. It shows a marked comparison of carbon intensity in 2050 after removing the tariff subsidies at different times. Results reveal that the early cancellation of subsidies in 2015 curbs carbon intensity reduction in the future, while the abolition of subsidies in 2025 makes little difference. The following reasons may provide a general understanding of this phenomenon. Although the cost of renewable energy is higher than that of fossil energy, renewable energy is the major means to reduce the emissions before 2015. Over the ten years, the development cost of renewable energy has dropped by more than 50%, in particular, the cost of photovoltaic power generation has decreased by 15% every year. The renewable electricity will be cheaper after 2015, which could be acceptable for some enterprises. However, the thermal power generation cost in China is likely to rise because of the increasing coal price. Moreover, in order to achieve the mitigation goals, enterprises are bound to use renewable energy even if it is more expensive because it is still cheaper than the introduction of new technology.

By comparing the scenario of removing subsidies in 2020 with that in 2025, it is found that the margin of carbon intensity reduction in all reduction measures is less than 0.2 t/1000 USD. In other words, the result coincides with experts' predictions that after 2020, the removal of subsidies is relatively appropriate at least from the perspective of reducing carbon emissions.

3.1.3. Limited Emission Trading and Carbon Intensity

Free emission trading implies that no trade credit and scope will be captive. However, emission trading may be restricted to a certain level. China's enterprises will be in trouble in free emission trading with other countries, such as the EU, stemming from the gap of production technology. China's enterprises are always at a disadvantage in market competition due to their technical level. In order to protect domestic firms, limited emission trading may play an important role in accelerating the development of native industries. Therefore, it is significant to explore the impacts of trade constraints on activity level and mitigation.

After a comprehensive analysis and calculation, results convincingly indicate the inappropriate expectations. For protectionist reasons, 50% of the trade is merely restricted to domestic sources but this is unable to produce the desired effect, and worse still, it prevents sectors from raising the level of output. Protectionism prevents enthusiasm for innovation, but is conducive to production technology development. The decline in GDP reaches 1.3% compared with that in the free trade level, unwilling to address this aspect, which means sector activity will be undoubtedly affected, partly due to the technical gap. Surprisingly, carbon emission intensity is not affected much by the carbon taxes deployed in the CM40ETC and CM50ETC scenarios, compared with the average increase of 0.15 t/1000 USD in other scenarios. Obviously, the emission trading for limiting the amount of the transaction does not lower the carbon emission intensity as expected, and instead it causes some adverse effects on economic development and technological progress.

3.1.4. Increasing Carbon Tax and Carbon Intensity

In this study, the carbon tax rate is held at 5 \$/t for simplicity sake, however, it is changeable in the practical economic and social context based on the government policies. Therefore, in this section, the carbon tax rate is assumed to increase from 2 \$/t to 10 \$/t within a decade, after which it remains unchanged and consistent with the current policy direction. The output in the industrial sector accounts for a large proportion of the GDP. As expected, a carbon tax inevitably has an adverse impact on the activity level in the industrial sector. However, the effects of carbon tax can be significantly reduced if a lower tax rate is followed initially in accordance with the deployment of government norms. In addition, the increase of marginal abatement costs is less pronounced. A lower carbon tax serves as a guide in improving the production technology and energy efficiency to some extent. With the increase of tax rate, enterprises will shoulder the pressure of reducing the energy consumption per unit of output to meet the cost increases due to energy consumption.

Simulation results indicate CO_2 emission intensity changes with the carbon tax rate gradually increasing from 2 \$/t to 10 \$/t according to the hypothesis. Compared with fixed rates, incrementing the tax rate contributes to a more substantial reduction in carbon emissions and the consequential drop in carbon intensity. From a practical point of view, higher carbon tax indeed can greatly reduce carbon emissions instead of carbon intensity. Based on the simulation results, emission intensity will see a big decline in CM50EC and CM50ETC, reducing to 0.51 t/1000 USD in CM50ETC, a drop of 0.2 t/1000 USD, however, a higher carbon tax rate has a huge impact on China's GDP. At the carbon tax rate of 10 \$/t, the GDP loss in CM40EC and CM50EC reaches 3.47% and 5.06%, respectively.

3.2. CO_2 Mitigation and CO_2 Prices

CO_2 prices are changeable with the increase of imposed CO_2 emission constraints, as presented in Table 4. In the CM40 and CM50 scenarios, it is assumed that carbon emissions per unit of GDP are reduced by 40% and 50%. The CO_2 prices in 2050 are consistent with emissions reduction.

Table 4. Carbon prices and the corresponding CO_2 mitigation.

Scenarios	CO_2 mitigation (%)	Carbon prices USD/t-CO_2
CM40	40	51.6
CM50	50	67.9
CM40E	56.6	98.1
CM50E	59.6	118.2
CM40ET	61.0	143.4
CM40EC	61.8	92.8
CM50ET	58.8	138.4
CM50EC	60.7	100.8
CM40ETC	57.9	84.75
CM50ETC	60.5	110.2

This occurs in the scenarios with renewable energy policies, *i.e.*, the CM40E and CM50E, in which CO_2 prices are much higher than those in the CM40 and CM50 scenarios. The reason is that higher emission reduction constraints result in higher cost of CO_2 mitigation procedures and high carbon prices. The CO_2 price is the most significant driving force of emission reduction. Therefore, higher prices help to drive more CO_2 emission reduction.

Emission trading results in high CO_2 prices and encourages more emission reductions. Emission trading brings about the highest CO_2 prices in the CM40ET and CM50ET scenarios. In the CM40ET scenario, the carbon emission reduction level is higher than that in the CM50ET, but with nearly the same CO_2 price. The key reason is that the relationship between the CERs supply and its price roughly shows an inverted U-shaped curve. When the CERs price is lower, the price increase will result in more CERs supply. The price of CERs is unlikely to go up unboundedly. Too high a CERs price is uneconomic or unacceptable for enterprises. In this case, the enterprises prefer to pay the penalty for uncompleted emissions reduction rather than take the initiative in mitigating the emissions. When all enterprises achieve their optimal reductions based on profit function, they lose the desire to cut the emissions unless the CERs prices fall. If not, mitigation target could not be achieved. Therefore, only a depressed CERs price would drive more emissions reduction. Therefore, it is possible that a certain price corresponds to two different emission reduction scenarios.

It can be seen that the carbon price in CM40E and CM50E is higher than that in CM40 and CM50. Generally, more renewable energy should reduce the demand for carbon permits and reduce their price. However, renewable energy availability promotes the output growth in sectors compared with mandatory emission reductions. The emissions increase due to production expansion is more than the decrease due to the use of renewable energy. In fact, the CO_2 emissions in the scenarios with renewable energy are relatively higher than in the mandatory reduction scenario, which would drive the rise of carbon price.

3.3. Macroeconomic Impacts

Deployment of emission mitigation policies will inevitably exert a far-reaching influence on Chinese macroeconomic operation. Mandatory carbon emission reduction requirements are equivalent to imposing carbon emission constraints on the macroeconomics. As a result, economic operation costs will increase, and economic agents will make necessary adjustments to concepts, technologies, business models and even consumption patterns. Affected by this mechanism, the government as organizers and managers have assigned emission limits to the enterprises. This would limit how much CO_2 can be emitted in production activities. The enterprises will be penalized for excessive emissions, which may severely disrupt production. Therefore, enterprises are bound to adjust their production plans or introduce low-carbon technology to meet the emission targets, which would raise the product price as likely as not. Besides, the decrease of fossil energy consumption is probably inevitable due to the law of conservation of carbon. The energy consumers must spend time to understand and find alternative energy sources. In this study, we focus on the direct influence on the yield of all sectors. By definition, the GDP would be calculated according to the output level and fixed base year price. Reasonable mechanism of emission reduction is helpful for promoting technological innovation and industrial upgrading, and attracting more investments, which contributes to the improvement of the trading environment, particularly for developing countries.

GDP is deemed to be changeable in line with the implementation of mitigation policies, which is usually used as the indicator of evaluating the economic impacts. Table 5 shows the effect of mitigation policies on GDP growth under the corresponding targets. Compared with the BAU scenario, cumulative GDP loss in 2020 is relatively mild in the CM40 and CM50 scenarios, *i.e.*, 2.54% in CM40 and 3.52% in CM50, respectively. Higher GDP loss in CM50 shows that when a higher mitigation target is set, the loss of GDP will increase greatly due to the rapid expansion of marginal abatement costs.

However, economic costs are relatively lower in the short run under the emission reduction scenario, which is ascribed to constant progress in technology and the improvement of labor productivity. However, GDP in CM40 and CM50 shows a considerably different growth rate in 2050. The loss of cumulative GDP is 3.61% in CM40, while a larger loss of 6.91% occurs in CM50, indicating that mandatory emission reductions will have substantial influence in the long run. This is probably due to an unreasonable adjustment of industry structure caused by exceeding emission reductions that makes economic growth flabby.

Table 5. GDP losses due to CO_2 mitigation policies (%).

Year	Emission reduction scenarios									
	CM40	CM50	CM40E	CM50E	CM40ET	CM50ET	CM40EC	CM50EC	CM40ETC	CM50ETC
2010	−0.83	−1.53	−1.22	−1.47	−1.22	−1.47	−1.22	−1.47	−1.22	−1.47
2015	−1.53	−2.01	−1.65	−1.83	−1.65	−1.83	−1.65	−1.83	−1.65	−1.83
2020	−2.54	−3.52	−2.11	−2.12	−1.82	−1.92	−2.11	−2.12	−1.82	−1.92
2025	−2.73	−3.85	−2.23	−2.36	−1.87	−2.16	−2.48	−3.11	−2.19	−2.54
2030	−3.25	−4.50	−2.36	−2.58	−1.95	−2.35	−2.81	−3.54	−2.37	−2.82
2040	−3.42	−5.42	−2.49	−2.71	−2.11	−2.51	−2.96	−4.07	−2.52	−3.10
2050	−3.61	−6.91	−2.57	−2.79	−2.23	−2.60	−3.09	−4.85	−2.63	−3.41

Note: The value of GDP is estimated based on commodity prices in 2005.

Expansion of renewable energy use plays an absolutely critical role in cutting emissions during the period from 2010 to 2020. However, it would have an adverse effect on GDP growth. As shown in Table 5, the measurable cumulative reduction of GDP is about 1.53% in CM40E, and 2.79% in CM50E by 2050. This is mainly because developing renewable energy sources is more costly than fossil energy in the short run. Moreover, the government will invest much in developing it. All these efforts increase the economic operation costs. However, in the long run, the costs will fall because of technology maturity and renewable energy will become as cheap as fossil energy. The relatively abundant energy decreases the cost of activity level, and consequently, boosts the economy.

Emission trading, expected to be fully implemented in 2016, significantly promotes economic growth. Table 5 shows that from 2015, the GDP cumulative loss compared with the BAU scenario has been continuously declining as time goes on. However, from 2018 in CM40ET and 2033 in CM50ET, GDP starts to exceed the BAU scenario, ultimately up to 1.85% in CM40ET and 1.04% in CM50ET compared to the BAU scenario by 2050. This gives rise to accumulative total GDP growth of 1.44% in CM40ET and 0.98% in CM50ET. Compared with CM40E and CM50E, the GDP loss in CM40ET and CM50ET is also lower. Under emission trading policy, CERs would be regarded as a kind of good to trade. According to the assumption and the definition of GDP, the GDP would increase compared with the scenarios without emission trading and this situation will continue.

A carbon tax means increasing the revenue for the government, which is stored before serving as production subsidies in various sectors. A carbon tax adversely affects the GDP in the short-term, leading to a large gap of 0.9% in CM40EC compared with CM40E in 2025. Energy costs in all sectors would increase if a carbon tax is introduced, which lowers the activity level in sectors and suppresses GDP growth. The analysis shows that the trend of GDP growth rates with carbon tax in the scenarios of 50% CO_2 mitigation is similar to that in CM40EC, the difference is that the GDP loss in CM50EC is much greater. In addition, the negative impacts of carbon tax on GDP decline in the long run. Table 5 shows that in CM40EC, CM50EC, CM40ETC and CM50ETC, the GDP loss continues to decline from 2020 to 2050, which shows that the impact of a carbon tax weakens gradually. Carbon taxes have an adverse impact on industrial structure optimization, which impedes economic development. Especially in the short term, the impact of carbon tax on economic growth is heavy. In the long run, the enterprises have enough time to decide where the money is going and then manufacture and expand the activity level. As a result, the impact of carbon tax is weakened. Comparatively speaking, emission trading have a positive impact which constantly drives GDP growth. On the contrary, the impact of carbon tax is negative and is significant in the short term.

We prefer to lower the GDP loss as much as possible when carrying out mitigation measures. In 2050, a GDP loss above 3% is seen in CM40, CM50, CM40EC, CM50EC and CM50ETC. However, in the scenario of BAU, GDP growth is assumed to be 3.05% in 2050. The above scenarios probably result in economy stagnation or negative growth and are unreasonable. A lower GDP loss is enjoyed in CM40E, CM40ET and CM40ETC. For this reason the above three policy combinations are acceptable.

Table 6. Change of activity level in various scenarios in 2050 compared with BAU (%).

Sectors	Scenarios									
	CM40	CM50	CM40E	CM50E	CM40ET	CM50ET	CM40EC	CM50EC	CM40ETC	CM50ETC
Agriculture	0.85	3.89	1.02	4.21	1.25	4.31	1.06	4.36	1.05	4.19
Mineral mining	−0.42	−2.63	−0.38	−2.44	−0.34	−2.01	−0.42	−2.65	−0.39	−2.43
Foods and Tobacco	0.81	5.26	0.96	5.32	1.00	6.47	0.99	5.38	0.94	5.61
Textiles and clothing	0.4	2.23	0.51	2.65	0.67	3.21	0.52	2.45	0.53	2.64
Wood and furniture	0.35	1.98	0.43	2.33	0.65	2.86	0.47	2.45	0.48	2.41
Paper products	−0.08	−0.64	−0.06	−0.54	−0.05	−0.45	−0.08	−0.62	−0.07	−0.56
Chemical	−0.32	−1.85	−0.25	−1.52	−0.18	−1.34	−0.32	−1.79	−0.27	−1.63
Cement	−0.93	−3.05	−0.76	−2.54	−0.72	−2.43	−0.88	−3.11	−0.82	−2.78
Non-metallic Mineral Products	−0.30	−1.67	−0.22	−1.54	−0.16	−1.41	−0.28	−1.98	−0.25	−1.65
Iron and steel	−1.06	−4.66	−0.94	−3.59	−0.88	−3.31	−1.26	−4.58	−1.04	−4.04
Nonferrous products	−0.56	−2.06	−0.43	−1.68	−0.39	−1.56	−0.55	−2.31	−0.48	−1.90
Water production	0.22	0.86	0.32	1.02	0.36	1.25	0.39	1.08	0.32	1.05
Machinery	−0.35	−2.34	−0.32	−2.14	−0.26	−2.04	−0.35	−2.38	−0.32	−2.23
Other manufacturing	0.12	0.96	0.15	0.98	0.18	1.25	0.89	0.15	0.34	0.84
Construction	0.02	−0.05	0.05	−0.03	0.08	0.01	0.03	−0.05	0.05	−0.03
Transportation	−0.21	−0.65	−0.16	−0.47	−0.16	−0.36	−0.12	−0.42	−0.16	−0.48
Service sector	0.2	0.78	0.35	1.14	0.46	1.45	0.38	1.32	0.35	1.17

3.4. Non-Energy Sector Output Impacts

The integrated policy of promoting emission reductions has different impacts at the sectoral level. With the mitigation measures being introduced, energy intensive enterprises will be faced with increased costs and shrinking profits [35]. Energy costs increases have an important impact on activity levels. The activity level in energy-intensive sectors, such as steelmaking, and cement production shall suffer the most (Table 6). For instance, the output in the steel sector drops by 1.06% and 4.66% in the CM40 and CM50 scenarios, respectively, suffering severely from mandatory emission reductions, followed by the cement sector with a decline in output by 0.93% in CM40 and 3.05% in CM50, respectively. The decrease of activity level means that CO_2 emissions from production will fall consequently.

The carbon tax increases enterprises' production costs due to the raising prices of fossil energy, which suppresses the activity level. The feed-in subsidies make renewable energy competitive in the energy market. Renewable energy becomes an significant force to drive the fossil energy price drop. Consequently, the production costs in all sectors will decline due to the deployment of renewable energy sources. The decline of production costs helps increase production, in particular for the energy-intensive industries such as the power, steel, chemical and transport sectors because the expenditures on energy use account for a large part of their total costs [15,36]. Under an emission trading mechanism, enterprises don't need to follow the specified emissions limits strictly and pay heavy penalties for excessive emissions. The emissions gap would be offset by purchasing CERs at lower cost. Compared with the mandatory emission reduction scenario, the production costs will fall when emissions exceed the limits. The enterprises can rearrange their production schedules and expand the production to obtain greater profits, therefore, emission trading would improve the activity level.

Nevertheless, Table 6 shows that mitigation measures cannot significantly alleviate the decline in output level. First, the implementation of a carbon tax increases the production costs both in the short term and long term, in particular to energy intensive sectors. The increased costs with hamper the production expansion. Second, renewable energy is more favored than fossil energy. Due to government feed-in tariff subsidies, the price of renewable electricity will be reduced which will make renewable energy highly competitive. However, given the assumption of subsidy removal in 2020, this high competitiveness would decrease. In addition renewable energy is in short supply and cannot fully meet the demand. Consequently, the motivation for energy intensive sectors is limited. Third, under a carbon trading mechanism, enterprises need to purchase CERs in the carbon market which increases their total production costs. Although the enterprises are able to meet emission reduction quotas, they cannot expand their production unconditionally due to the emission limits and only do so if the government forecasts their output and allocates reasonable emissions. Unfortunately, the carbon trading would then become meaningless.

Energy efficient sectors, including the service, agriculture and food industry shall enjoy a drastic increase. The energy costs accounts for a small share of costs in energy efficient sectors and the impact of energy cost changes on output is limited. On the contrary, the introduction of new technology improves production efficiency. The most striking discovery is that the output levels in energy efficient sectors go up in pace with the rise in emission reductions For example, the output of service production increases the most under any scenario, in particular, in CM40ETC (3.11%) and CM50ETC (4.59%). Even so, only a small amount of the increase in emissions comes from service, foods and agriculture sectors.

3.5. Fossil Energy Supply and Final Consumption

The impacts of carbon abatement on fossil energy supply in all the scenarios in 2050 are illustrated in Figure 4. Despite the mitigation measures, coal supply by 2030 is expected to suffer a blow, which is due to the fact that coal combustion releases more CO_2 than other fossil energies.

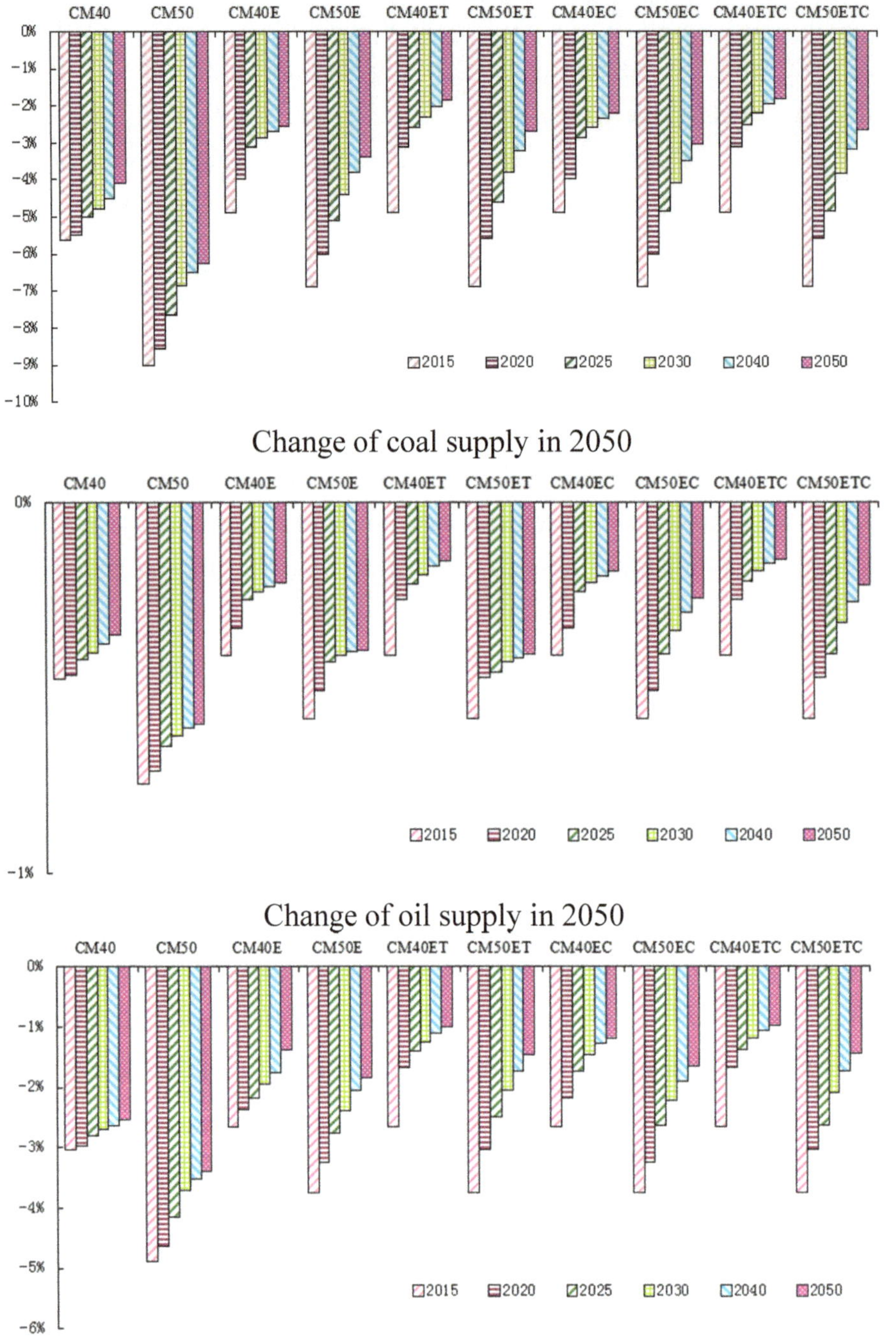

Change of gas supply in 2050

Figure 4. Change of fossil energy supply in the varied scenarios compared with BAU in 2050.

Coal supply will be 3158 million tonnes coal equivalent (Mtce) in CM40ETC and 2960 Mtce in CM50ETC, accounting for 45.2% and 43.2% respectively in the total energy supply in 2050, while oil production will be almost unaffected, *i.e.*, a slight drop of 0.78% in CM50, whose share will roughly remain unchanged in all scenarios. Limited resources and production capacities result in China's heavy dependence on imported counterparts. The share of oil consumption remains steady, but oil consumption will severely depend on imports in 2050, with a degree of dependence of up to 80%, which is bound to increase energy security risks. China's renewable energy development of wind, solar, and non-traditional biomass will replace fossil energy use. Meanwhile, renewable electricity generation increases significantly with the decreasing cost of renewable electricity. Renewable electricity in both the CM40 and CM50 scenarios sees relatively moderate increases, which indicates binding emission reduction commitments promote renewable electricity development to a certain extent (Figure 5). Current emission reduction policies result in significant growth in renewable electricity generation. Furthermore, the total and renewable electricity in general are higher under the CO_2 emission reduction target of 50% than that at 40%. It is strongly noticeable that renewable electricity increases by approximately six times in CM50ET and the total electricity generation reaches 2584.52 Mtce, including 555.80 Mtce in thermal power, 730.02 Mtce in hydro power, 848.70 Mtce in nuclear power and 450.00 Mtce in renewable power, respectively. In contrast to CM40ET, the renewable energy contribution grows faster than that in CM40EC and the same situation occurs in CM50EC and CM50ET. In addition, the contributions of nuclear and hydro power in the scenarios with emission trading increase significantly that with a carbon tax. In comparison, emission trading is more efficient than a carbon tax in term of promote renewable energy development.

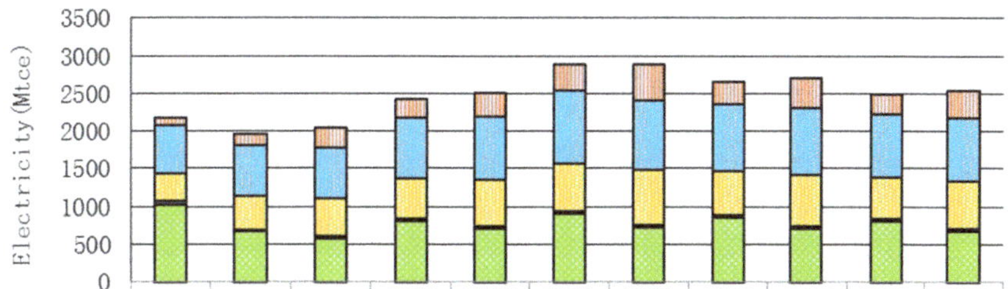

	BAU	CM40	CM50	CM40 E	CM50 E	CM40 ET	CM50 ET	CM40 EC	CM50 EC	CM40 ETC	CM50 ETC
Renewables	100	150	261	251	321	351	481	301	401	263	366
Nuclear	641	673	686	808	844	969	928	889	886	835	836
Hydro	368	441	497	530	611	636	733	583	672	547	628
Oil	7	5	4	6	5	7	6	6	5	6	5
Natural gas	36	23	20	28	25	33	30	31	27	29	25
Coal	1025	666	574	800	706	895	713	847	710	802	676

Coal, Natural gas, Oil, Hydro, Nuclear, Renewables

Figure 5. The total electricity generation mix through 2050.

Figures 6 and 7 show the total primary energy consumption by sector and by type in 2050. Primary energy consumption decreases dramatically in all countermeasure scenarios. Compared with the BAU scenario, primary energy consumption sees a decrease of 3.01% and 5.50% in CM40 and CM50 respectively. A substantial fall in primary energy consumption is observed in the scenario with emission reduction measures, whose trend is more obvious when a higher emission reduction proportion is adopted. For instance, in the scenario, the renewable and emission trading and the primary energy consumption drop significantly by 17.35% in CM40ET, 37.12% in CM40ETC, 27.13% in CM50ET, and 41.61% in CM50ETC, respectively. Meanwhile, the total primary energy consumption reduces to 4835.20 Mtce in CM40ETC and 4490 Mtce in CM50ETC, in which the share of non-fossil energy is 29.19% and 32.33%, respectively. The primary energy structure in the CM40 and CM50 scenarios is not significantly different from that in the BAU scenario, implying that compulsory emission reduction measures do not improve the energy mix. The oil and natural gas consumption remain rather stable in CM40 and CM50, while the consumption in all countermeasure scenarios declines by 10%–40%. Coal consumption experiences dramatic decreases with the amount being 44.45% and 46.67% of BAU's level in CM40ETC and CM50ETC. Cumulative energy savings is estimated to be 27,625 Mtce in CM50ETC and 15,301 Mtce in CM50ETC from 2020 to 2050.

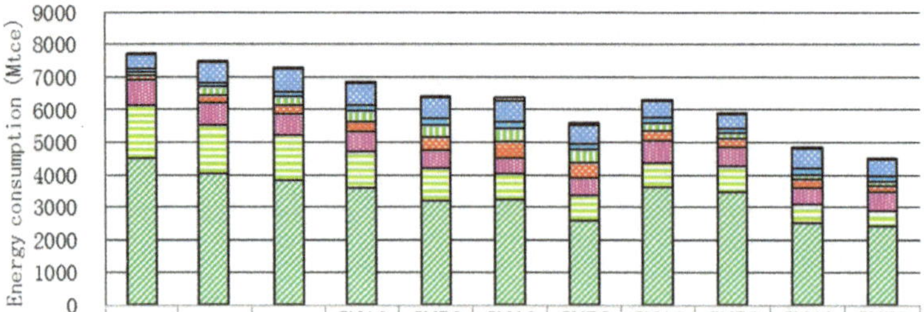

	BAU	CM40	CM50	CM40 E	CM50 E	CM40 ET	CM50 ET	CM40 EC	CM50 EC	CM40 ETC	CM50 ETC
Biomass	4	7	13	11	14	70	84	42	34	35	30
Hydro	430	645	710	645	613	643	553	526	421	600	500
Nuclear	95	114	137	190	228	228	194	167	133	190	150
Solar	112	258	283	336	370	407	386	216	173	150	130
Wind	150	225	259	300	390	507	482	328	262	260	230
Natural gas	750	675	638	600	540	450	540	648	583	500	550
Crude oil	1650	1485	1403	1122	1010	825	784	745	819	600	500
Coal	4500	4050	3825	3600	3204	3225	2580	3644	3462	2500	2400

Coal Crude oil Natural gas Wind Solar Nuclear Hydro Biomass

Figure 6. The total energy consumption by fuel types through 2050.

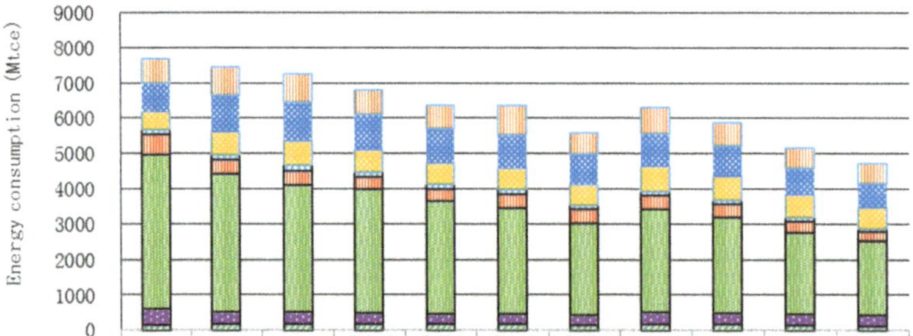

	BAU	CM40	CM50	CM40 E	CM50 E	CM40 ET	CM50 ET	CM40 EC	CM50 EC	CM40 ETC	CM50 ETC
Others	680	789	806	708	673	830	611	750	675	575	550
Residential	827	1060	1100	980	950	930	860	930	860	750	700
Service	508	650	700	630	620	620	600	700	680	650	600
Construction	130	130	150	146	135	135	132	135	132	120	100
Electricty	575	420	400	360	340	400	380	390	350	320	250
manufacture	4384	3900	3600	3500	3200	3000	2600	2900	2700	2300	2100
Mining	438	350	340	320	300	290	280	360	340	320	300
Agriculture	149	160	170	160	150	150	140	150	150	130	120

Agriculture Mining manufacture Electricty
Construction Service Residential Others

Figure 7. The total energy consumption by sector through 2050.

In the countermeasure scenarios, the energy use in energy-intensive sectors declines significantly. It is visible that manufacturing, whose energy consumption decreases by 31.5% in CM40ETC and 35.2% in CM50ETC, respectively, is the highest energy consuming sector which suffers the greatest decline. The energy consumption in the service sector increases from 18%–34%.

3.6. Sensitivity Analysis

The sensitivity analysis of this AIM/CGE model should be tested to identify whether the result will be heavily influenced by parameter assumption. In this study, a sensitivity analysis is conducted with respect to the elasticities between the production factors labor and capital (σ_{KL}), fossil energy and electricity (σ_E). We focus on the impacts of a 5% increase or drop from the base value on GDP and carbon intensity in CM40E and CM40ET.

Higher elasticity of σ_{KL} means higher flexibility for producers to choose the factors of labor and capital when fulfilling their production quotas. Consequently, production cost drops away relatively, which results in higher activity levels and GDP growth. Results show that GDP loss is reduced by 4.36% from the base value. However, the carbon intensity do not change dramatically.

The σ_E have a significant impact on carbon intensity and GDP. A 5% increase of σ_E leads to a 11.49% drop of carbon intensity and a 5% drop of σ_E leads to a 6.38% increase. Higher σ_E put pressure on energy and carbon prices. Energy consumers have greater choice and flexibility to choose which kind of energy to consume. Hence, lower cost energy will be chosen, *i.e.*, natural gas and electricity. The lower energy price stimulates production and drives GDP growth. Meanwhile, the carbon intensity will drop due to habit changes.

4. Conclusions

This study introduces the AIM/CGE model to investigate the effects of alternative mitigation policies, including renewable energy, emission trading and a carbon tax during 2010–2050. The simulation results show the effect on carbon intensity, activity level, GDP, energy system, and carbon prices in detail.

The emission reduction rate in the BAU scenario is approximately 28.68% in 2020 and 46.8% in 2050, based on the level of 2005. Absolute emission limits show negative impacts on future economic growth and activity level in energy intensive sectors. Activity level in non-energy intensive sectors increases slightly. In addition, the impact is larger when a higher emission reduction target level is set. In contrast, mitigation measures will curb CO_2 emissions, promote economic growth and optimize the energy mix. More importantly, energy-intensive sectors will improve energy utilization technology as far as possible to lower their costs. The cumulative energy consumption reduction is estimated to be 27,625 Mtce in CM50ETC from 2020 to 2050. Moreover, climate policies lead to lower emissions, reduced energy intensity, and improved energy and industrial mixes. It is necessary to take all the measures into account to reduce emissions.

Among the mitigation measures, renewable energy makes a critical difference in abating emissions during the period from 2010 to 2020, reducing emissions per unit of GDP by 33.33% in CM40E and 36.43% in CM50E. After 2020, according to the hypotheses, abolishing subsidies on renewable energy will slow down the pace of development of renewable energy, which means that energy efficiency improved by renewable energy will affect the accompanying emission mitigation. Nevertheless, the renewable introduced reduces about 36% of the emissions per GDP, but still fails to meet the goals. Therefore, it is necessary to pay more attention to the development of renewable energy and the determination of the appropriate subsidy rates for renewable energy.

The emission trading introduced in 2016 will bring about high carbon prices, *i.e.*, 143.4 USD/t in CM40ET, which encourages emissions reduction. Limited emission trading does not lower the carbon emission intensity as expected. Free emission trading must be integrated with mitigation measures to achieve the emission reduction targets. The government should try its best to plan the implementation of emission trading to avert unfair and unreasonable factors.

Carbon taxes would reduce carbon emissions, however, they does not work very well in the long term. Most noticeably, carbon taxes have a considerable negative impact on the sustainability, optimization of energy and economy system in the short term. It is necessary to give perfect rein to carbon tax by setting different tax rates.

Acknowledgments

This research was financially supported by the National Natural Science Foundation of China (NSFC) (71,471,061), Hebei Province Science and Technology Plan Project (14,457,694D) and the Fundamental Research Funds for the Central Universities (2014ZP11).

Author Contributions

Wei Li designed this research and gave the whole guidance; Hao Li largely wrote the research, including data simulation, result display and analysis; Shuang Sun. collected all the data. All authors read and approved the final manuscript.

Appendix: Definition of Symbols Used in this Model

The variable symbols and meanings used in this study are as follows:

F_i	the fossil energy aggregate in i th sector
A_i^f	the efficiency coefficient of fossil energy input in i th sector
$\alpha_{k,i}$	the share coefficient of k th energy in i th sector
$EC_{k,i}$	the input of k th energy in i th sector
$\sigma_{f,i}$	the elasticity of substitution of various fossil energy in i th sector
c_k	the CO_2 emissions factor of k th energy
EM_i	the CO_2 emissions in i th sector
EL_i	the renewable electricity input in i th sector
A_i^e	the efficiency coefficient of energy input in i th sector
$\alpha_{F,i}$	the share coefficient of fossil energy in i th sector
$\sigma_{E,i}$	the elasticity of substitution between fossil energy and electricity in i th sector
E_i	the energy aggregate in i th sector
K_i	the capital input in i th sector
L_i	the labor input in i th sector
A_i^{kl}	the efficiency coefficient of capital and labor input in i th sector
$\sigma_{KL,i}$	the elasticity of substitution between capital and labor in i th sector
$\alpha_{L,i}$	the share coefficient of labor input in i th sector
KL_i	the capital-labor aggregate in i th sector
$\sigma_{KLE,i}$	the elasticity of substitution between energy and capital-labor in i th sector
$\alpha_{KLE,i}$	the share coefficient of capital-labor input in i th sector
A_i^{kle}	the efficiency coefficient of capital-labor and energy input in i th sector
KLE_i	the capital-labor and energy aggregate in i th sector
$\alpha_{i,j}$	the direct consumption coefficients of j th intermediate inputs in i th sector

$X_{i,j}$	the jth intermediate inputs per unit of outputs in ith sector
M_i	the intermediate inputs aggregate in ith sector
A_i^q	the efficiency coefficient of output in ith sector
$\sigma_{KLEM,i}$	the elasticity of substitution between capital-labor-energy and intermediate inputs in ith sector
$\alpha_{KLEM,i}$	the share coefficient of capital-labor-energy input in ith sector
Q_i	the output in ith sector
Y_h	the gross income of household sector
W_i	the wage rate in ith sector
r_i	the rate of return on capital in ith sector
HK	the capital stock in household sector
K_i	the capital input in ith sector
$\overline{TR_{gov}}$	the transfer payment from government to household and enterprise
$\overline{TR_{eni}}$	the transfer payment from enterprise
β	the share of transfer payment from government to household
γ	the share of transfer payment from government to enterprise
Y_{gov}	the gross income of government
Y_{ent}	the gross income of enterprise
τ_i	the ad valorem rate of duty in ith sector
P_i	the product prices in ith sector
τ_h	the rate of individual income tax
τ_{ent}	the rate of corporate income tax
$TARIFF$	tariff revenue
TAX_{carbon}	the carbon tax revenue
QC_i	the aggregate supply in domestic market in ith sector
A_l^m	the scale parameter of the aggregate supply in domestic market of lth commodity
$\alpha_{m,l}$	the share parameter of imported commodities of lth commodity
$\sigma_{m,l}$	the elasticity of substitution between imported commodities and domestic production of lth commodity
QM_l	the imports of lth commodity
QAM_l	the domestic production of lth commodity
PC_l	the consumer price of lth commodity
PM_l	the imports price of lth commodity
PAM_l	domestic production price of lth commodity
$\tau_{m,l}$	the import tariffs of lth commodity
$\overline{pwm_l}$	the world price of lth commodity
$\overline{EXR}$	the exchange rate

Table A1. The description of sector classifications used in this model.

Non-Energy Sectors		Energy Sectors	
1	Agriculture	1	Coal mining
2	Mineral mining	2	Extractionof petroleum
3	Foods and Tobacco	3	Natural gas
4	textiles and clothing	4	Coking
5	Wood and furniture	5	Petroleum and nuclear fuel processing
6	Paper products	6	Gas production and supply
7	Chemical	7	Electricity
8	Cement		
9	Non-metallic mineral products		
10	Iron and steel		
11	Nonferrous products		
12	Water production		
13	Machinery		
14	Other manufacturing		
15	Construction		
16	Transportation		
17	Service sectors		

Table A2. The estimated feed-in subsidies for the renewable during the study period (Billion USD).

Year	CM40E			CM50E			CM40ET			CM50ET		
	Wind	Solar	Biomass	Wind	Solar	Biomass	Wind	Solar	Biomass	Wind	Solar	Biomass
2010	0.75	0.02	0.13	0.83	0.17	0.22	0.75	0.02	0.13	0.83	0.17	0.22
2015	2.42	0.84	0.31	2.78	1.25	0.56	2.42	0.84	0.31	2.78	1.25	0.56
2020	4.89	1.21	0.74	5.71	1.88	1.17	5.70	2.47	1.60	6.30	2.98	2.16

Table A3. The estimated revenues of carbon tax during 2020–2050 (Billion USD).

Year	CM40EC	CM50EC	CM40ETC	CM50ETC
2020	8.39	8.06	6.84	6.59
2025	9.32	8.98	7.27	7.14
2030	10.07	9.73	7.61	7.52
2035	10.79	10.34	7.89	7.85
2040	11.45	11.04	8.36	8.14
2045	11.93	11.42	8.57	8.43
2050	12.37	11.88	8.70	8.32

Table A4. The government receipts in all scenarios during 2010–2050 (Billion USD).

Year	CM40	CM50	CM40E	CM50E	CM40ET	CM50ET	CM40EC	CM50EC	CM40ETC	CM50ETC
2010	1336.03	1315.59	1299.54	1280.43	1264.81	1246.22	1231.02	1212.92	1198.12	1180.51
2015	2229.65	2184.83	2148.78	2109.46	2074.65	2036.69	2003.08	1966.43	1933.98	1898.59
2020	3091.58	2982.76	2919.82	2857.92	2805.91	2752.03	2693.96	2636.85	2588.86	2539.16
2025	3766.12	3621.13	3540.38	3456.82	3392.18	3318.91	3236.60	3135.94	3067.26	2989.36
2030	4922.17	4700.67	4589.74	4471.32	4384.13	4281.11	4160.81	4013.51	3918.39	3807.89
2040	8055.88	7619.25	7429.53	7228.19	7075.68	6898.08	6693.89	6421.45	6259.63	6065.58
2050	11,091.77	10,325.33	10,059.97	9779.29	9561.21	9312.62	9024.86	8587.16	8361.31	8076.19

Table A5. Change of coal supply in 2050.

Year	CM40	CM50	CM40E	CM50E	CM40ET	CM50ET	CM40EC	CM50EC	CM40ETC	CM50ETC
2015	−5.62%	−9.00%	−4.90%	−6.90%	−4.90%	−6.90%	−4.90%	−6.90%	−4.90%	−3.74%
2020	−5.50%	−8.56%	−4.00%	−6.00%	−3.10%	−5.60%	−4.00%	−6.00%	−3.10%	−3.04%
2025	−5.00%	−7.66%	−3.10%	−5.10%	−2.60%	−4.60%	−2.85%	−3.82%	−2.53%	−2.63%
2030	−4.80%	−6.85%	−2.85%	−4.40%	−2.30%	−3.80%	−2.58%	−3.30%	−2.20%	−2.09%
2040	−4.50%	−6.50%	−2.68%	−3.80%	−2.03%	−3.20%	−2.36%	−2.88%	−1.96%	−1.73%
2050	−4.10%	−6.25%	−2.55%	−3.40%	−1.85%	−2.70%	−2.20%	−2.61%	−1.82%	−1.44%

Table A6. Change of oill supply in 2050.

Year	CM40	CM50	CM40E	CM50E	CM40ET	CM50ET	CM40EC	CM50EC	CM40ETC	CM50ETC
2015	−0.47%	−0.76%	−0.41%	−0.58%	−0.41%	−0.58%	−0.41%	−0.58%	−0.41%	−0.58%
2020	−0.46%	−0.72%	−0.34%	−0.51%	−0.26%	−0.47%	−0.34%	−0.51%	−0.26%	−0.47%
2025	−0.42%	−0.65%	−0.26%	−0.43%	−0.22%	−0.39%	−0.24%	−0.41%	−0.21%	−0.41%
2030	−0.40%	−0.58%	−0.24%	−0.37%	−0.19%	−0.32%	−0.22%	−0.35%	−0.19%	−0.33%
2040	−0.38%	−0.55%	−0.23%	−0.32%	−0.17%	−0.27%	−0.20%	−0.29%	−0.17%	−0.27%
2050	−0.35%	−0.53%	−0.21%	−0.29%	−0.16%	−0.23%	−0.19%	−0.26%	−0.15%	−0.22%

Table A7. Change of gas supply in 2050.

Year	CM40	CM50	CM40E	CM50E	CM40ET	CM50ET	CM40EC	CM50EC	CM40ETC	CM50ETC
2015	−3.05%	−4.88%	−2.66%	−3.74%	−2.66%	−3.74%	−2.66%	−3.74%	−2.66%	−3.74%
2020	−2.98%	−4.64%	−2.17%	−3.25%	−1.68%	−3.04%	−2.17%	−3.25%	−1.68%	−3.04%
2025	−2.71%	−4.15%	−1.68%	−2.76%	−1.41%	−2.49%	−1.55%	−2.63%	−1.37%	−2.63%
2030	−2.60%	−3.71%	−1.55%	−2.39%	−1.25%	−2.06%	−1.40%	−2.22%	−1.19%	−2.09%
2040	−2.44%	−3.52%	−1.45%	−2.06%	−1.10%	−1.73%	−1.28%	−1.90%	−1.06%	−1.73%
2050	−2.22%	−3.39%	−1.38%	−1.84%	−1.00%	−1.46%	−1.19%	−1.65%	−0.99%	−1.44%

Table A8. CO_2 emission intensity from 2020 to 2050.

Year	BAU	CM40	CM50	CM40E	CM50E	CM40ET	CM50ET	CM40EC	CM50EC	CM40ETC	CM50ETC
2020	0.92	0.77	0.68	0.86	0.82	0.70	0.68	0.78	0.75	0.78	0.73
2025	0.90	0.74	0.65	0.82	0.78	0.64	0.63	0.72	0.71	0.73	0.69
2030	0.88	0.71	0.62	0.78	0.75	0.60	0.58	0.67	0.64	0.69	0.65
2040	0.86	0.65	0.58	0.68	0.63	0.56	0.54	0.62	0.59	0.63	0.58
2050	0.83	0.61	0.54	0.59	0.55	0.53	0.52	0.56	0.54	0.57	0.54

Conflicts of Interest

The authors declare no conflict of interest.

References

1. Yuan, H.; Zhou, P.; Zhou, D. What is low-carbon development? A conceptual analysis. *Energy Procedia* **2011**, *5*, 1706–1712.
2. European Union (EU). Roadmap for a Low-Carbon Economy in 2050. Available online: http://ec.europa.eu/clima/consultations/articles/0005_en.htm (accessed on 8 December 2010).

3. Japan-U.S. Summit Meeting. Japan-U.S. Joint Message on Climate Change Negotiations. Available online: http://www.mofa.go.jp/region/n-america/us/pv0911/summit.html (accessed on 13 October 2010).
4. National Development and Reform Commission (NDRC). National Climate Change Plan (2014–2020). Available online: http://www.sdpc.gov.cn/zcfb/zcfbtz/201411/t20141104_642612.html (accessed on 19 September 2014).
5. Dai, H.; Masui, T.; Matsuoka, Y. Assessment of China's climate commitment and non-fossil energy plan towards 2020 using hybrid AIM/CGE model. *Energy Policy* **2011**, *39*, 2875–2887.
6. Bandyopadhyay, G.; Bagheri, F.; Mann, M. Reduction of fossil fuel emissions in the USA: A holistic approach towards policy formulation. *Energy Policy* **2007**, *35*, 950–965.
7. Hammar, H.; Sjöström, M. Accounting for behavioral effects of increases in the carbon dioxide (CO_2) tax in revenue estimation in Sweden. *Energy Policy* **2011**, *39*, 6672–6676.
8. Lin, B.; Li, X. The effect of carbon tax on per capita CO_2 emissions. *Energy Policy* **2011**, *39*, 5137–5146.
9. Szabó, L.; Hidalgo, I.; Ciscar, J.C.; Soria, A. CO_2 emission trading within the European Union and Annex B countries: The cement industry case. *Energy Policy* **2006**, *34*, 72–87.
10. Lee, C.F.; Lin, S.J.; Lewis, C. Analysis of the impacts of combining carbon taxation and emission trading on different industry sectors. *Energy Policy* **2008**, *36*, 722–729.
11. Sugino, M.; Arimura, T.H.; Morgenstern, R.D. The effects of alternative carbon mitigation policies on Japanese industries. *Energy Policy* **2013**, *62*, 1254–1267.
12. Deetman, S.; Hof, A.F.; Pfluger, B.; van Vuuren, D.P.; Girod, B.; van Ruijven, B.J. Deep greenhouse gas emission reductions in Europe: Exploring different options. *Energy Policy* **2013**, *55*, 152–164.
13. Hübler, M.; Löschel, A. The EU Decarbonisation Roadmap 2050—What way to walk? *Energy Policy* **2013**, *55*, 190–207.
14. Thepkhun, P.; Limmeechokchai, B.; Fujimori, S.; Masui, T.; Shrestha, R.M. Thailand's Low-Carbon Scenario 2050: The AIM/CGE analyses of CO_2 mitigation measures. *Energy Policy* **2013**, *62*, 561–572.
15. Qi, T.; Zhang, X.; Karplus, V.J. The energy and CO_2 emissions impact of renewable energy development in China. *Energy Policy* **2014**, *68*, 60–69.
16. Wang, Y.; Liang, S. Carbon dioxide mitigation target of China in 2020 and key economic sectors. *Energy Policy* **2013**, *58*, 90–96.
17. Li, Y.P.; Huang, G.H.; Li, M.W. An integrated optimization modeling approach for planning emission trading and clean-energy development under uncertainty. *Renew. Energy* **2014**, *62*, 31–46.
18. Hermeling, C.; Löschel, A.; Mennel, T. A new robustness analysis for climate policy evaluations: A CGE application for the EU 2020 targets. *Energy Policy* **2013**, *55*, 27–35.
19. Orlov, A.; Grethe, H. Carbon taxation and market structure: A CGE analysis for Russia. *Energy Policy* **2012**, *51*, 696–707.
20. Matsumoto, K.; Masui, T. Economic impacts to avoid dangerous climate change using the AIM/CGE model. *Procedia Environ. Sci.* **2010**, *6*, 162–168.

21. Okagawa, A.; Masui, T.; Akashi, O.; Hijioka, Y.; Matsumoto, K.; Kainuma, M. Assessment of GHG emission reduction pathways in a society without carbon capture and nuclear technologies. *Energy Econ.* **2012**, *34*, S391–S398.
22. Xu, Y.; Masui, T. Local air pollutant emission reduction and ancillary carbon benets of SO_2 control policies: Application of AIM/CGE model to China. *Eur. J. Oper. Res.* **2009**, *198*, 315–325.
23. Fujimori, S.; Tu, T.T.; Masui, T.; Matsuoka, Y. *AIM/CGE [Basic] Manual*; National Institute for Environmental Studies: Tsukuba, Japan, 2011.
24. China Input-Output Association (CIOA): 2005 Input-Output Table. Available online: http://www.iochina.org.cn/Download/xgxz.html (accessed on 1 January 2006).
25. National Bureau of Statistics (NBS). *China Statistical Yearbook 2010*; China Statistics Press: Beijing, China, 2010.
26. IPCC. IPCC Guide Lines for National Green House Gas Inventories. 2006. Available online: http://www.ipcc-nggip.iges.or.jp/public/2006gl (accessed on 1 April 2007).
27. Peng, Z. *In-Output Tables of China*; China Statistics Press: Beijing, China, 2007.
28. Wei, S.; Huang, Y. Keqiang economics. *Western Development.* **2013**, *9*, 22–25.
29. Wang, K.; Wang, C.; Chen, J. Analysis of the economic impact of different Chinese climate policy options based on a CGE model incorporating endogenous technological change. *Energy Policy* **2013**, *37*, 2930–2940.
30. The Central People's Government of the People's Republic of China. National Population Development Strategy Research Report. Available online: http://www.gov.cn/gzdt/2007-01/11/content_493677.htm (accessed on 11 January 2007).
31. National Bureau of Statistics of China. *China Statistical Yearbook*; China Statistics Press: Beijing, China, 2008–2010. (In Chinese)
32. National Energy Administration (NEA). Energy Development "Twelfth Five Year Plan". Available online: http://www.nea.gov.cn/2013–01/28/c_132132808.htm (accessed on 1 January 2013).
33. Jiang, K.; Hu, X. China's low carbon development scenario in 2050. In *2050 China Energy and CO_2 Emissions Report. Study Team of 2005 China Energy and CO_2 Emissions Report*; Science Press: Beijing, China, 2009.
34. National Development and Reform Commission (NDRC). Interim Measures on Management Carbon Trading. Available online: http://www.in-en.com/finance/html/energy07360736802226927.html (accessed on 22 December 2014).
35. Yang, Y.; Qiu, W.; He, D. The game analysis between the government and the enterprises under mandatory emission reduction mechanism. *Syst. Eng.* **2012**, *2*, 110–114.
36. Wang, X.; Pang, J. The challenges, opportunities and recommendations for Chinese energy intensive enterprises to cope with climate change. *Clim. Chang. Res. Prog.* **2009**, *2*, 110–116.

A Novel Hybrid-Flux Magnetic Gear and Its Performance Analysis Using the 3-D Finite Element Method

Yiduan Chen and Weinong Fu *

Department of Electrical Engineering, the Hong Kong Polytechnic University, Kowloon 999077, Hong Kong, China; E-Mail: duanduan_happy90@163.com

* Author to whom correspondence should be addressed; E-Mail: eewnfu@polyu.edu.hk;

Academic Editor: Paul Stewart

Abstract: This paper presents a novel hybrid-flux magnetic gear, which integrates a transverse-flux magnetic gear and an axial-flux magnetic gear into a single unit. Compared to its conventional counterparts, the proposed magnetic gear transmits a relatively high torque density. When compared to the transverse-flux magnetic gear, this new structure employs an extra iron segment between the low-speed rotor and high-speed rotor to modulate the magnetic field and contribute to the transmission of additional torque. A three-dimensional (3-D) finite element method (FEM) is used for the analysis of the magnetic field. In the paper a variables-decoupling method based on the sensitivity analysis of the design parameters is also presented to accelerate the optimization process of the proposed machine.

Keywords: design method; finite element method; hybrid-flux; magnetic field; magnetic gear; optimization; torque transmission

1. Introduction

Magnetic gears (MGs), as an emerging technology, are now competing with mechanical gears in many applications as the former have no mechanical contacts, less mechanical loss and noise, as well as less maintenance [1,2]. Radial-flux MG is the most basic type of topology which combines a highly competitive torque transmission capability with very high efficiency [3–5]. The choices of design parameters have a significant influence on the torque transmission capability. The permanent magnets (PMs) in the rotors can be made of spoke type instead of surface mounted type, which has a flux focusing function to achieve higher performance.

However, if the outer diameter of the whole machine system is to be bound by a strict limit, the MG operating with transverse magnetic flux or axial magnetic flux is an attractive alternative as the space required is only along the axial direction [6]. A transverse-flux MG which has two rotors arranged side by side along the axial direction which interact by the magnetic flux flowing transversely across outer iron segments has been proposed. The magnets are magnetized along the azimuthal direction, which allows a flux focusing effect [7,8]. It is relatively easy to fabricate and has unique merits in applications which requires airproof devices and low vibration. However, the performance of transverse-flux MG is very sensitive to the materials of the iron segments. To obtain a high torque density iron segment materials with high saturation flux density and high permeability are required. An axial-flux MG has been described, which is particularly suitable for the applications requiring flat outside shapes and hermetic isolation between the input and output shafts [9]. Another axial-flux MG has been investigated, in which the ferrite magnets are arranged in a flux focusing configuration in order to increase the air-gap flux density [10]. The two rotors at the two sides interact by the magnetic flux flowing axially across the iron segments. If the ratio of the outer diameter to its axial length is large, the axial-flux MG has the potential of high torque density.

In this paper a novel hybrid-flux MG, which has a higher electromagnetic torque density than either the transverse-flux MG or the axial-flux MG, is presented. A global optimal design method based on three-dimensional (3-D) finite element method (FEM) of magnetic field and mechanical motion coupled computation for sizing the proposed gear is developed. The performance of the proposed structure is quantitatively compared to those of conventional configurations. The numerical method of FEM presents the advantage of taking into account the real geometry of the MGs as well as the magnetic saturation of the iron parts.

2. Basic Principle

The PM rotor with p pole pairs rotates at a rotational velocity of Ω_r and the magnetic field produced by PMs is modulated by the stationary iron segments, then appropriate space harmonics having the requisite number of poles as the associated PM rotor result. The radial component of the flux density distribution at a radius distance r and a space angle θ produced by either PM rotor is given by [11,12]:

$$B_r(r,\theta)=\left(\sum_{m=1,3,5\ldots} b_{rm}(r)\cos\left(mp(\theta-\Omega_r t)\right)+mp\theta_0\right)\times\left(\lambda_{r0}(r)+\sum_{j=1,2,3\ldots}\lambda_{rj}(r)\cos(jn_s\theta)\right) \quad (1)$$

Similarly, the circumferential component of the flux density distribution is given by:

$$B_\theta(r,\theta)=\left(\sum_{m=1,3,5\ldots} b_{\theta m}(r)\sin\left(mp(\theta-\Omega_r t)\right)+mp\theta_0\right)\times\left(\lambda_{\theta 0}(r)+\sum_{j=1,2,3\ldots}\lambda_{\theta j}(r)\cos(jn_s\theta)\right) \quad (2)$$

where b_{rm} and $b_{\theta m}$ are the Fourier coefficients for the radial and circumferential components of the flux density distribution, without the iron segments, respectively. λ_{rj} and $\lambda_{\theta j}$ are the Fourier coefficients for the modulating function associated with the radial and circumferential components of the flux density distribution, resulting from the introduction of the iron segments, respectively. t is time variable, θ_0 is the initial spatial phase angle, and n_s is the number of iron segments. λ_{r0} and $\lambda_{\theta 0}$ are for the radial and circumferential components of the average permeability of the iron segments.

The magnetic flux density can be rewritten as:

$$\begin{aligned} B_r(r,\theta) &= \lambda_{r0} \sum_{m=1,3,5...} b_{rm}(r)\cos\left(mp(\theta-\Omega_r t)+mp\theta_0\right) \\ &+\frac{1}{2}\sum_{m=1,3,5...}\sum_{j=1,2,3...}\lambda_{rj}(r)b_{rm}(r)\cos\left(\left(mp+jn_s\right)\left(\theta-\frac{mp\Omega_r}{mp+jn_s}t\right)+mp\theta_0\right) \\ &+\frac{1}{2}\sum_{m=1,3,5...}\sum_{j=1,2,3...}\lambda_{rj}(r)b_{rm}(r)\cos\left(\left(mp-jn_s\right)\left(\theta-\frac{mp\Omega_r}{mp-jn_s}t\right)+mp\theta_0\right) \end{aligned} \tag{3}$$

$$\begin{aligned} B_\theta(r,\theta) &= \lambda_{\theta 0} \sum_{m=1,3,5...} b_{\theta m}(r)\sin\left(mp(\theta-\Omega_r t)+mp\theta_0\right) \\ &+\frac{1}{2}\sum_{m=1,3,5...}\sum_{j=1,2,3...}\lambda_{\theta j}(r)b_{\theta m}(r)\sin\left(\left(mp+jn_s\right)\left(\theta-\frac{mp\Omega_r}{mp+jn_s}t\right)+mp\theta_0\right) \\ &+\frac{1}{2}\sum_{m=1,3,5...}\sum_{j=1,2,3...}\lambda_{\theta j}(r)b_{\theta m}(r)\sin\left(\left(mp-jn_s\right)\left(\theta-\frac{mp\Omega_r}{mp-jn_s}t\right)+mp\theta_0\right) \end{aligned} \tag{4}$$

One can find that the coefficients of the space angle θ in the cosine or sine function indicate different order of harmonics, so the number of pole-pairs in the space harmonic flux density distribution produced by either PM rotor can be given by:

$$\begin{aligned} p_{m,k} &= \left|mp+kn_s\right| \\ m &= 1, 3, 5, \ldots, \infty \\ k &= 0, \pm 1, \pm 2, \pm 3, \ldots, \pm\infty \end{aligned} \tag{5}$$

Similarly, the coefficients of the time variable t in the cosine or sine function indicate the rotational velocities of the space harmonic flux density distribution, so the different rotational velocities of the harmonic magnetic fields can be given by:

$$\begin{aligned} \Omega_{m,k} &= \frac{mp}{mp+kn_s}\Omega_r \\ m &= 1, 3, 5, \ldots, \infty \\ k &= 0, \pm 1, \pm 2, \pm 3, \ldots, \pm\infty \end{aligned} \tag{6}$$

When $k = 0$, it is the basic harmonic magnetic field that the rotational velocity of the harmonic field is equal to that of the PM rotor where there is no modulation. When $k \neq 0$, they are modulation harmonic magnetic fields that the rotational velocities of the harmonic fields and the PM rotor are different, due to the introduction of the iron segments. What's more, the asynchronous space harmonic is the highest when $m = 1$ and $k = -1$. The harmonic order and the rotational velocity of the harmonic field are:

$$p_{1,-1} = \left|p-n_s\right| \tag{7}$$

$$\Omega_{1,-1} = \frac{p}{p - n_s}\Omega_r \tag{8}$$

To transmit energy steadily at a different rotational velocity, the pole pairs of the two magnetic fields should be the same. Therefore, the number of pole pairs of the other PM rotor should be $|n_s - p|$. So the gear ratio is given by:

$$G_r = \frac{\Omega_r}{\frac{p}{p - n_s}\Omega_r} = \frac{p - n_s}{p} \tag{9}$$

3. Proposed Configuration

The magnetic field distribution of the proposed MG is determined using 3-D FEM. The electromagnetic torque is computed using virtual work method in the post processing of FEM [13].

In order to evaluate the torque capability of the constructed MG, the torque-load angle characteristic is investigated by computing the electromagnetic torque on the low-speed rotor which rotates at a speed of 20 rpm when the high-speed rotor is locked [14–16].

The configuration of the magnetic gear proposed in this paper is shown in Figure 1. It has one high-speed rotor and one low-speed rotor arranged in parallel along the axial direction. The high-speed rotor and low-speed rotor have, respectively, p_{high} and p_{low} pairs of rare-earth PM poles. It has specific stationary iron segments with $n_s = p_{high} + p_{low}$ pole pairs, outside of the two rotors, which constitutes a transverse-flux magnetic gear. It also has stationary iron segments with p_{Fe} pole pairs between the two rotors, which constitutes an axial-flux magnetic gear.

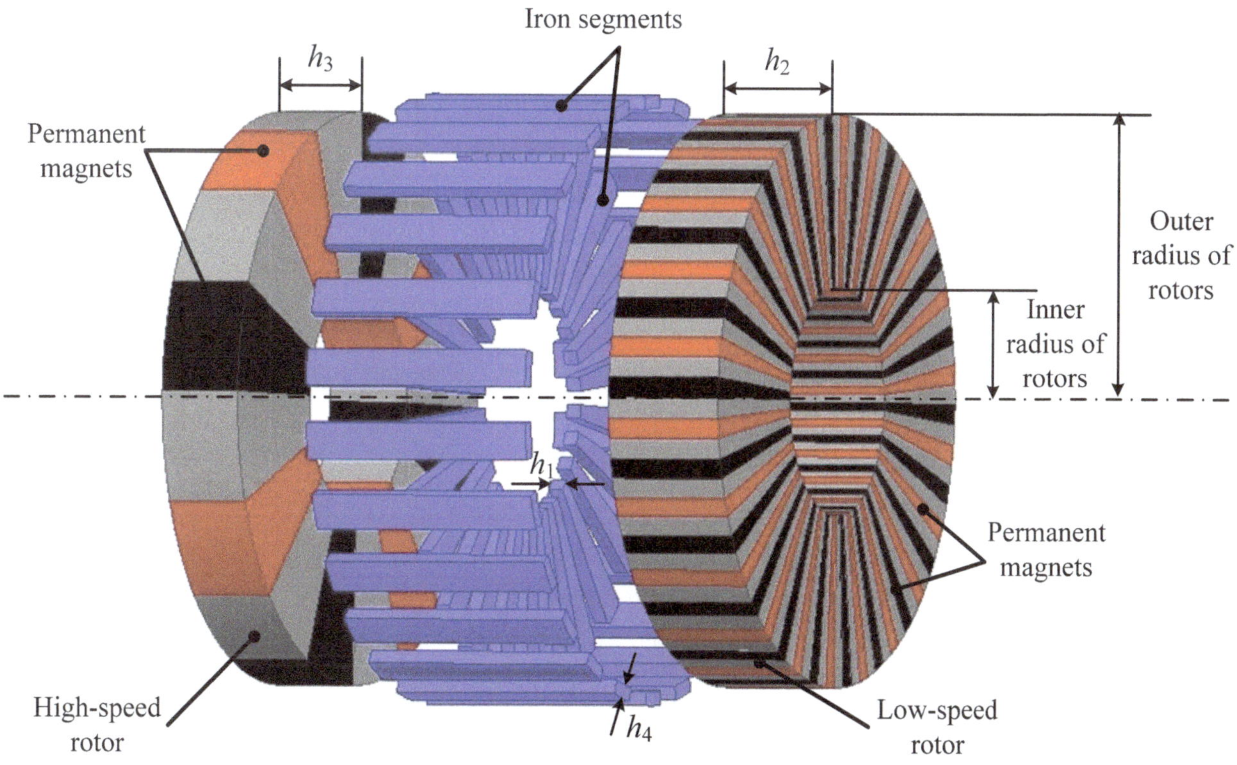

Figure 1. Proposed hybrid-flux magnetic gear.

The proposed structure is a combination of these two aforementioned conventional topologies and the space can be properly and fully utilized. When the rotors rotate, the iron segments will modulate the transverse and axial magnetic fields at the same time, and the capacity of the torque being transmitted can be enhanced significantly.

For the case in this paper, the high-speed rotor has p_{high} = 4 pole pairs, the low speed rotor has p_{low} = 22 and the number of stationary iron segments is n_s = 26. After the pole combination is chosen, the gear ratio is then determined as $G_r = -22/4 = -5.5$. The geometric parameters of this hybrid-flux MG are listed in Table 1. With the help of the 3-D finite element analysis the performance of the MG is numerically analyzed.

Table 1. Initial design parameters of the proposed MG.

Design parameters	Values
Number of pole pairs of high-speed rotor p_{high}	4
Number of pole pairs of low-speed rotor p_{low}	22
Number of pole pairs of iron segments n_s	26
Outer radius of rotors [mm]	150
Inner radius of rotors [mm]	60
Thickness of middle iron segments h_1 [mm]	8
Axial length of low-speed rotor h_2 [mm]	65
Axial length of high-speed rotor h_3 [mm]	45
Thickness of outer iron segments h_4 [mm]	10
Air-gaps [mm]	0.5

4. Analysis of Design Parameter

To find the effect of geometric parameters on the output torque capacity, eight design variables are studied. They are the axial length of the two rotors and the corresponding ratios of PM to pole pitch, the thickness of the two sets of iron segments and the corresponding ratios of iron to pole pitch.

Golden Section (GS) Search is a method that shrinks continually the known range of the optimum value of a monotonic function in finding its optimal value [17,18]. The algorithm keeps the spacing between three points with golden section properties. Here, the axial length of low-speed rotor h_2 and the axial length of high-speed rotor h_3 are optimized using GS search method. That is assuming h_3 is infinite and one finds the minimum h_2 for maximum torque by changing the value of h_2 using GS method. After h_2 is determined, the value of h_3 is changed in a similar way to find the minimum h_3 at which the maximum torque is reached.

Figure 2 shows the maximum torque versus the axial length of the two rotors. It can be found that if the axial length of the rotors is larger, a higher torque can be achieved. But when the rotor length reaches a certain value, the torque will not be greatly improved even if the length continues to increase. On the contrary, it will increase the cost of this MG as excessive PM is used. Therefore, it is very necessary to find the most appropriate rotor length to ensure the maximum torque is realized with minimum PM usage.

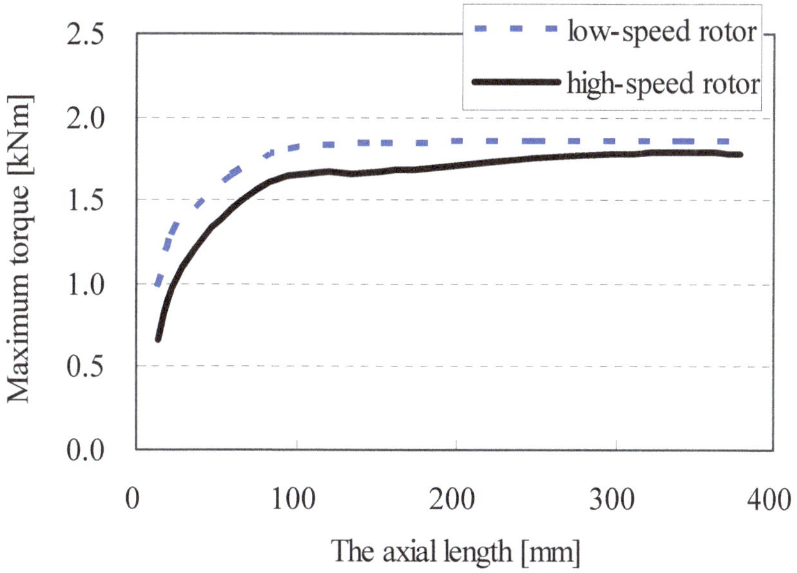

Figure 2. Maximum torque *versus* the axial length of rotors.

Figure 3 shows the maximum torque versus the thickness of two sets of iron segments. It can be seen that the maximum torque increases when the thickness of the middle iron segments is increased appropriately. But lower torque will be obtained if the thickness is too large due to the appearance of unduly long magnetic path. The effect of the thickness of the outer iron segments is that there is an increase in the maximum torque as the thickness increases. However, oversize thickness is needless because excessive PM is of little use to improve the torque. In this case, using materials with high saturation flux density is required.

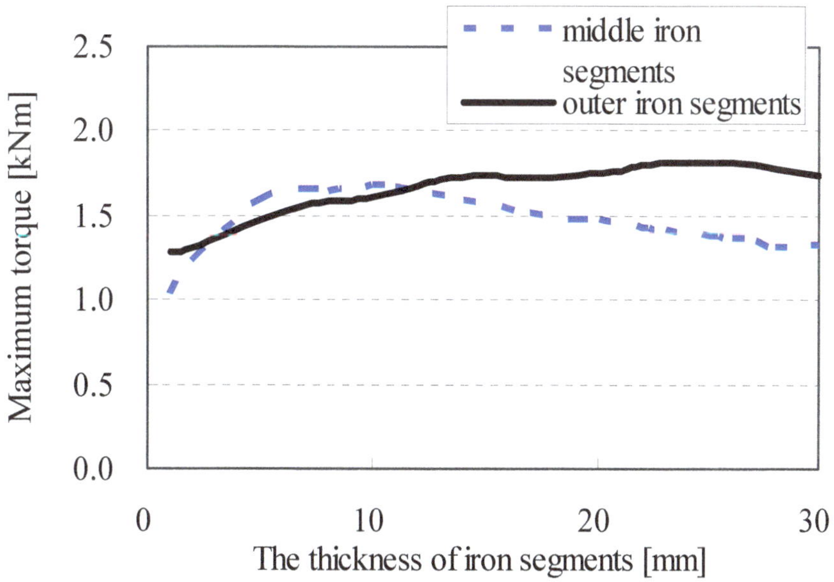

Figure 3. Maximum torque *versus* the thickness of iron segments.

The effect of the ratios of PM to pole pitch of the two rotors on the maximum torque is shown in Figure 4.

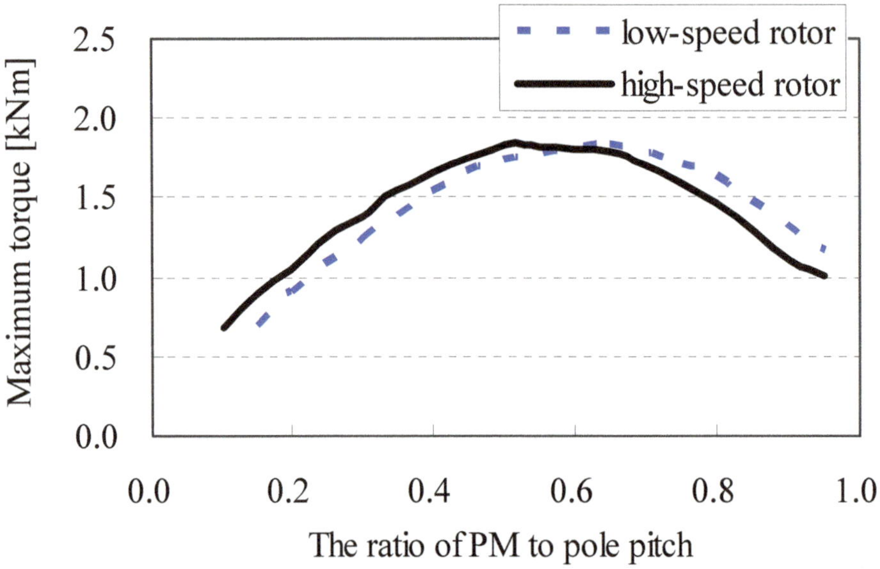

Figure 4. Maximum torque *versus* the ratio of PM to pole pitch of rotors.

Figure 5 shows the maximum torque *versus* the ratios of iron to pole pitch of two sets of iron segments.

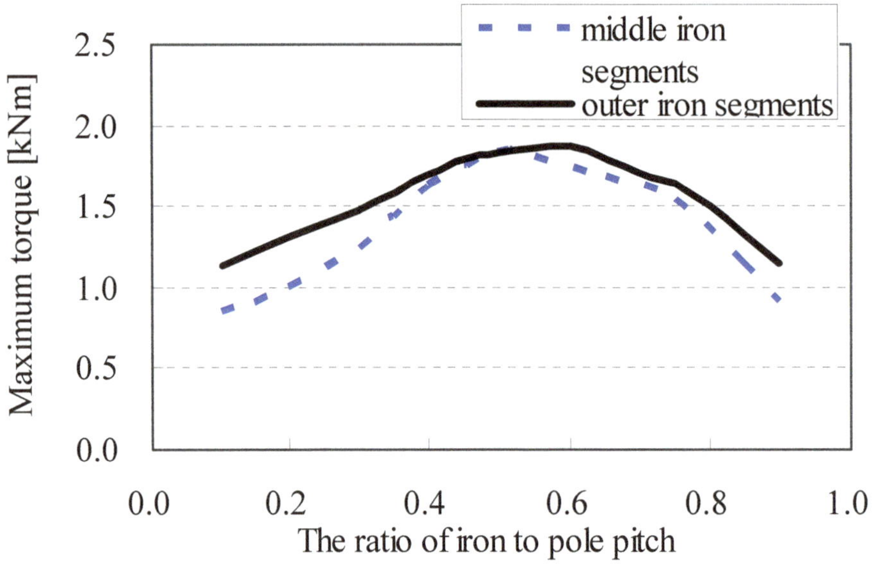

Figure 5. Maximum torque versus the ratio of iron to pole pitch of iron segments.

From Figures 4 and 5 one can find that choosing appropriate ratios of PM to pole pitch and iron to pole pitch can produce preferable maximum torque. The analysis above illustrates that all the eight parameters have influence on the torque capacity of the MG. During the study, the correlations among the parameters have not been taken into consideration while they are studied simply by the sweeping method. So the structure obtained cannot achieve optimal torque performance. However, if one tries to optimize all the parameters at the same time, it will take a lot of computing time as 3-D FEM is required for the torque computation.

5. Design Method and Results

A global optimization method is used to determine the sizes of the gear. As 3-D FEM is required to compute the electromagnetic torque, it is essential to reduce the computing time. In this paper the design parameters are divided into several groups based on the sensitivity analysis among them [19]. The optimization of these parameter groups can be processed in series, which can significantly reduce the number of objective function computation.

The sensitivity analysis is to study the correlation between the design parameters. If they are not related, they will be optimized separately using the sweep method directly. Otherwise, they should be optimized simultaneously using the response surface methodology (RSM).

In this embodiment, the specific analysis process is as follows. As shown in Table 2, it is assumed that there are two variables x_1 and x_2, and the corresponding maximum torque is T. If x_1 has a small increment while x_2 remains unchanged, the obtained maximum torque is T_1. Similarly if x_2 has a small increment while x_1 stays the same, the maximum torque is T_2. When there are little increments in both x_1 and x_2 the torque is T_{12}. By calculating the difference between the three torques and T respectively, the relationship among them can be analyzed. If the value $|\Delta T_{12}-\Delta T_1-\Delta T_2|$ is within the margin of error, then the two variables are uncorrelated. On the contrary, they are correlated. By using this method, the correlation between two parameters can be worked out.

Table 2. Sensitivity analysis.

Design parameters	Torque	Torque increment
$x_1\ x_2$	T	
$x_1 + \Delta x_1\ x_2$	T_1	$\Delta T_1 = T_1 - T$
$x_1\ x_2 + \Delta x_2$	T_2	$\Delta T_2 = T_2 - T$
$x_1 + \Delta x_1\ x_2 + \Delta x_2$	T_{12}	$\Delta T_{12} = T_{12} - T$

Response surface methodology explores the relationships between several explanatory variables and one or more response variables [20]. The method was introduced by G. E. P. Box and K. B. Wilson in 1951. It allows individual effects of the k-factors to be estimated independently without (or with minimal) confounding. Also the orthogonality provides minimum variance estimates of the model coefficient so that they are uncorrelated.

An easy way to estimate a first-degree polynomial model is to use a factorial experiment or a fractional factorial design. This is sufficient to determine which explanatory variables have an impact on the response variable(s) of interest. Once it is suspected that only significant explanatory variables are left, then a more complicated design, such as a central composite design can be implemented to estimate a second-degree polynomial model, which is still only an approximation at best. However, the second-degree model can be used to optimize (maximize, minimize, or attain a specific target for) [21–25].

The parameter design flow chart is shown in Figure 6. For the case studied in this paper, the specific parameter optimization flow chart is shown in Figure 7. By using the parameter analysis described in the previous Section, it can be seen that the axial length of the rotors and the corresponding ratios of PM to pole pitch are correlated and they should be optimized simultaneously.

The thickness of the two iron segments can be optimized separately and sequentially. In order to make the optimization more comprehensive, several iterations are done and a respond surface method is used.

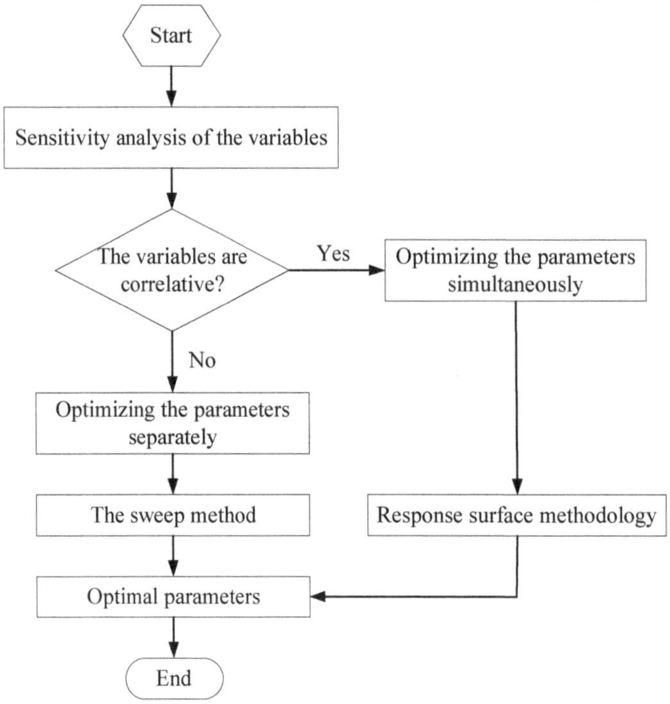

Figure 6. Parameter design flow chart.

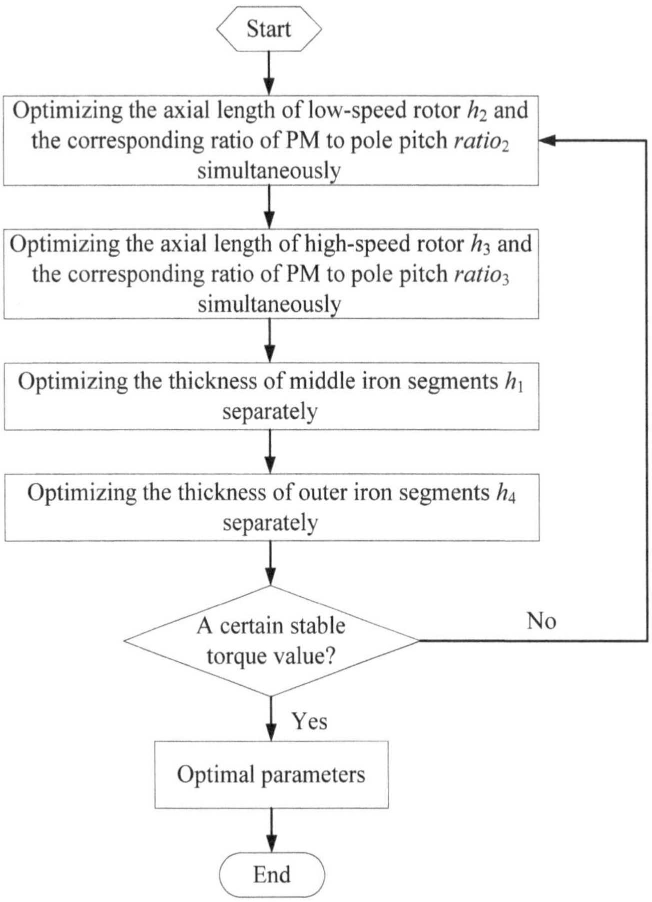

Figure 7. Specific parameter optimization flow chart.

The effect of the axial length of the rotors and the corresponding ratios of PM to pole pitch on the maximum torque are shown in Figures 8 and 9. The results are obtained using simultaneous analysis.

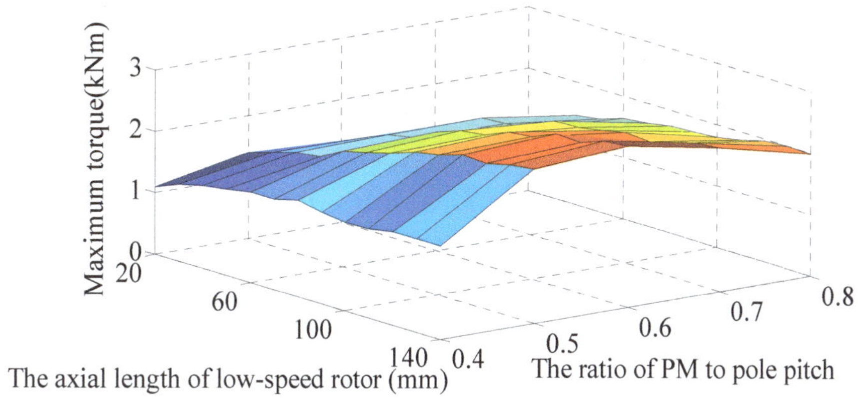

Figure 8. The effect of the axial length of low-speed rotor and the corresponding ratio of PM to pole pitch on the maximum torque.

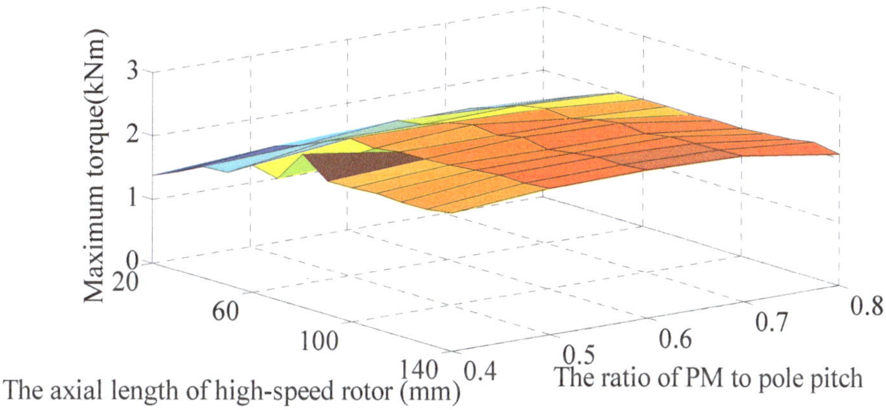

Figure 9. The effect of the axial length of high-speed rotor and the corresponding ratio of PM to pole pitch on the maximum torque.

By the aforementioned analysis, the final design parameters of the proposed hybrid-flux MG are listed in Table 3.

Table 3. Final design parameters of the proposed MG.

Design parameters	Values
Axial length of low-speed rotor [mm]	110
The ratio of PM to pole pitch of low-speed rotor	0.6
Axial length of high-speed rotor [mm]	90
The ratio of PM to pole pitch of high-speed rotor	0.6
Thickness of middle iron segments [mm]	10
The ratio of iron to pole pitch of middle iron segments	0.5
Thickness of outer iron segments [mm]	15
The ratio of iron to pole pitch of outer iron segments	0.6

6. Performance Analysis and Comparison

The proposed novel hybrid-flux MG has a higher torque capacity. To make the performance prominent, it is analyzed by comparing its performance with those of other MGs. As an important index, torque density (electromagnetic torque per unit volume) is computed for evaluating the performance of MGs with different structures. In this paper, three types of configurations are considered, these are transverse-flux MG, axial-flux MG and hybrid-flux MG.

The torque-load characteristics of MGs with different structures are shown in Figure 10. In Table 4, the maximum torques and the corresponding torque density are compared. It can be seen that the torque of the proposed MG has the largest value among the three configurations. Quantitative analysis using 3-D FEM shows that it has a better performance.

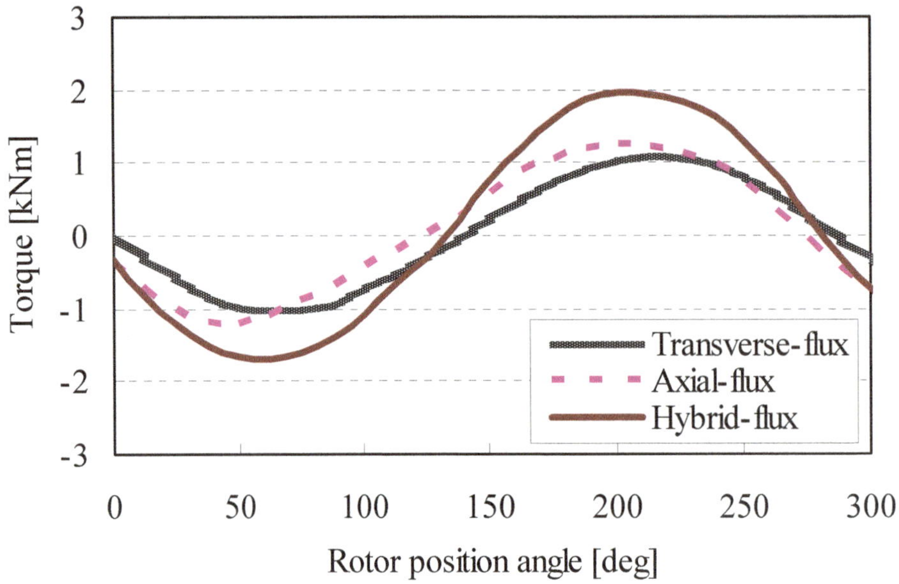

Figure 10. The torque-load characteristics of MGs with different structures.

Table 4. Torque comparison of three configurations.

Configuration	Maximum torque (kNm)	Torque density (kNm/m^3)
Transverse-flux magnetic gear	1.12	61
Axial-flux magnetic gear	1.25	69
Hybrid-flux magnetic gear	1.97	108

The flux density distribution of the proposed MG on the low-speed rotor side and high-speed rotor side are shown in Figures 11 and 12 respectively. The flux density distribution of the proposed MG on the iron segments is shown in Figure 13. One PM rotor rotates at a certain rotational velocity and the magnetic field produced by PMs is modulated by the two set of stationary iron segments, then appropriate space harmonics at the two air gaps result. From Figure 13 we can find that there are some saturation problems on the iron segments. It may be caused by the thickness and materials of the iron segments which should be study further.

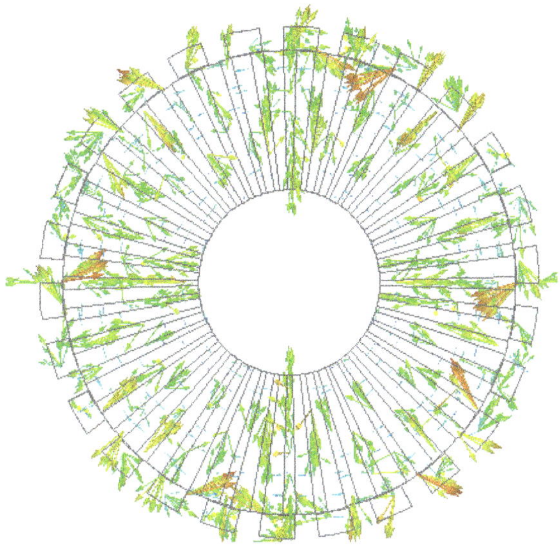

Figure 11. The flux density distribution of the MG on the low-speed rotor side.

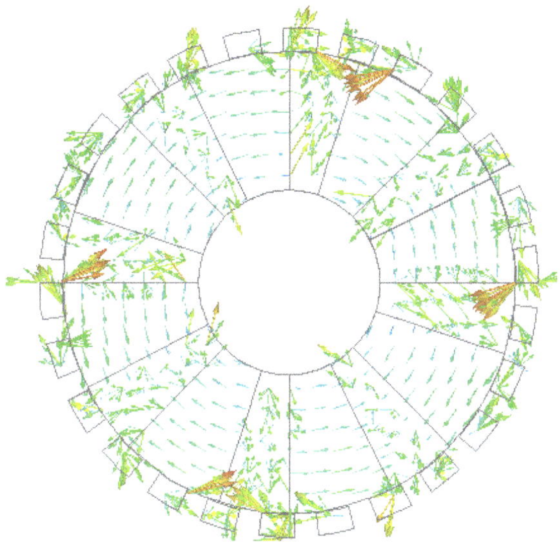

Figure 12. The flux density distribution of the MG on the high-speed rotor side.

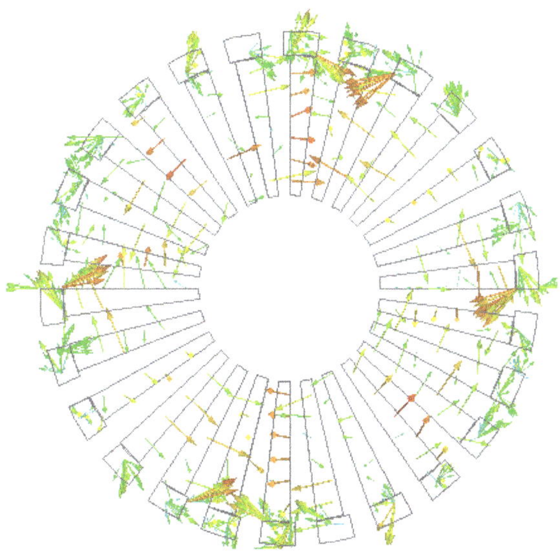

Figure 13. The flux density distribution of the MG on the iron segments.

The performance analysis and comparison in this paper show the higher torque density of the proposed hybrid-flux MG. However, there are several aspects without consideration, such as core loss, materials, manufacturing process, and so on. The more complicated structure will bring some problems that should be solved in the further study.

7. Conclusions

In this paper, a novel hybrid-flux MG is presented. It integrates a transverse-flux MG and an axial-flux MG into a single unit which produces a relatively higher electromagnetic torque density when compared to pure transverse-flux or axial-flux counterparts. The magnetic field of the proposed MG is analyzed using 3-D finite element method. The structure is designed based on a global optimal design method with sensitivity analysis of the parameters being used for variables decoupling. It accelerates the optimization process of the proposed MG. Finally, the torque characteristics of the proposed configuration and other conventional configurations are compared quantitatively. The comparison reveals the prominent advantageous performance of the proposed hybrid-flux MG, as reported in this paper.

Acknowledgments

This work was supported by the Research Grant Council of the Hong Kong SAR Government under projects PolyU 5153/13E and G-YM14.

Author Contributions

W.N. Fu conceived the idea of research, provide guidance and supervision; Yiduan Chen implemented the research, performed the analysis and wrote the paper. All authors have contributed significantly to this work.

Conflicts of Interest

The authors declare no conflict of interest.

References

1. Chau, K.T.; Chan, C.C.; Liu, C. Overview of permanent-magnet brushless drives for electric and hybrid electric vehicles. *IEEE Trans. Ind. Electron.* **2008**, *55*, 2246–2257.
2. Chau, K.T.; Zhang, D.; Jiang, J.Z.; Liu, C. Design of a magnetic-geared outer-rotor permanent-magnet brushless motor for electric vehicles. *IEEE Trans. Magn.* **2007**, *43*, 2504–2506.
3. Rasmussen, P.O.; Andersen, T.O.; Jorgensen, F.T.; Nielsen, O. Development of a high-performance magnetic gear. *IEEE Trans. Ind. Appl.* **2005**, *41*, 764–770.
4. Uppalapati, K.; Bird, J. A flux focusing ferrite magnetic gear. In Proceedings of the 6th IET International Conference on Power Electronics, Machines and Drives, Bristol, UK, 27–29 March 2012.
5. Chen, Y.D.; Fu, W.N.; Ho, S.L.; Liu, H.J. A quantitative comparison analysis of radial-flux, transverse-flux and axial-flux magnetic gears. *IEEE Trans. Magn.* **2014**, *50*, 8104604.

6. Hirata, K.; Yamamoto, M.; Muramatsu, M. Proposal of an axial gap magnetic harmonic gear. *IEEE J. Trans. Sens. Micromach.* **2009**, *130*, 802–807.
7. Li, Y.; Xing, W.; Lu, Y.P.; Yin, Z.J. Torque analysis of a novel non-contact permanent variable transmission. *IEEE Trans. Magn.* **2011**, *47*, 4465–4468.
8. Bomela, W.; Bird, J.Z.; Acharya, V.M. The performance of a transverse flux magnetic gear. *IEEE Trans. Magn.* **2014**, *50*, 4000104.
9. Mezani, S.; Atallah, K.; Howe, D. A high-performance axial-field magnetic gear. *J. Appl. Phys.* **2006**, *99*, 08R303.
10. Acharya, V.M.; Bird, J.Z.; Calvin, M. A flux focusing axial magnetic gear. *IEEE Trans. Magn.* **2013**, *49*, 4092–4095.
11. Atallah, K.; Calverley, S.; Howe, D. Design, analysis and realization of a high-performance magnetic gear. *Proc. Inst. Elect. Eng.* **2004**, *151*, 135–143.
12. Lubin, T.; Mezani, S.; Rezzoug, A. Development of a 2-D analytical model for the electromagnetic computation of axial-field magnetic gears. *IEEE Trans. Magn.* **2013**, *49*, 5507–5521.
13. Fu, W.N.; Zhou, D.; Lin, D.; Stanton, S. Magnetic force computation in permanent magnets using a local energy coordinate derivative method. *IEEE Trans. Magn.* **2004**, *40*, 683–686.
14. Ho, S.L.; Niu, S.X.; Fu, W.N.; Zhu, J.G. A mesh-insensitive methodology for magnetic force computation in finite-element analysis. *IEEE Trans. Magn.* **2012**, *48*, 287–290.
15. Ho, S.L.; Niu, S.X.; Fu, W.N. Design and analysis of a novel axial-flux electric machine. *IEEE Trans. Magn.* **2011**, *47*, 4368–4371.
16. Ho, S.L.; Niu, S.X.; Fu, W.N. Design and comparison of Vernier permanent magnet machines. *IEEE Trans. Magn.* **2011**, *47*, 3280–3283.
17. Sehoon, O.; Hori, Y. Parameter optimization for nc machine tool based on golden section search driven PSO. In Proceedings of the IEEE International Symposium on Industrial Electronics, Vigo, Spain, 4–7 June 2007.
18. Li, X.; Jiang, Y.Y. Golden-section peak search in fractional fourier domain. In Proceedings of the International Conference on Electric Information and Control Engineering (ICEICE), Wuhan, China, 15–17 April 2011.
19. Varella, H.; Guerif, M.; Buis, S. Global Sensitivity Analysis (GSA) measures the quality of parameter estimation. In Proceedings of the IEEE International Geoscience and Remote Sensing Symposium (IGARSS), Boston, MA, USA, 7–11 July 2008.
20. Box, G.E.P.; Wilson, K.B. On the experimental attainment of optimum conditions. *J. R. Stat. Soc.* **1951**, *XIII*, 1–45.
21. Osborne, D.M.; Armacost, R.L.; Pet-Edwards, J. State of the art in multiple response surface methodology. In Proceedings of the IEEE International Conference on Systems, Man, and Cybernetics, Orlando, FL, USA, 12–15 October 1997.
22. Gonda Neddermeijer, H.; van Oortmarssen, G.J.; Piersma, N.; Dekker, R. A framework for response surface methodology for simulation optimization. In Proceedings of the 2000 Winter Simulation Conference Proceedings, Orlando, FL, USA, 10–13 December 2000.
23. Quarantelli, M.; Daldoss, L.; Gubian, P.; Guardiani, C. Predictive tolerance and sensitivity analysis based on parametric response surface methodology. In Proceedings of the 3rd International Workshop on Statistical Metrology, Honolulu, HI, USA, 7 June 1998.

24. Versele, C.; Deblecker, O.; Lobry, J. A surface response methodology approach to study the influence of specifications in the multiobjective optimal design of power converters. In Proceedings of the 2011–14th European Conference on Power Electronics and Applications (EPE 2011), Birmingham, UK, 30 August–1 September 2011.
25. Koyamada, K.; Saki, K.; Itoch, T. Parameter optimization technique using the response surface methodology. In Proceedings of the 26th Annual International Conference of the IEEE, Engineering in Medicine and Biology Society (IEMBS '04), San Francisco, CA, USA, 1–5 September 2004.

Magnetic Field Equivalent Current Analysis-Based Radial Force Control for Bearingless Permanent Magnet Synchronous Motors

Huangqiu Zhu and Hui Li *

School of Electrical and Information Engineering, Jiangsu University, Zhenjiang 212013, China; E-Mail: zhuhuangqiu@ujs.edu.cn

* Author to whom correspondence should be addressed; E-Mail: lihui090232@163.com

Academic Editor: Enrico Sciubba

Abstract: Bearingless permanent magnet synchronous motors (BPMSMs), with all advantages of permanent magnet motors (PMSMs) and magnetic bearings, have become an important research direction in the bearingless motor field. To realize a stable suspension for the BPMSM, accurate decoupling control between the electromagnetic torque and radial suspension force is indispensable. In this paper, a concise and reliable analysis method based on a magnetic field equivalent current is presented. By this analysis method, the operation principle is analyzed theoretically, and the necessary conditions to produce a stable radial suspension force are confirmed. In addition, mathematical models of the torque and radial suspension force are established which is verified by the finite element analysis (FEA) software ANSYS. Finally, an experimental prototype of a 2-4 poles surface-mounted BPMSM is tested with the customized control strategy. The simulation and experimental results have shown that the motor has good rotation and suspension performance, and validated the accuracy of the proposed analysis method and the feasibility of the control strategy.

Keywords: bearingless permanent magnet synchronous motor; magnetic field equivalent current; decoupling control; radial suspension force

1. Introduction

Bearingless motors combine the characteristics of conventional motors and magnetic bearings. Compared with magnetic bearings, bearingless motors possess the advantages of shorter shaft length, higher critical speed, and lower cost [1]. Among different types of bearingless motors, bearingless permanent magnet synchronous motors (BPMSMs) are highly valued around the world due to their virtues of small size, light weight, non-contact, no wear, no lubrication, high efficiency, high power factor, and so on [2]. They have broad application prospects in centrifugal machines, aeronautics and astronautics, life science, *etc.* Especially due to their non-contact and no lubrication characteristics, BPMSMs are employed successfully in high purity environments, such as the semiconductor, pharmaceutical and medical industries.

Until now, the advanced technologies of the BPMSMs, including mathematical models, motor topologies and control strategies, have been investigated and discussed extensively by researchers. The complete mathematical models of the surface mounted BPMSMs are studied in [3] by using the Maxwell stress tensor method, which can effectively compensate for the effect of rotor eccentricity and Lorentz force. In [4], a new modeling method based on mechanical to electrical coordinate transformation is put forward, which proves an effective modeling method. For the motor topologies, Ooshima proposed a rotor design method considering demagnetization for a 2-DOF BPMSM [5]. Besides, a 5-DOF BPMSM, consisting of a 3-DOF axial-radial magnetic bearing and a 2-DOF BPMSM, is presented in [6]. At the same time, Kascak provided another novel 5-DOF BPMSM, only consisting of two conical BPMSMs to realize the 5-DOF active control [7]. For the control strategies, the decoupling control algorithm based on the rotor flux orientation control (FOC) has been widely used in recent decades [4]. The direct radial displacement control strategy, based on the relationship between the rotor radial displacement and radial suspension force, is pointed out in [8]. In addition, many artificial-intelligence-based control strategies to effectively handle the nonlinearities of the BPMSM system are also presented, such as sliding mode variable structure control [9], artificial neural network inverse control [10] and genetic algorithm control.

The double-winding BPMSM has two sets of coupling windings in the stator slot, namely torque windings (pole pair is P_M) and suspension force windings (pole pair is P_B). Thus, the precise decoupling control between them is indispensable to realize the suspension operation. In [11], by marking the distribution situation of the torque winding flux linkage and suspension force winding flux linkage, the generation principle of radial suspension forces is studied. Then, the corresponding performance of the induction type bearingless motor can be verified by the 2D finite element analysis (FEA) model. Based on these methods, analyses of the bearingless motor suspension characteristics have become simple, direct and operable. However, as the calculated coefficients are hard to get, the mathematical models are inaccurate. In [12], the mathematical model of the surface-mounted BPMSM is established based on the principle of virtual displacement. The derived idea is as follows: firstly, according to the equivalent magnetic circuit principle, the inductance matrixes are deduced. Then the expressions of electromagnetic energy based on inductance matrixes are derived. Finally, the mathematical expressions of radial suspension forces can be deduced by partial derivation with respect to the electromagnetic energy expressions. This method makes the deduction of radial suspension forces easier and more visual. The inductance parameters, however, are difficult to obtain precisely through experimental measurements, therefore, the precise calculation of radial suspension force also cannot be achieved for the BPMSM.

In this paper, an accurate and reliable analysis method based on the magnetic field equivalent current is presented. Based on this analysis method, the practical motor windings are replaced by equivalent windings existing in different material spaces. Distributions of these equivalent windings only depend on air gap magnetic fields, and they are not subject to the mechanical structure features. Then, the interactions among the air gap magnetic fields can be converted to that of the carrying-current equivalent windings. Thus, some complex analyses of magnetic fields can be avoided and the difficulty of system analysis can be effectively reduced, which provides a systemic theoretical foundation for the control of radial suspension forces.

2. Magnetic Field Equivalent Current-Based Analysis of Radial Suspension Forces

2.1. Fundamental Theory of the Magnetic Field Equivalent Current Analysis Method

Assuming that the motor windings are designed as spatially symmetric distributions, the permeability in the air gap is equal everywhere, the leakage inductances of permanent magnets and magnetic saturation are ignored. The air gap magnetic fields are thought to be produced by a series of carrying-current equivalent windings. The pole pairs of equivalent windings are equal to that of the original air gap magnetic fields. The distributions of the equivalent windings are flexibly designed according to the actual air gap magnetic fields. The equivalent full pitch windings are employed in this paper and they can rotate with the change of the air gap magnetic fields. The current in equivalent windings is considered as direct current. Then, the interaction relationships between torque winding air gap magnetic field and suspension force winding air gap magnetic field can be obtained indirectly from that of the corresponding equivalent currents. Figure 1 shows the schematics of magnetic field equivalent current analysis method.

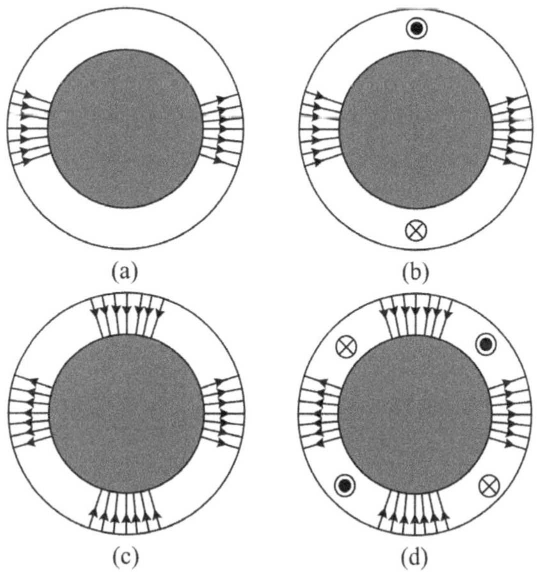

Figure 1. Schematic diagrams of magnetic field equivalent current analysis method. (**a**) 2-pole air gap magnetic field; (**b**) Equivalent winding currents of 2-pole air gap magnetic field; (**c**) 4-pole air gap magnetic field; (**d**) Equivalent winding currents of 4-pole air gap magnetic field.

In this paper, the analysis method based on carrying-current equivalent windings is defined as the magnetic field equivalent current analysis method. The equivalent windings produced by the torque winding air gap magnetic field are defined as the equivalent torque windings, and the equivalent windings generated by the suspension force winding air gap magnetic field are defined as the equivalent suspension force windings. In addition, on the basis of field distributions, many structural forms can be composed of equivalent torque windings and equivalent suspension force windings. Among them, intensifying factor and softening factor are two key structural forms, as shown in Figure 2. The upper layer of windings corresponds to equivalent suspension force windings, while the lower layer of windings corresponds to equivalent torque windings. Intensifying factor and softening factor are formed by two sets of equivalent windings with their centre line overlapped. The enhanced resultant air gap magnetic fields can be produced by the intensifying factor, and the maximum magnetic motive force (MMF) can be formed. The reduced resultant air gap magnetic fields can be produced by the softening factor, and the minimum MMF can be generated. If the centre lines of the two sets of equivalent windings are no longer overlapped, but there exists an inclusion relation in space, the magnetic field superposition effect between windings is the same as that of intensifying factor and softening factor. Then, these structure forms are defined as partial-intensifying factor and partial-softening factor.

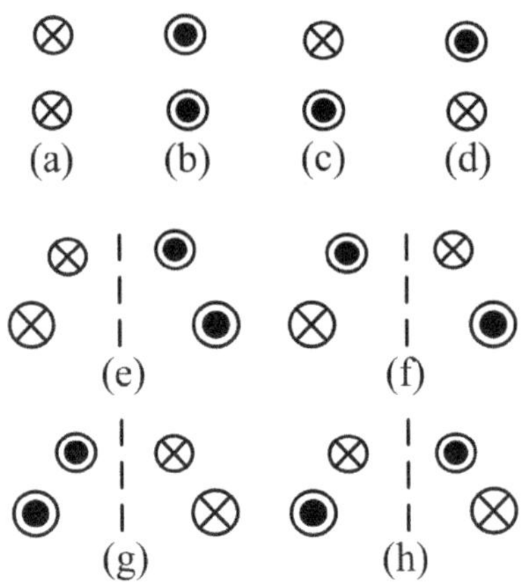

Figure 2. Basic current structure formed by two sets of windings. (**a**), (**b**), (**e**) and (**g**) are intensifying factors. (**c**), (**d**), (**f**) and (**h**) are softening factors.

2.2. The Principle of Radial Suspension Force Generation for the BPMSM

Figure 3 shows the stator configuration and the 2D FEA model of a surface-mounted BPMSM. To orientate and start easily, it is usual to bring the A phase winding axes of the torque windings in line with that of the suspension force windings. The suspension force windings are embedded in the outer layer of the stator slot, and the phase sequence arrangement is $N_{BA} \rightarrow N_{BB} \rightarrow N_{BC}$, $P_B = 2$. The torque windings are embedded in the inner layer of the stator slot, and the phase sequence arrangement is $N_{MA} \rightarrow N_{MB} \rightarrow N_{MC}$, $P_M = 1$. The motor structure diagram of the 2-4 poles BPMSM in the rotational coordinates is illustrated in Figure 4. Due to the fact that the pole pair of the permanent magnet air gap magnetic field is the same as that of the torque windings, their resultant magnetic field can similarly be considered as the products of currents in the torque windings [13]. The diagrams of the motor structure and operation principle for the 2-4 poles BPMSM, based on the conventional analysis method and on the magnetic field equivalent current analysis method, are depicted in Figure 4a,b, respectively. The windings N_M stand for equivalent torque windings, and windings N_B represent equivalent suspension force windings. Then the principles of radial suspension force generation are compared below.

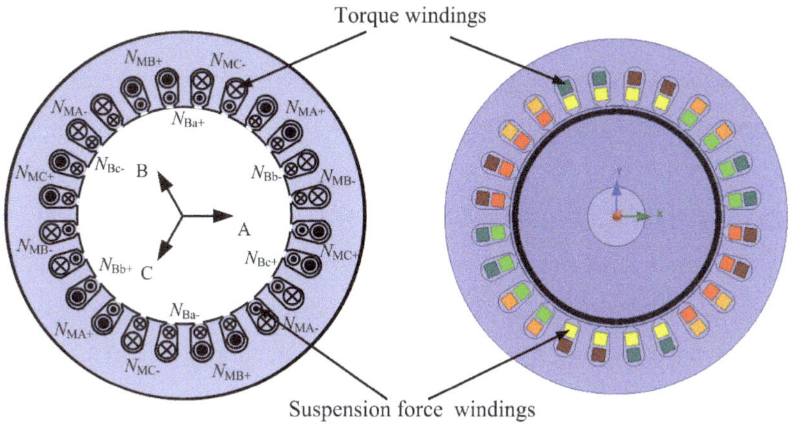

(a) (b)

Figure 3. Structure diagram of the bearingless permanent magnetic synchronous motor (BPMSM). (**a**) Stator winding distribution. (**b**) Finite element analysis model.

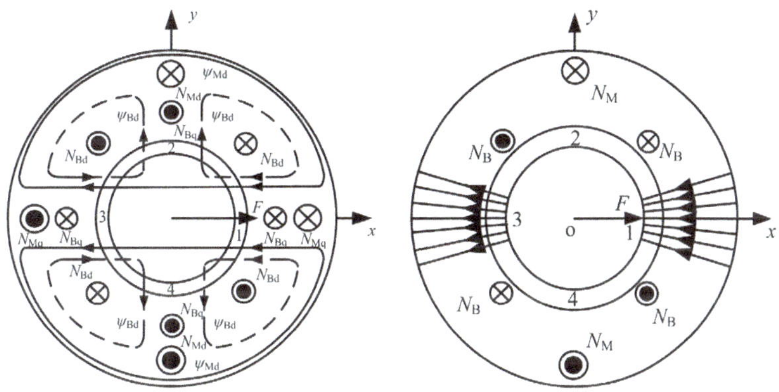

(a) (b)

Figure 4. Structure and operation principle diagram of the 2-4 poles BPMSM. (**a**) Based on conventional analysis method. (**b**) Based on the magnetic field equivalent current analysis method.

Figure 4a illustrates the principle based on the conventional method for the BPMSM [14,15]. The symmetrical 2-pole excitation flux ψ_{Md} is generated by the d-axis component N_{Md} of the torque windings. When a suspension force winding current I_{Bd} is provided, the symmetrical 4-pole flux ψ_{Bd} is generated in the air gap. The directions of the torque winding fluxes ψ_{Md} and the torque winding fluxes ψ_{Bd} are same in the air gap 1, and thus, the resultant air gap flux density is increased. In the air gap 3, their flux directions are opposite, so the resultant air gap flux density is decreased. Then, according to Maxwell's stress tensor, the radial suspension force F is produced and points to the positive x-axis. The opposite-direction radial suspension force in the x-axis can be produced with a negative current I_{Bd}. Moreover, the vertical radial suspension force F in the y-axis can be produced by providing the suspension current I_{Bq} in the suspension winding N_{Bq}. In short, the directions of the radial suspension forces are affected by the phase of the 4-pole suspension force winding current.

Figure 4b depicts the principle based on the proposed magnetic field equivalent current analysis method. In the air gap 1, an intensifying factor is formed and a softening factor is formed at the position of air gap 3. The direction of the composite MMF is the positive x-axis. Thus, a radial suspension force F is generated which is oriented in the positive x-axis. The analysis result of the proposed method agrees with that of conventional method. In the same way, if the equivalent suspension force windings N_B rotate a 90° electrical angle in the counter-clockwise direction, the vertical radial suspension force F in the positive y-axis is generated. The radial suspension force F in the negative x-axis can be produced through rotating N_B by 180° electrical angle in the counter-clockwise direction. In conclusion, the direction of radial suspension force rotation has relations with that of equivalent suspension force winding rotation, and the rotating angular velocity is P_B times to that of equivalent suspension force windings, as can be seen in [16]. The description of the principle for the 2-4 poles BPMSM also applies to other pole pair compositions and various types of bearingless motors.

3. Analysis on Producing Stable Radial Suspension Force for the BPMSM

Based on the magnetic field equivalent current analysis method, the necessary condition to produce a stable radial suspension force in a certain direction can be derived. In this section, the counter-clockwise rotation direction is defined as the rotation positive direction. It is supposed that torque winding air gap magnetic field rotates in the positive direction with mechanical angular velocity ω_M. Then, to obtain a stable radial suspension force in a certain direction, the absolute angular velocity of the radial suspension force should be set as 0 rad/s, and its direction should be fixed. Treating the torque winding air gap magnetic field as a reference, the relative angular velocity ω_F of radial suspension force is $-\omega_M$. The relative angular velocity ω of suspension force winding air gap magnetic field can be obtained as [17]:

$$\begin{cases} \omega = -\omega_F / P_B = -(-\omega_M)/P_B = \omega_M / P_B, \quad P_B = P_M - 1 \\ \omega = \omega_F / P_B = (-\omega_M)/P_B = -\omega_M / P_B, \quad P_B = P_M + 1 \end{cases} \tag{1}$$

The absolute angular velocity ω_B of suspension force winding air gap magnetic field can be written as:

$$\begin{cases} \omega_B = \omega_M + \omega = \omega_M + \omega_M / P_B = P_M\omega_M / P_B, \quad P_B = P_M - 1 \\ \omega_B = \omega_M + \omega = \omega_M + (-\omega_M / P_B) = P_M\omega_M / P_B, \quad P_B = P_M + 1 \end{cases} \tag{2}$$

Likewise, when the angular velocity of the torque winding air gap magnetic field is $-\omega_M$, (that is, rotating in the negative direction), the absolute angular velocity of suspension force winding air gap magnetic field can be expressed as $-P_M\omega_M/P_B$.

Therefore, the necessary condition to generate a stable radial suspension force is that the angular velocity ω_B of suspension force winding air gap magnetic field should be P_M/P_B times that of the torque winding air gap magnetic field, and their directions should remain the same. In other words, the suspension force winding air gap magnetic field and the torque air gap winding magnetic field have the same rotation electric angular velocity and rotation direction:

$$P_B \cdot \omega_B = P_B \cdot (P_M\omega_M / P_B) = P_M \cdot \omega_M \tag{3}$$

The verification of the necessary condition by means of the magnetic field equivalent current analysis method is as follows: the relative motion analysis situations are shown in Figures 5 and 6.

Among them, large dimension and small dimension of windings represent equivalent suspension force windings and equivalent torque windings, respectively.

If $P_B = P_M + 1$ (that is $P_B = 2$, $P_M = 1$), when the equivalent torque windings rotate $\pi/2$ rad mechanical angle in positive direction, according to the necessary condition, the equivalent suspension force windings must rotate $\theta_B = P_M\theta_M/P_B = \pi/4$ rad mechanical angle in the same direction, as can be seen from Figure 5b. There is an intensifying factor at $\alpha = 0$ rad and a softening factor at $\alpha = \pi$ rad. Thereby, radial suspension force is produced towards the direction of 0 rad. In this situation, the rotation electrical angle of the equivalent suspension force windings $P_B{\cdot}\theta_B = 2{\cdot}\pi/4 = \pi/2$ rad is the same as that of the equivalent torque windings $P_M{\cdot}\theta_M = 1{\cdot}\pi/2 = \pi/2$ rad. When the equivalent torque windings rotate π rad mechanical angle in positive direction, the equivalent suspension force windings must rotate $\theta_B = P_M\theta_M / P_B = \pi/2$ rad mechanical angle in positive direction, as shown in Figure 5c. There also is an intensifying factor at $\alpha = 0$ rad and a softening factor at $\alpha = \pi$ rad, so the direction of the radial suspension force remains 0 rad. The rotation electrical angle of the equivalent suspension force windings $P_B{\cdot}\theta_B = 2{\cdot}\pi/2 = \pi$ rad is equal to that of the equivalent torque windings $P_M{\cdot}\theta_M = 1{\cdot}\pi = \pi$ rad. In Figure 5d, when the equivalent torque windings rotate $3\pi/2$ rad mechanical angle in the positive direction, the same conclusion could be obtained. In consequence, when $P_B = P_M + 1$, to generate a stable and reliable radial suspension force, the rotation electrical angular of the equivalent suspension force windings $P_B{\cdot}\theta_B$ is equal to that of equivalent torque windings $P_M{\cdot}\theta_M$. Besides, the rotation direction of equivalent suspension force windings is the same as that of equivalent torque windings, and the converse is also true.

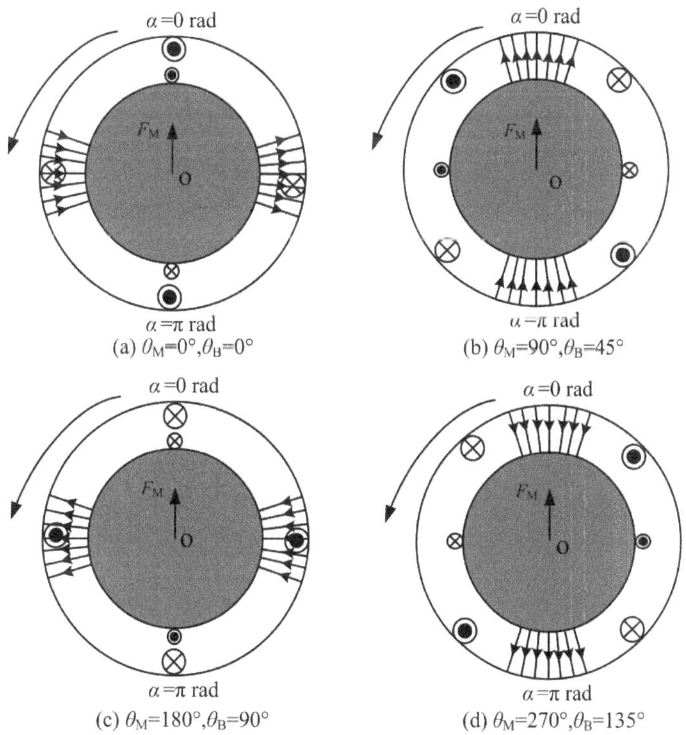

Figure 5. Schematic diagram of stable radial suspension force in a single direction ($P_B = P_M + 1$). (**a**) Initial state. (**b**) Rotation of equivalent torque windings with the mechanical angular 90°. (**c**) Rotation in the positive direction with the mechanical angular 180°. (**d**) Rotation in the positive direction at the mechanical angular 270°.

If $P_B = P_M - 1$ (that is $P_B = 1$, $P_M = 2$), when the equivalent torque windings rotate π/4 rad mechanical angle in the positive direction, the equivalent suspension force windings must rotate $\theta_B = P_M\theta_M / P_B = \pi/2$ rad mechanical angle in the same direction to hold the radial suspension force in a stationary direction, as plotted in Figure 6b. The intensifying factor is at $\alpha = 0$ rad and a softening factor is at $\alpha = \pi$ rad. Thereby, the radial suspension force points to the direction of 0 rad. Then, the rotation electrical angle of the equivalent suspension force windings $P_B \cdot \theta_B = 1 \cdot \pi/2 = \pi/2$ rad is equal to that of the equivalent torque windings $P_M \cdot \theta_M = 2 \cdot \pi/2 = \pi$ rad. When the equivalent torque windings rotate π/2 rad mechanical angle in positive direction, the equivalent suspension force windings must rotate π rad mechanical angle in positive direction, as shown in Figure 6c. The rotation electrical angle of the equivalent suspension force windings $P_B \cdot \theta_B = 1 \cdot \pi = \pi$ rad is also equal to that of equivalent torque windings $P_M \cdot \theta_M = 2 \cdot \pi/2 = \pi$ rad. In Figure 6(d), when the equivalent torque windings rotate 3π/4 rad mechanical angle in positive direction, the same conclusion could be obtained. Thus, when $P_B = P_M - 1$, the conclusion is consistent with that of the condition of $P_B = P_M + 1$.

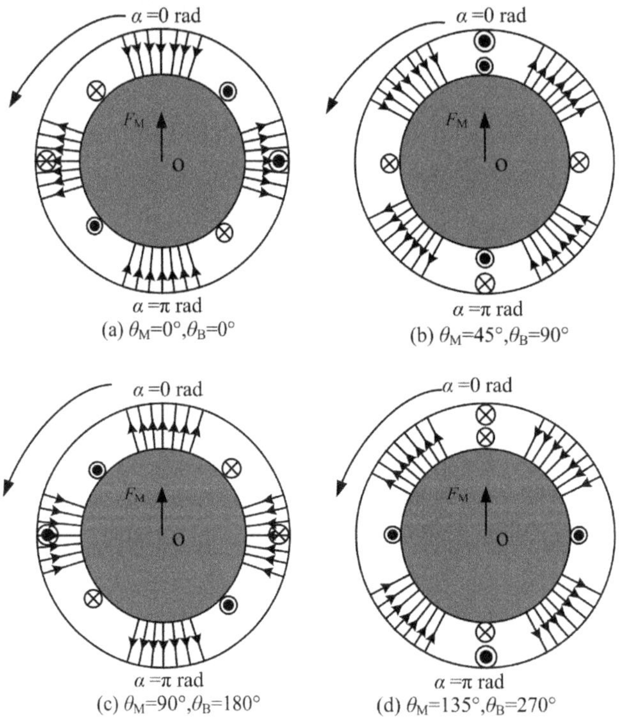

Figure 6. Schematic diagram of a stable radial suspension force in a single direction ($P_B = P_M - 1$). (**a**) Initial state. (**b**) Rotation of equivalent torque windings with the mechanical angular rotation of 45°. (**c**) Rotation in the positive direction with the mechanical angular rotation of 90°. (**d**) Rotation in the positive direction at the mechanical angular rotation of 135°.

In brief, the electrical angular velocity of torque winding air gap magnetic field $P_M \cdot \omega_M$ is in complete agreement with that of suspension force winding air gap magnetic field $P_B \cdot \omega_B$. The current frequency of suspension force windings and torque windings can be expressed as:

$$f_B = f_M = P_M \cdot \omega_M / 2\pi \quad (4)$$

Therefore, the necessary condition to generate a stable and reliable radial suspension force is that the suspension force winding air gap magnetic field rotates in the same direction as the torque winding air gap magnetic field with P_M/P_B times to the angular velocity ω_M of the torque winding air gap magnetic field.

4. Mathematical Model of Radial Suspension Force for the BPMSM

4.1. Analytical Model of Radial Suspension Force for the BPMSM

According to Maxwell's stress tensor method, the radial suspension force vector F as shown in Figure 5b can be expressed as [3,18]:

$$F = \begin{pmatrix} F_n \\ F_t \end{pmatrix} = \frac{1}{2\mu_0} \oint_A \begin{pmatrix} B_n^2 - B_t^2 \\ 2B_n B_t \end{pmatrix} \mathrm{d}A = \frac{lr}{2\mu_0} \int_0^{2\pi} \begin{pmatrix} B_n^2 - B_t^2 \\ 2B_n B_t \end{pmatrix} \mathrm{d}\theta \tag{5}$$

with μ_0 as the air permeability, B_n and B_t being the radial and tangential components of air gap flux density for the BPMSM, respectively, A being the stator inner surface area, r as the stator inner diameter, l representing the motor length and θ being the space angle.

Taking only the fundamental components of current into account, the air gap flux density B_1 caused by a permanent magnet and torque winding current loading A_1, and the air gap flux density B_2 excited by a suspension force winding current loading A_2 can be described as:

$$\begin{aligned} B_1(t,\theta) &= B_{1\mathrm{m}} \cos(\omega_1 t - P_\mathrm{M}\theta + \lambda_1) \\ B_2(t,\theta) &= B_{2\mathrm{m}} \cos(\omega_2 t - P_\mathrm{B}\theta + \lambda_2) \end{aligned} \tag{6}$$

while the current loadings of torque windings and suspension force windings A_1, A_2 can be expressed as follows:

$$\begin{aligned} A_1(t,\theta) &= -A_{1\mathrm{m}} \sin(\omega_1 t - P_\mathrm{M}\theta + \lambda_1) \\ A_2(t,\theta) &= -A_{2\mathrm{m}} \sin(\omega_2 t - P_\mathrm{B}\theta + \lambda_2) \end{aligned} \tag{7}$$

where: $B_{1\mathrm{m}} = \frac{\mu_0}{\delta}\left(\frac{3}{2}\frac{4}{\pi}\frac{I_{1\mathrm{m}}}{2}\frac{N_1 g_\mathrm{M}}{P_\mathrm{M}}\right)$, $B_{2\mathrm{m}} = \frac{\mu_0}{\delta}\left(\frac{3}{2}\frac{4}{\pi}\frac{I_{2\mathrm{m}}}{2}\frac{N_2 g_\mathrm{B}}{P_\mathrm{B}}\right)$, $A_{1\mathrm{m}} = \frac{mN_1 I_{1\mathrm{m}} g_\mathrm{M}}{\pi r}$, $A_{2\mathrm{m}} = \frac{mN_2 I_{2\mathrm{m}} g_\mathrm{B}}{\pi r}$. $\delta_0 = h_\mathrm{m} + g$ is the equivalent air gap length without rotor eccentricity, h_m is the permanent magnet thickness, g is the mechanical air gap length, g_M and g_B correspond to the fundamental winding factors of torque windings and suspension force windings, respectively, P_M and P_B are their pair poles. $I_{1\mathrm{m}} = I_\mathrm{M} + I_f$ is the torque winding resultant current magnitude, I_M and I_f represent as the magnitude of the torque winding current and the equivalent current of the permanent magnets, respectively. $I_{2\mathrm{m}}$ is the suspension force winding current value. N_1, N_2 are the turn numbers of torque windings and suspension force windings, respectively, ω_1 and ω_2 correspond to their electrical angular frequencies, λ_1 and λ_2 are their initial phase angle.

The radial flux density component B_n and tangential flux density component B_t in the air gap are given by:

$$\begin{aligned} B_n(t,\theta) &= B_1(t,\theta) + B_2(t,\theta) \\ B_t(t,\theta) &= \mu_0 A(t,\theta) \end{aligned} \tag{8}$$

Supposing $P_M = P_B \pm 1$ and $\omega_1 = \omega_2$, the radial suspension force components in the x- and y-direction are obtained as:

$$\begin{aligned} F_x &= \pi \cdot l \cdot r\left(\frac{B_{1m}B_{2m}}{2\mu_0} \pm \left(\frac{B_{1m}A_2}{2} - \frac{B_{2m}A_1}{2}\right)\right)\cos(\gamma_1-\gamma_2) = (k_m \pm k_l)I_{1m}I_{2m}\cos(\gamma_1-\gamma_2) \\ F_y &= \pi \cdot l \cdot r\left(\pm\frac{B_{1m}B_{2m}}{2\mu_0} + \left(\frac{B_{1m}A_2}{2} - \frac{B_{2m}A_1}{2}\right)\right)\sin(\gamma_1-\gamma_2) = (\pm k_m + k_l)I_{1m}I_{2m}\sin(\gamma_1-\gamma_2) \end{aligned} \tag{9}$$

where $k_m = \frac{9}{2\pi}\frac{\mu_0 lrN_1N_2g_Mg_B}{P_MP_B\delta_0^2}$, $k_l = \frac{3}{2\pi}\frac{m\mu_0 lN_1N_2g_Mg_B}{P_MP_B\delta_0}$ and m is the phase number for the BPMSM.

The first part is the Maxwell controllable suspension forces influenced by the air gap flux density B_1 and B_2, whereas the second part is the radial component of the Lorentz forces generated by the interaction of the air gap flux density B_1 with the current loading A_2 and the interaction of the air gap flux density B_2 with the current loading A_1. It can be seen from (9) that the amplitude of the controllable radial suspension forces is linear with the suspension force winding current value I_{2m}. And the direction depends only on phase difference between the torque winding air gap flux density B_1 and suspension force winding air gap flux density B_2, given by $\lambda_1-\lambda_2$. Due to $r >> \delta_0$ for the BPMSM, k_l is much less than k_m, thus the radial component of Lorentz forces can be ignored.

Therefore, under dq synchronous rotating reference frame, the radial suspension force F_d and F_q can be expressed as:

$$\begin{pmatrix} F_d \\ F_q \end{pmatrix} = k_m \begin{pmatrix} I_{Md}+I_f & I_{Mq} \\ I_{Mq} & -(I_{Md}+I_f) \end{pmatrix}\begin{pmatrix} I_{Bd} \\ \pm I_{Bq} \end{pmatrix} \tag{10}$$

with the I_{Md} and I_{Bd} as the d-axis components of torque and suspension force winding currents, the I_{Mq} and I_{Bq} corresponding to their q-axis components, respectively.

Except for currents in the suspension force windings, the rotor eccentricity e will also cause a radial suspension force, namely, unilateral magnetic force. When rotor eccentric displacement occurs, the motor air gap length δ will change which can be approximately expressed as:

$$\delta(\theta,t) \approx \delta_0 - e\cos(\theta-\varphi) = \delta_0 - x\cos\theta - y\sin\theta \tag{11}$$

with $x = e\cos\varphi$, $y = e\sin\varphi$, and φ being the deviation angle between the eccentricity orientation and x-axis. Thus, the radial suspension force can be expressed by [19,20]:

$$\begin{pmatrix} F_x \\ F_y \end{pmatrix} = \begin{pmatrix} F_d \\ F_q \end{pmatrix} + \begin{pmatrix} F_{ed} \\ F_{eq} \end{pmatrix} = k_m \begin{pmatrix} I_{Md}+I_f & I_{Mq} \\ I_{Mq} & -(I_{Md}+I_f) \end{pmatrix}\begin{pmatrix} I_{Bd} \\ \pm I_{Bq} \end{pmatrix} + k_{ecc}\begin{pmatrix} x \\ y \end{pmatrix} \tag{12}$$

Here, $k_{ecc} = \frac{9}{2}\frac{\mu_0 lrN_1^2g_M^2I_f^2}{\pi P_M^2\delta_0^3}$.

In order to support the stable suspension, it is necessary to compensate the unilateral magnetic forces by a reasonable adjustment of the suspension force winding currents.

4.2. Analytical Model of Electromagnetic Torque for the BPMSM

Because the d- and q-axis components of the torque winding inductances are approximately equal ($L_d = L_q$) for the surface-mount BPMSM, the mathematical formula of electromagnetic torque has the same form as the conventional PMSM in a dq synchronous rotating reference frame which can be given by [21]:

$$T_{em} = P_M \psi_{1d} I_{1q} + P_M (L_d - L_q) I_{1d} I_{1q} = P_M \psi_{1d} I_{1q} \tag{13}$$

where, $\psi_{1d} = \psi_{Md} + \psi_{fd}$, $\psi_{1q} = \psi_{Mq} + \psi_{fq}$, ψ_{1d} and ψ_{1q} are the d- and q-axis components of the equivalent torque winding air gap flux linkage, respectively. ψ_{Md} and ψ_{Mq} are d- and q-axis components of the torque winding air gap flux linkage, respectively. ψ_{fd} and ψ_{fq} are d- and q-axis components of the air gap flux linkage excited by the permanent magnet.

4.3. FEA Validation of the Model

To verify radial suspension force generation, the 2D FEA simulation model is developed by means of the software-ANSYS (ANSYS Inc., Cecil Township, PA, USA) for the BPMSM, as shown in Figure 3b. The main specifications of the BPMSM simulation model are listed in Table 1.

Table 1. Specification of the prototype machine.

Parameter	Symbol	Vaule
Motor unit geometry and torque system data		
Rated mechanical power	P_N	500 W
Stator slot count	Q	24
Stator inner diameter	d_{si}	75 mm
Stator outer diameter	d_{so}	120 mm
Rotor outer diameter	d_{ro}	73 mm
Auxiliary bearing thickness	h_a	0.7 mm
Pole pair count	P_M	1
Torque winding wire diameter	Φ	0.71 mm
Turns of torque winding per one slot	N_M	22
Stacking coefficient	k_s	0.95
Number of winding parallel ways	a_M	1
Winding connection	Y	-
Mechanical air gap length	g	1 mm
Permanent magnetic material	-	NdFeB
Coercive force	H_c	−890 kA/m
Magnet thickness	h_m	2 mm
Magnet remanence at 20°C	B_r	1.1 T
Suspension force system data		
Pole pair count	P_B	2
Turns of suspension force winding per one slot	N_B	22
Number of winding parallel ways	a_B	1
Winding connection	Y	-
Force/current coefficient	k_m	122.325 N/A
Force/displacement coefficient	k_{ecc}	568.02 N/mm

In Figures 7–9, the FEA results of the radial suspension force are compared with the mathematical model calculation values. This leads to the following conclusions: the amplitudes of the radial suspension force increase with the suspension force winding current I_{2m}, as illustrated in Figure 7a. When the suspension force winding current I_2 is small (less than 3 A), the radial suspension forces show a linearly increasing trend. The FEA results are in good agreement with the mathematical model calculations and the error between them is less than 3 percent, as depicted in Table 2. When the current has reached 3 A, the motor core enters a saturated state. Thus the FEA results are significantly less than the values of the mathematical model calculation, and the error between them is more than 10 percent, so the rated suspension force current is set at 2 A in this paper, and the maximum radial suspension force available in this case is 244.65 N. Besides, from Figure 7b, it can be seen that the direction of the radial suspension force is proportional to the initial phase angle λ_2 of the suspension force winding current. It is almost unaffected by saturation of the magnetic circuit which is completely consistent with the mathematical model calculation values. In conclusion, when the motor operates in a linear state, the amplitude and direction of the radial suspension force change with the magnitude and phase angle of the suspension force winding current, respectively. On the basis of this rule, the orientation control of radial suspension force can be realized, which can build a foundation for the BPMSM.

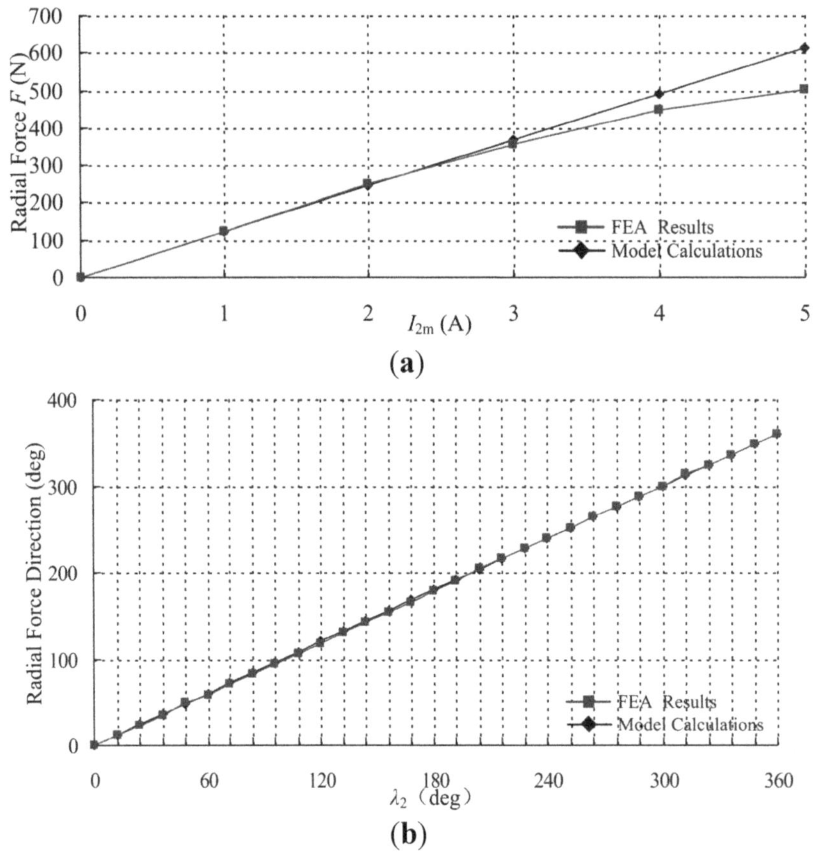

Figure 7. Relationships between radial suspension forces and suspension currents. (**a**) Radial suspension force *versus* I_{2m} (I_{1m} = 0 A, λ_2 = 0 deg), (**b**) Radial suspension force *versus* λ_2 (I_{1m} = 0 A, I_{2m} = 2 A). FEA—Finite Element Analysis.

In Figure 8, the radial suspension force component in the *x*-direction changes in cosines with initial phase angle λ_2 of suspension force winding current, and the *y*-direction component is dominated by the sines rule with initial phase angle λ_2. The variation tendencies of the FEA results and mathematical model calculations are essentially identical. When the suspension force winding current is zero, the related FEA for a unilateral magnetic force is carried out, as shown in Figure 9.

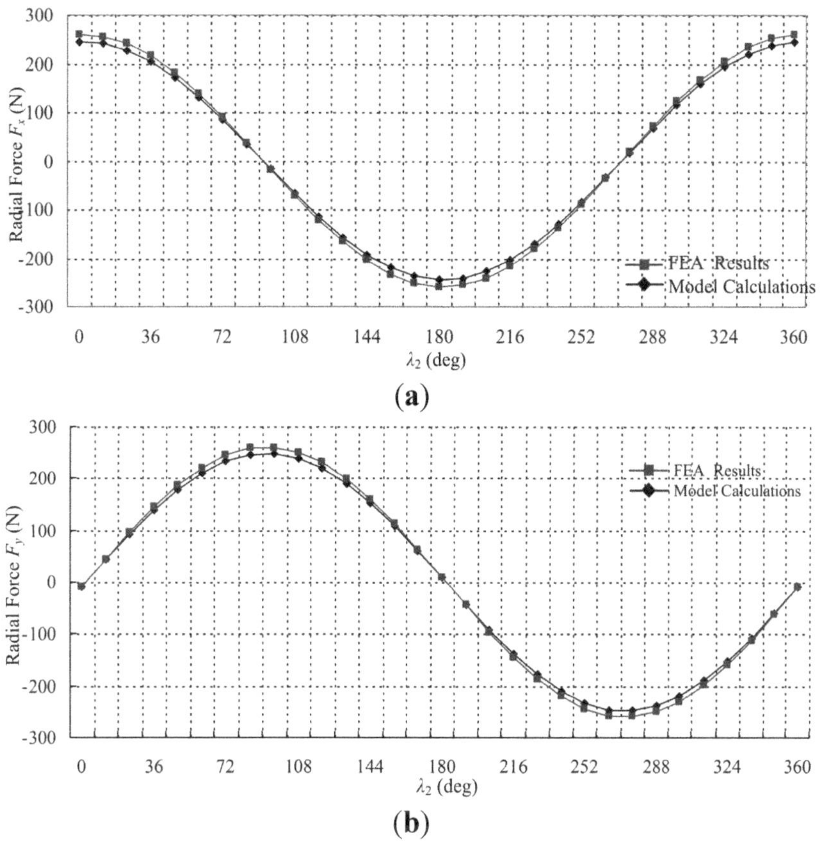

Figure 8. Effects of suspension force winding spatial position angle λ_2 on suspension force. (**a**) Radial suspension force F_x *versus* λ_2 ($I_1 = 0$ A, $I_2 = 2$ A), (**b**) Radial suspension force F_y *versus* λ_2 ($I_1 = 0$ A, $I_2 = 2$ A).

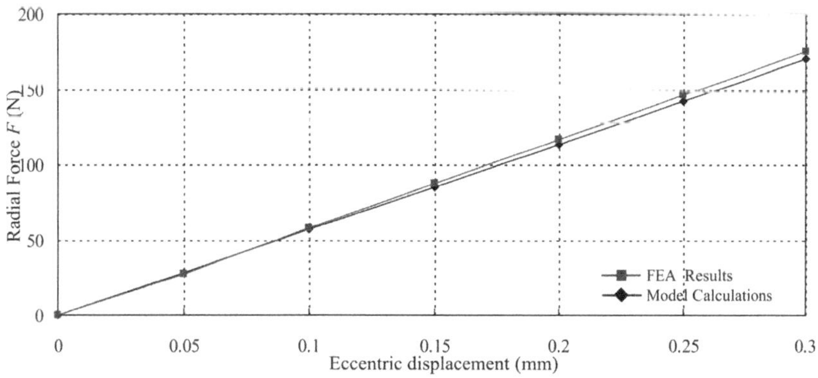

Figure 9. Relationship between unilateral magnetic force and eccentric displacement of rotor ($I_1 = 0$ A, $I_2 = 0$ A).

Table 2. Comparison of FEA and mathematical models in controllable radial suspension forces.

I_{Bm} (A)	The FEA results (N)	Mathematical model values (N)	Error
1	123.5032	122.325	0.96%
2	247.6507	244.65	1.23%
3	365.9582	366.975	2.73%
4	448.0844	489.3	8.42%
5	501.6163	611.625	17.99%

It's concluded that the FEA results basically agree well with the mathematical model calculation values, and the error between them is less than 2.7 percent (Table 3). Due to the assembly auxiliary bearing, the maximum rotor radial deviation ranges from −0.3 mm to 0.3 mm, and the amplitude of maximum unilateral magnetic force is 175 N.

Table 3. Comparison of FEA and mathematical model in unilateral magnetic force.

Eccentric displacement (mm)	The FEA results (N)	Mathematical model values (N)	Error
0.1	28.4	27.34	1.99%
0.2	113.604	116.462	2.51%
0.3	170.46	174.953	2.67%

5. Experimental System for the BPMSM

5.1. Prototype Machine Structure

The 5-DOF suspension control for the BPMSM is an essential prerequisite to realize its high-speed and stable operation. The prototype BPMSM, consisting of a 3-DOF axial-radial magnetic bearing and a 2-DOF BPMSM, has been built, as shown in Figure 10.

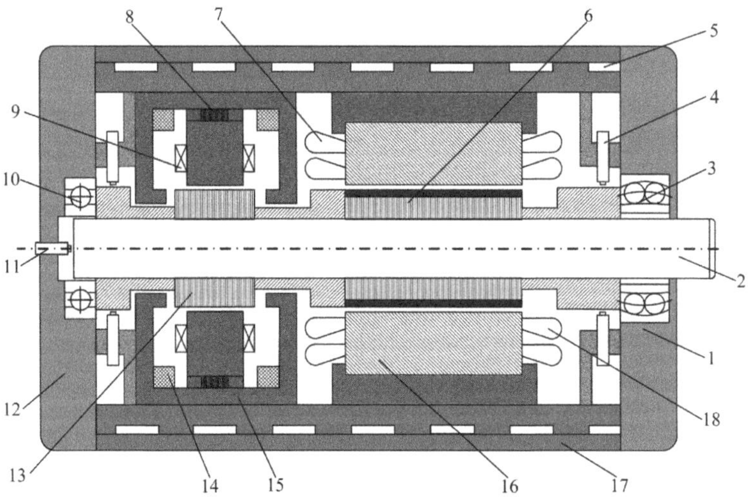

Figure 10. Schematic drawing of a test machine: 1-Front cover, 2-Shaft, 3-Radial-axial auxiliary bearing, 4-Radial displacement sensor, 5-Internal thread cooling pipe, 6-Permanent magnet rotor, 7-Suspension force winding overhang, 8-Permanent magnet, 9-Radial control windings, 10-Radial auxiliary bearing, 11-Axial displacement sensor, 12-Bake cover, 13-Rotor, 14-Axial control windings, 15-Axial stator lamination, 16-Stator lamination, 17-Aluminium housing, 18-Torque winding overhang.

The 2-DOF on the right end of the high-speed machine and the rotor rotation are controlled by the BPMSM unit, while the other 2-DOF on the left end and the axial direction are controlled by the 3-DOF axial-radial magnetic bearing unit. Among them, the BPMSM unit is the key technology to realize the stable operation of the high-speed machine. Thus, this paper mainly focuses on the 2-DOF BPMSM unit.

The stator and rotor iron core are wire cut from DW465_50 silicon steel sheet and neodymium-iron-boron (NdFeB) is selected as the permanent magnet material. The rotor radial positions are measured by eddy-current displacement sensors and the rotation speed is measured by a photoelectric encoder. If there is no external disturbance or load imposed on the rotor, the radial suspension force just supports the rotor in the balance position.

5.2. Control of Torque and Radial Suspension Force for the BPMSM

The air gap magnetic field of the BPMSM is composed of the suspension force winding air gap magnetic field, torque winding air gap magnetic field and permanent magnet air gap magnetic field. The torque production mechanism is the same as that of PMSM, namely, the product of the interactions of the permanent magnet and torque winding air gap magnetic field. In addition, because of its long effective air gap length for the surface-mounted BPMSM, the torque winding air gap magnetic field can be ignored compared with the permanent magnet air gap magnetic field. Then, the radial suspension force could be considered as the interactions between the suspension force winding air gap magnetic field and permanent magnet air gap magnetic field. Thus, the decoupling control between torque and radial suspension force can be realized by adoption of the traditional FOC [22].

Figure 11 shows the relative position between torque windings and suspension force windings. In order to facilitate the control of the radial suspension force, the *xoy* coordinate is set up along the displacement sensors. It is supposed that the angle between the *x*-axis and A phase axes of the torque windings ($P_M = 1$) or the nearest A phase axes relative to the positive *x*-axis ($P_M > 1$) is θ_M, and the angle between the *x*-axis and A phase axes of suspension force windings ($P_B = 1$) or the nearest A phase axes relative to the positive *x*-axis ($P_B > 1$) is θ_B. If $\theta_M \neq \theta_B$, defining $\Delta\theta = \theta_B - \theta_M$, (see Figure 11a). An advanced mechanical angle $\Delta\theta$ must be compensated by the suspension force winding currents. Based on the relative position relations between the suspension force winding air gap magnetic field and the torque winding air gap magnetic field, the result is equivalent to $\Delta\theta = 0°$, (see Figure 11b). If $P_B = P_M + 1$, the advanced electrical angle θ_1 must be compensated by the suspension force winding currents. If $P_B = P_M - 1$, the lag electrical angle θ_1 must be compensated by the suspension force winding currents. Thus, the suspension force winding air gap magnetic field and torque winding air gap magnetic field have the same direction of rotation and electric angular velocity. In the control system, the suspension force winding current command i_{Bd} will directly control the component of the radial suspension force in the *x*-axis. Then, due to the current component i_{Bq} advancing i_{Bd} with mechanical angle $90°/P_B$, if $P_B = P_M - 1$, the current component i_{Bq} will generate a radial suspension force lagging 90° with respect to the *x*-axis. If $P_B = P_M + 1$, the current component i_{Bq} will produce a radial suspension force advancing 90° with respect to the *x*-axis. Consequently, the 2-DOF radial suspension force control for the BPMSM can be realized.

When the BPMSM runs at load states, the torque winding current component i_{Mq} will produce an air gap magnetic field in the q-axis. Then, an advancing electrical angle of θ_1 (< 90°) is produced by the torque winding air gap magnetic field. In other words, the suspension force winding air gap magnetic field lags a mechanical angle θ_1/P_M with respect to the torque winding air gap magnetic field. If $P_B = P_M + 1$, the angle of suspension force is $\theta_1/P_M + (-P_B\theta_1/P_M) = -\theta_1$. To obtain a stable suspension force, an advancing mechanical angle θ_1/P_B must be compensated by the suspension force winding currents. If $P_B = P_M - 1$, the angle of radial suspension force is $\theta_1/P_M + P_B\theta_1/P_M = \theta_1$. Similarly, an advancing mechanical angle θ_1/P_B must be compensated by the suspension force winding currents. The advanced angle θ_1 of the torque winding current can be obtained by the proportional relation the air gap magnetic field of torque winding current component i_q and the original torque winding air gap magnetic field.

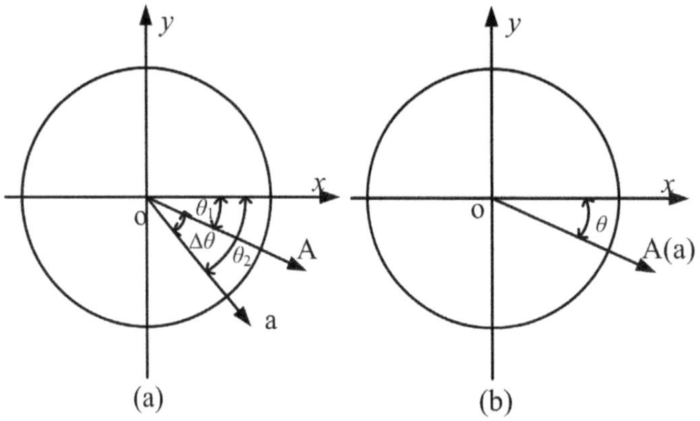

Figure 11. Relative positions of torque windings and suspension force windings.

Based on the rotor flux-orientation control, it is concluded that:

$$\begin{aligned} &I_{Md} = 0,\ I_{1d} = I_f,\ I_{1q} = I_{Mq} \\ &\psi_{fd} = \psi_f, \psi_{fq} = 0 \end{aligned} \tag{14}$$

Thus, Equations (12) and (13) can be expressed respectively as:

$$\begin{pmatrix} F_x \\ F_y \end{pmatrix} = k_m \begin{pmatrix} I_f & I_{Mq} \\ I_{Mq} & -I_f \end{pmatrix} \begin{pmatrix} I_{Bd} \\ \pm I_{Bq} \end{pmatrix} + k_{ecc} \begin{pmatrix} x \\ y \end{pmatrix} \tag{15}$$

$$T_{em} = P_M \psi_f I_{Mq} \tag{16}$$

Hence, the radial suspension forces and torque are controlled by changing the suspension force winding current (I_{Bd}, I_{Bq}) and torque winding current (I_{Mq}), respectively. Moreover, compared with CRPWM, the space vector pulse width modulation (SVPWM) method can reduce the inverter switching frequency and decrease the harmonic content of the inverter output current. Hence, a control strategy based on the SVPWM method is proposed to realize the FOC in this section. Figure 12 shows the vector control block diagram, which consists of two basic schemes, namely, torque control and radial suspension force control.

In torque control, the rotational speed signal n is obtained by an optical-electricity encoder in the torque control system. The error between the speed command n^* and measurement n is amplified in the proportional-integral (PI) controllers, and the torque winding current command i_{Mq}* can be generated by the torque control algorithm. Due to the adoption of field orientation control, the torque current commands i_{Md}* is defined as zero. Then, the reference voltage vectors $V_{M\alpha}$* and $V_{M\beta}$* can be generated through the current PI controllers and PARK inverse transformation. Finally, through the SVPWM modulation module, six PWM signals are generated to drive the voltage source inverter.

In radial suspension force control, the radial displacement closed-loop control is adopted as well as the current closed-loop control. The radial positions x and y of the rotor are measured by eddy current sensors. The error between the displacement commands and measurements is amplified in the proportional-integral-differential (PID) controllers. Thus, radial suspension force commands F_x* and F_y* are generated. Through the radial suspension force control algorithm, the F_x* and F_y* are transformed into the current commands i_{Bd}* and i_{Bq}* under the d-q rotating coordinate system. Then, the reference voltage vectors $V_{B\alpha}$* and $V_{B\beta}$* can be obtained through the current PI controllers and PARK inverse transformation. Finally, through the SVPWM modulation module, 6 PWM signals are produced to drive the voltage source inverter.

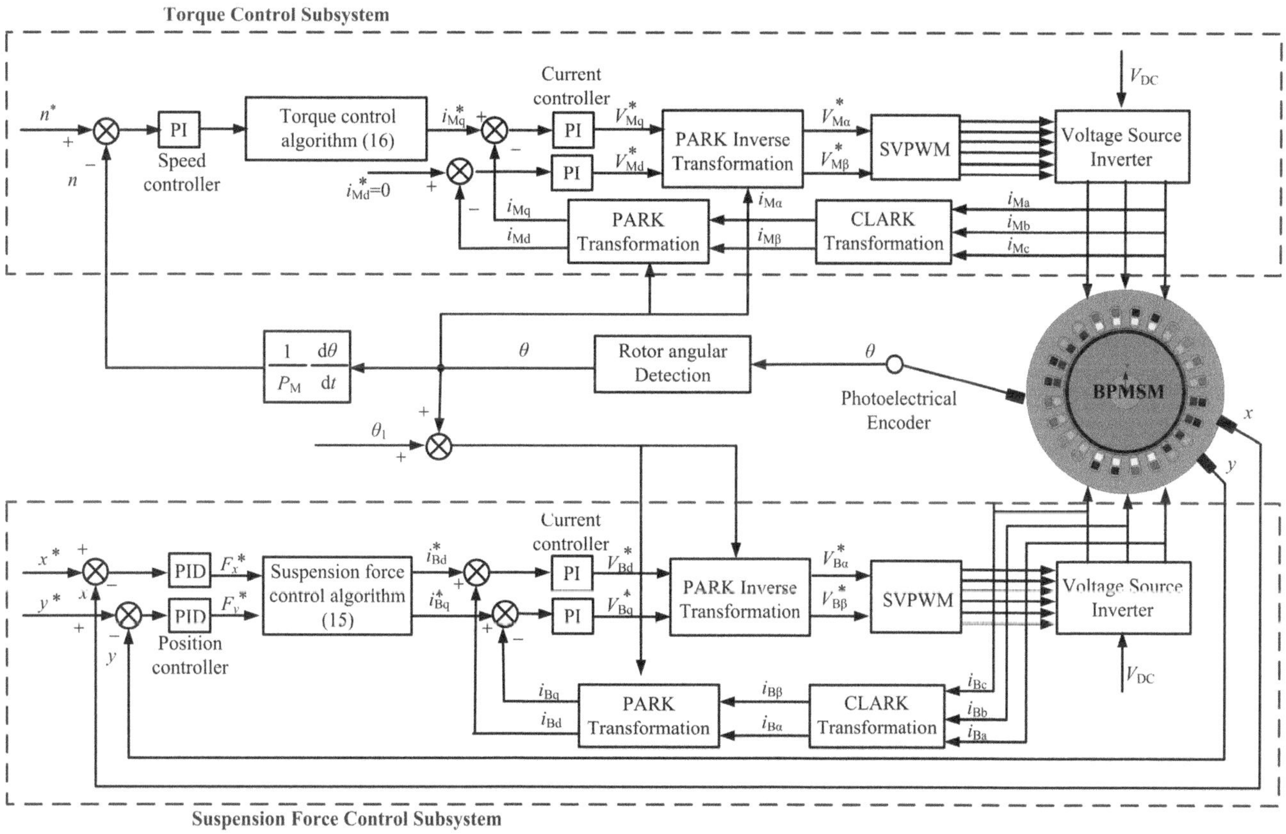

Figure 12. A vector control block diagram of BPMSM.

On the basis of the above analyses, the double closed control circuits are used both in the torque control and radial suspension force control systems. Thereinto, PID controller tuning is one of the key and more difficult points for stable suspension operation. In this paper, the engineering tuning method is adopted, due to its advantages of simplicity, valid and easy operation. Below, the specific setting steps are given. Firstly, set the integral gain k_i and differential gain k_d to zero, increase the proportional gain k_p until the output of the loop starts to oscillate. Then the k_p should be set as 0.6–0.7 times the current value. Increase the integral gain k_i until the loop oscillates as well, and the k_i should be set as 1.5–1.8 times the current value. Then, keep k_i unchanged and change the value of k_p for tracking control. If the condition is improved, continue to adjust the k_p until a satisfactory result is achieved. Otherwise increase the original k_p a bit, adjust the k_i to improve the control effect. Repetition goes on until the best coefficients of the integral gain k_i and the proportional gain k_p are achieved. Follow the same steps above, the appropriate differential gain k_d can also be obtained.

6. Analysis on Experimental

To verify the validity of the proposed analysis method, relevant experiments are designed and conducted for the surface-mounted BPMSM. The basic parameters of the experimental prototype are as follows: rated power P_N is 500 W, rotor outer diameter d_{ro} is 73 mm, permanent magnet thickness h_m is 2 mm, air gap length g is 1 mm, torque winding resistance is 2.07 Ω, the d- and q-axis inductance of the torque windings are 8 mH, suspension force winding resistance is 1.13 Ω, the d- and q-axis inductance of the suspension force windings are 3.6 mH, the mutual inductance between torque and suspension force windings is 0.16 mH, air gap length g_1 with the auxiliary bearing is 0.3 mm.

Figure 13 shows an experiment system which takes a TMS320F2812 digital signal processor as the BPMSM drive controller, and an intelligent power module (IPM) as its power driver [23,24]. Besides, the range of analog input voltage is 0–3V in the digital converter of DSP board, but the output voltage of the displacement sensor QH8500 ranges from −2 V to −18 V and the output range of the CNSE151-100 Hall sensor is −24 mA–24 mA. Thus, the corresponding signals detected by the sensors must be transferred to the DSP board through the corresponding interface circuit. After that, when the rotor is located in the equilibrium position, the output voltage is set as 1.5 V. The output voltage should be 3 V when the rotor is located at the maximum displacement with the positive x-axis and y-axis, while the output voltage should be 0 V when the rotor is located at the minimum displacement of the negative x-axis and y-axis.

Figure 13. Experimental system of the prototype motor. DC — direct current; DSP—Digital Signal Processor.

The voltage signal is linear with the rotor displacement in the other position. Based on this method, the radial position of rotor can be detected accurately. Besides, the current interface circuit must be employed to transform the sensor output signal to the range of 0–3 V. Finally, the PID controller parameters can be achieved by the method above: in the torque control system, the current closed loop k_p is 0.6, and k_i is 0.04, the speed closed loop k_p is 0.6, and k_i is 0.00005; in the suspension force control system, the current closed loop k_p is 0.90, and k_i is 0.005, the displacement closed loop k_p is 1.8, k_i is 0.006 2, and k_d is 0.85.Then, the suspension operation tests are carried out to verify the effectiveness of the proposed equivalent winding current analysis method. Figure 14 shows the radial displacement waveforms in the *x*- and *y*-direction with the start of suspension.

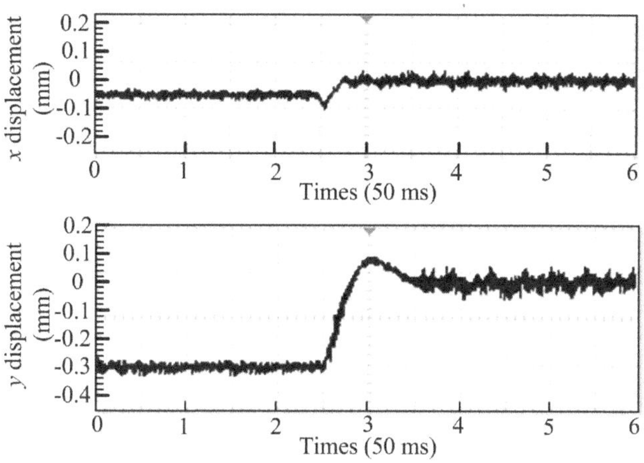

Figure 14. Radial displacement waveforms in the *x*- and *y*-direction when the start of levitation.

When the radial suspension force control system is not active, the rotor stops on a random position of the auxiliary bearing due to the influence of gravity. From Figure 14, it can be found that the original position of the rotor is (−0.05 mm, −0.3 mm). When the radial suspension force control system is active, the rotor comes rapidly back to the equilibrium position (0 mm, 0 mm). At this time, the rise time in the *y*-direction is 0.3 s, the settling time is 1 s, and the maximum deviation is 67 μm, thus the maximum overshoot is less than 22.67%, which is much smaller than the air gap at the equilibrium point (g_0 = 0.3 mm). In the *x*-direction, the vibration amplitude is about 20 μm, and the vibration amplitude is almost 32 μm in the *y*-direction. The deviations of radial displacements, which are mainly caused by mass unbalance of the rotor, are acceptable. The deviations can be reduced by more precise control schemes with an unbalance compensation circuit. The frequency of the radial displacement vibration is dependent on the rotating speed of the rotor. Figures 15–17 show the rotor position angle and rotation speed waveforms, the radial displacement waveforms in the *x*- and *y*-direction, and the waveform of relationship between them, respectively, when the rated currents are applied. In Figure 15, the rotor position angle changes periodically with time in the range of 0°–360°. The corresponding frequency is 50 Hz, which is in agreement with the rotational speed calculation formula, and the oscillation of the rotational speed is slight. Thereby, the reasonable operating characteristic of speed closed loop is verified at the speed of 3000 r/min.

From Figure 16 and Figure 17 the maximal displacement in the positive *x*-axis is about 16 μm and 12 μm in the negative *x*-axis. The maximal displacement in the positive *y*-axis is about 18 μm and 21 μm in the negative *y*-axis, when the rotor runs steadily at 3000 r/min. They are significantly less than the length of air gap at the equilibrium point. It is found that the steady suspension characteristics of proposed 2-4 BPMSM can be obtained at 3000 r/min. Due to the effects of gravity, the displacements in the *y*-direction are larger than that in the *x*-direction. Thus, the eccentric displacement track diagrams are nearly-circular or elliptical, as shown in Figure 17.

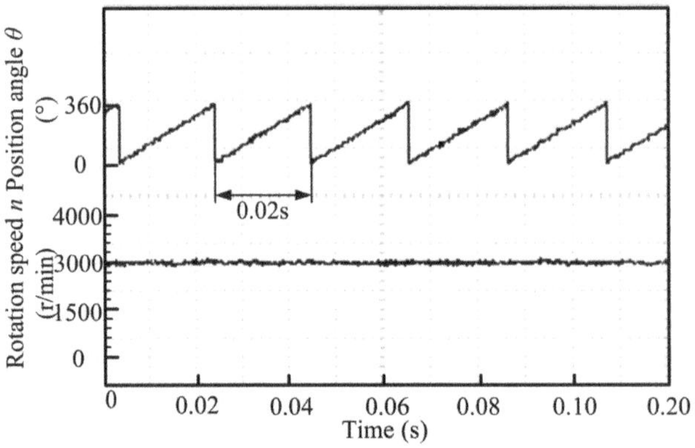

Figure 15. Waveform of the rotor position angle and rotation speed when the stable suspension operation is realized.

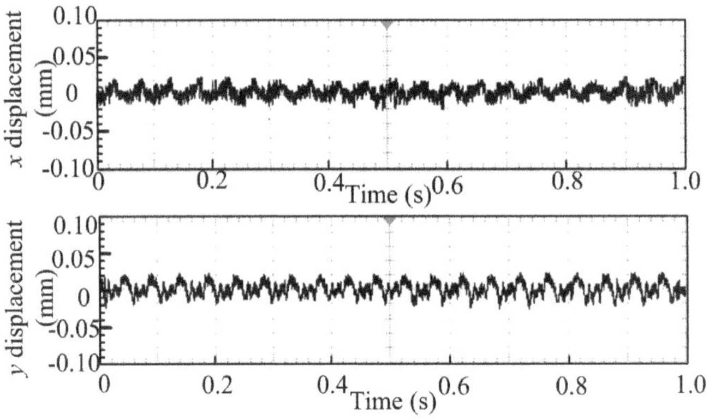

Figure 16. Radial displacement waveforms in the *x*- and *y*-direction when the start of levitation.

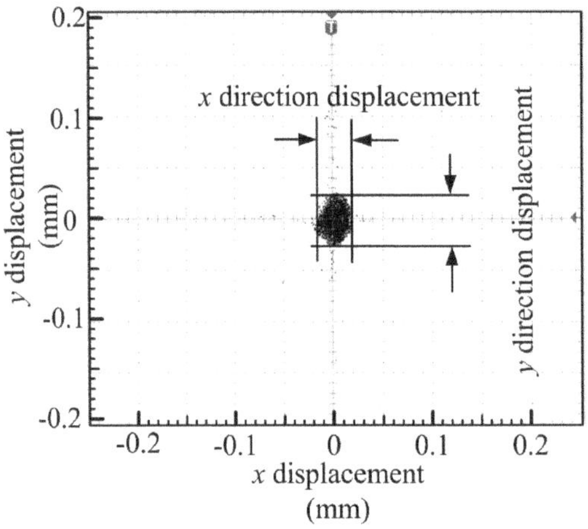

Figure 17. The relationships between radical displacements in the *x*- and *y*-direction for BPMSM.

Figure 18 describes the responses of radical displacement in the *x*- and *y*-direction when the rotation speed is changed from 1200 r/min to 3000 r/min. It can be found that the rotational speed command n^*

is accelerated suddenly from 1200 r/min to 3000 r/min at 0.4 s. After a 0.2 s settling time, the stable suspension operation is realized and the overshoot is 0. The short working time of the proposed experimental prototype shows that the dynamic response of the BPMSM is good. At this point, there is a little vibration in the *x*- and *y*-direction when the speed has been changed. But the changes of vibration amplitudes for the radical displacements which may originate from the motor's unbalance mass are slight. The vibration peak-to-peak value is about 28 μm in the *x*-direction, and the peak-to-peak value is almost 39 μm in the *y*-direction. The radial displacement deviations are within acceptable ranges, which can be reduced further by more precise control schemes with an unbalance compensation circuit, so the independent control between radial displacement and rotating speed can be realized. The validity and feasibility of the control strategy is confirmed, which establishes a solid foundation for further deliberation and research, and has some theoretical and practical reference value.

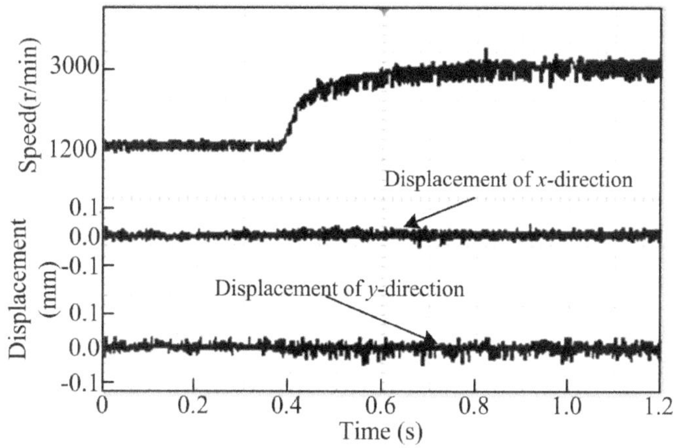

Figure 18. The responses of speed and radical displacement in the *x*- and *y*-direction when the rotation speed is changed.

7. Conclusions

This paper proposes an analysis method based on equivalent winding currents. On that basis, the validity of the necessary conditions to produce a stable radial suspension force for bearingless motors is proved. The simulation model is established by the ANSYS simulation software and the simulation results verify the theoretical analysis. In addition, good rotation and suspension performance are obviously shown by the simulation and experimental results, as well as the accuracy of the proposed analysis method and the feasibility of the control strategy. Besides, the decoupling control between electromagnetic torque and radial suspension force is also successfully achieved. Thus, the new analysis method of the BPMSM has some theoretical significance and practical reference value for further research compared with other methods.

Acknowledgments

This work was sponsored by National Natural Science Foundation of China (60974053, 61174055), Jiangsu Province University Achievements in Scientific Research Industrial Production Advancement Project (JHB2012-39), Jiangsu Province "Qinglan Project" (2014), and Jiangsu Province "333 Project" (2014).

Author Contributions

Huangqiu Zhu proposed the analysis method, performed simulation analysis, and drafted the manuscript. Hui Li designed the control strategy and helped refinement the manuscript. All authors discussed the results and approved the final manuscript.

Conflicts of Interest

The authors declare no conflict of interest.

References

1. Baumgartner, T.; Burkart, R.M.; Kolar, J. Analysis and Design of a 300-W 500 000-r/min Slotless Self-Bearing Permanent-Magnet Motor. *IEEE Trans. Ind. Electron.* **2014**, *61*, 4326–4336.
2. Nguyen, Q.D.; Ueno, S. Modeling and control of Salient-Pole Permanent Magnet Axial-Gap Self-Bearing Motor. *IEEE Trans. Mechatron.* **2011**, *16*, 516–528.
3. Qiu, Z.J.; Deng, Z.Q.; Wang, X.L.; Meng, X.L. Study on the Modeling and Control of a Permanent-magnet-type Bearingless Motor Considering Rotor Eccentricity and Lorenz Force. *Proc. Chin. Soc. Electr. Eng.* **2007**, *27*, 64–70.
4. Zhu, H.Q.; Cheng, Q.L.; Wang, C.B. Modeling of bearingless permanent magnet synchronous motor based on mechanical to electrical coordinates transformation. *Sci. China Ser. E* **2009**, *52*, 3736–3744.
5. Ooshima, M.; Miyazawa, S.; Chiba, A.; Nakamura, F.; Fukao, T. A rotor design of a permanent magnet-type bearingless motor considering demagnetization. In Proceedings of the Power Conversation, Nagaoka, Japan, 3–6 August 1997; pp. 655–660.
6. Sun, X.D.; Zhu, H.Q.; Zhang, T. Nonlinear decoupling control for 5 degrees-of-freedom bearingless permanent magnet synchronous motor. In Proceedings of the Power Electronics and Motion Control Conference, Wuhan, China, 17–20 May 2009.
7. Kascak, P.; Jansen, R.; Dever, T.; Loparo, K.; Nahorny, A. Motoring performance of a conical pole-pair separated bearingless electric machine. In *Energyteck*; IEEE: New York, NY, USA, 2011; pp. 1–6.
8. Zhang, S.; Luo, F.L. Direct control of radial displacement for bearingless permanent-magnet-type synchronous motors. *IEEE Trans. Ind. Electron.* **2009**, *56*, 542–552.
9. Sun, X.D.; Zhu, H.Q.; Zhang, T. Sliding mode variable structure control for radial suspension forces of bearingless permanent magnet synchronous motor based on inverse system method. In Proceedings of the Power Electronic and Motion Control Conference, Wuhan, China, 17–20 May 2009.
10. Sun, X.D.; Zhu, H.Q.; Pan, W. Decoupling control of bearingless permanent magnet-type synchronous motor using artificial neural networks-based inverse system method. *Int. J. Model. Ident. Control* **2009**, *8*, 114–121.
11. Khoo, W.K.S. Bridge configured winding for polyphase self-bearing machines. *IEEE Trans. Magn.* **2005**, *41*, 1289–1295.

12. Zhang, S.R.; Luo, F.L. Direct Control of Radial Displacement for Bearingless Permanent-Magnet-Type Synchronous Motors. *IEEE Trans. Ind. Electron.* **2009**, *56*, 542–552.
13. Amrhein, W.; Silber, S.; Nenninger, K. Levitation forces in bearingless permanent magnet motors. *IEEE Trans. Magn.* **1999**, *35*, 4052–4054.
14. Chiba, A.; Fukao, T.; Ichikawa, O.; Oshima, M.; Takemoto, M.; Dorrell, D.G. *Magnetic Bearings and Bearingless Drives*; Elserver: Amsterdam, The Netherland, 2005; p. 400.
15. Chiba, A.; Hanazawa, M.; Fukao, T.; Rahman, M.A. Effects of magnetic saturation on radial suspension force of bearingless synchronous reluctance motors. *IEEE Trans. Ind. Appl.* **1996**, *32*, 354–362.
16. Li, H.; Zhu, H.Q. Bearingless motor's radial suspension force control based on virtual winding current analysis. In Proceedings of the Electrical Machines and Systems (ICEMS), Hangzhou, China, 22–25 October 2014; pp. 2141–2146.
17. Pillay, P.; Krishnan, R. Modeling of permanent magnet motor drives. *IEEE Trans. Ind. Election.* **2002**, *35*, 537–541.
18. Sun, X.D.; Chen, L.; Yang, Z.B.; Zhu, H.Q.; Zuo, W.Q.; Shi, K. Modeling of a Bearingless Permanent Magnet Synchronous Motor Considering Rotor Eccentricity and Coupling Relationship of Windings. *Trans. CES* **2013**, *28*, 63–70.
19. Zhu, H.Q.; Wei, J. Radial suspension force modeling on interior bearingless permanent magnet synchronous motors. *Electr. Mach. Control* **2013**, *5*, 45–50.
20. Zhu, H.Q.; Wu, L.; Li, T.H.; Sun, Y.K. Suspension principle of rotor and decoupling control for bearingless permanent magnet-type synchronous motors. *J. Southeast Univ. (Nat. Sci. Edit.)* **2005**, *35*, 34–38.
21. Inagaki, K.; Chiba, A.; Rahman, M.A.; Fukao, T. Performance characteristics of inset-type permanent magnet bearingless motor drives. In Proceedings of the Power Engineering Society Winter Meeting, Singapore, 23–27 January 2000; pp. 202–207.
22. Deng, Z.Q.; Qiu, Z.J.; Wang, X.L.; Yan, Y.G. Study on rotor flux orientation control of permanent magnet bearingless synchronous motors. *Proc. Chin. Soc. Electr. Eng.* **2005**, *25*, 48–52.
23. Muntcanu, G.; Binder, A.; Schneider, T.; Funieru, B. No-load tests of a 40 kW high-speed bearingless permanent magnet synchronous motor. In Proceedings of the 2010 International Symposium on Power Electronics Electrical Drives Automation and Motion (SPEEDAM), Pisa, Italy, 14–16 June 2010; IEEE: New York, NY, USA; pp. 1460–1465.
24. Yajima, S.; Takemoto, M.; Tanaka, Y.; Chiba, A.; Tadashi, F. Total efficiency of a deeply buried permanent magnet type bearingless motor equipped with 2-pole motor windings and 4-pole suspension windings. In Proceedings of the Power Engineering Society General Meeting, Tampa, FL, USA, 24–28 June 2007; IEEE: New York, NY, USA; pp. 1–7.

6

Residual Mix Calculation at the Heart of Reliable Electricity Disclosure in Europe—A Case Study on the Effect of the RE-DISS Project

Markus Klimscheffskij [1,*], Thierry Van Craenenbroeck [2,†], Marko Lehtovaara [1,†], Diane Lescot [3,†], Angela Tschernutter [4,†], Claudia Raimundo [5,†], Dominik Seebach [6,†] and Christof Timpe [6,†]

[1] Grexel Systems Oy, Lautatarhankatu 6, FI-00580 Helsinki, Finland; E-Mail: marko.lehtovaara@grexel.com

[2] Vlaamse regulator van de elektriciteits-en gasmarkt, 1000 Brussels, Belgium; E-Mail: thierry.vancraenenbroeck@vreg.be

[3] Observatoire des énergies renouvelables, 146 rue de l'Université, 75007 Paris, France; E-Mail: diane.lescot@energies-renouvelables.org

[4] Energie-Control Austria, Rudolfsplatz 13a, 1010 Vienna, Austria; E-Mail: angela.tschernutter@e-control.at

[5] IT Power Consulting Ltd., St. Brandon's House 29 Great George Street, Bristol BS1 5QT, UK; E-Mail: claudia.raimundo@itpowergroup.com

[6] Oeko-Institut e.V., Merzhauser Strasse 173, PO Box 17 71, 79017 Freiburg, Germany; E-Mails: d.seebach@oeko.de (D.S.); c.timpe@oeko.de (C.T.)

† These authors contributed equally to this work.

* Author to whom correspondence should be addressed; E-Mail: markus.klimscheffskij@grexel.com

Academic Editor: Vincenzo Dovì

Abstract: In the EU, electricity suppliers are obliged to disclose to their customers the energy origin and environmental impacts of sold electricity. To this end, guarantees of origin (GOs) are used to explicitly track electricity generation attributes to individual electricity consumers. When part of a reliable electricity disclosure system, GOs deliver an important means for consumers to participate in the support of renewable power. In order to be considered reliable, GOs require the support of an implicit disclosure system, a residual mix, which prevents once explicitly tracked attributes from being double counted in a default energy mix. This article outlines the key problems in implicit electricity disclosure: (1) uncorrected generation statistics used for implicit disclosure; (2) contract-based tracking; (3) uncoordinated calculation within Europe; (4) overlapping regions for implicit disclosure; (5) active GOs. The improvements achieved during the RE-DISS project (04/2010-10/2012) with regard to these problems have reduced the total implicit disclosure error by 168 TWh and double counting of renewable generation attributes by 70 TWh, in 16 selected countries. Quantitatively, largest individual improvements were achieved in Norway, Germany and Italy. Within the 16 countries, a total disclosure error of 75 TWh and double counting of renewable generation attributes of 36 TWh still reside after the end of the project on national level. Regarding the residual mix calculation methodology, the article justifies the implementation of a shifted transaction-based method instead of a production year-based method.

Keywords: guarantee of origin; electricity disclosure; residual mix

---

1. Introduction

Electricity flows to our houses and businesses from a mix of sources: from all the power stations that are connected to our power system [1]. Hence, in the physical sense, the origin of the energy that lights up our living rooms is just about as traceable as the origin of the dust on our shoes.

Nonetheless, Directive 2009/72/EC, Article 3(9) [2], requires electricity suppliers to disclose in electricity bills and in promotional materials the energy origin of the electricity they are selling. This process is commonly referred to as *electricity disclosure*, which constitutes of both explicit tracking and implicit allocation of electricity generation attributes (Information about the characteristics of the power production which need to be tracked, most importantly the "energy source and technology used for power production, the related CO_2 emissions and radioactive waste produced" [3]) from production to consumption.

In the EU, guarantees of origin (GO) as defined by Article 15 of the Directive 2009/28/EC, are the main mechanism for explicit tracking of electricity generation attributes. As long as not all electricity consumption is explicitly tracked to certain generation attributes, explicit tracking mechanisms always require the support of an implicit disclosure system, a *residual mix*, in order to avoid double counting. This is important, because consumers expect the tracking system to be reliable and their willingness to purchase renewable power can be significantly reduced if the attributes are double counted [4]. Residual mix is the main topic of this article.

This article presents the improvements in the implicit electricity disclosure of 16 (10 were chosen in the analysis due to their participation in the RE-DISS Project (Austria, Belgium, Denmark, Finland, Italy, Luxembourg, The Netherlands, Norway, Sweden and Switzerland) and 6 due to their relevance in terms of the European GO and/or electricity market (France, Germany, Ireland, Portugal, Slovenia and Spain)) selected European countries, achieved during the RE-DISS (Reliable Disclosure Systems in Europe, Berlin, Germany) project (4/2010–10/2012). These improvements are two-fold. First, RE-DISS has helped competent bodies of electricity disclosure in setting up a reliable disclosure system in their countries through a series of workshops, bilateral meetings, tailor-made recommendations and commenting of legal texts. The effect of these actions, and actions taken independently by Competent Bodies during the project, is quantified in Section 4.1, after comparing implicit disclosure errors before (2010) and after (2012) the project in Section 3.2. Second, RE-DISS has generated significant enhancements in the residual mix calculation methodology as developed by the E-Track [5] project. These enhancements, and their effects in the avoidance of double counting, are elaborated in Section 4.2 and later discussed in Section 5.

The article is structured as follows. Section 2.1 explains the importance of reliable electricity disclosure information in today's electricity markets. Sections 2.2 and 2.3 aim at elaborating how the current explicit and implicit (respectively) tracking and disclosure systems work in Europe and what is their role in renewable energy support. Section 3.1 presents five types of implicit disclosure problems discovered in the 16 selected countries before the RE-DISS project, and Section 3.2 describes how these problems have been overcome during the RE-DISS project, highlighting the major advancements. Section 4.1 quantifies the implicit disclosure error avoided due to these improvements. In Section 4.2, the residual mix calculation methodology presented in Section 2.3 is further studied. Also, the major deficiency of the methodology, corrected during the RE-DISS project, is depicted and new improvements are proposed. Section 5, shows what remains to be done in the field of implicit disclosure in the 16 countries as well as discusses the options for the further improvements of the residual mix calculation methodology.

The scientific scope of this study is the reliable implementation of the guarantee of origin and electricity disclosure mechanisms in the 16 selected European countries. Therefore critical assessment of the effects of GOs and disclosure on the RE market is left out of the research; the focus being on how well the mechanisms are operating. Furthermore, the European countries besides the 16 were not included due to restrictions of available data and small relevance to the international GO system.

Since the article focuses on the implementation of the mechanism itself, it also doesn't consider the sustainability of different energy sources. GO is a mechanism of consumer choice and does not in itself promote sustainable energy production and therefore detailed analysis of various fuels (including their Life-Cycle Assessment) is left out of the study.

The accuracy of the analysis is also limited in how well it considers so-called non-Reliable Tracking Systems. The main problem of these systems is the availability of data and therefore some volumes require estimation. Where estimations are used, this is clearly elaborated in the text.

Finally an important limitation regarding the results is that they are calculated at a national level. It is highly likely that problems beyond this analysis exist at individual supplier level, but these could not be assessed in the scope of this study, because data collection was made on a country rather than individual supplier level.

2. Electricity Tracking and Disclosure in Modern Electricity Markets

2.1. The Role of Electricity Disclosure in the Modern Electricity Markets

The cornerstone of a well-functioning internal electricity market is a deliberate choice of a supplier by the consumers. This choice is founded on general commercial aspects such as price, quality and reliability of service, but it also relates to the generation characteristics of the electricity supplied.

As mentioned, through 2009/72/EC, Article 3(9) [2], customers have the right to know the generation attributes of the electricity supplied to them, meaning from which energy source(s) the electricity was generated and what impacts this production had on the environment (CO_2 and radioactive waste generated). A reliable and transparent electricity disclosure system leads to trust in the energy market [6]. Also references [7,8] point out the importance of reliable disclosure information not only for consumers, but for energy companies as well.

Over the past years, energy regulators have put customer empowerment at the heart of their working program [9]. This development is vital, because according to [10,11] customers' trust in the supplier and the disclosure information increases when a governmental body is involved in checking this information. Also the European Commission (EC) has established the Citizens' Energy Forum in 2007 as a new regulatory platform based on the experiences gained in the Florence and Madrid Forums [12]. "The aim of the Forum is the implementation of competitive, energy-efficient and fair retail energy markets for consumers" [12], and this starts from empowered customers.

Trustworthy and well-presented ([6] highlight that because the electricity bills already tend to be confusing and complex, a clear structured, easy to understand disclosure information is essential. The information is most effective when placed on the bill, whereas placed in an attachment or appendix turns out to be less effective [7,11]. However the placing and appearance of disclosure information is not the topic of this article, but rather the reliability of this information) electricity disclosure information empowers the consumer to opt for desired sources of energy production and leads to a shift in the relevance of parameters for a decision to contract a supplier. Based on their survey, [6] see a possible link between transparent disclosure information available and the switching of electricity suppliers by consumers, but stress that this will only happen if the disclosure system is considered reliable and worthy of paying a premium. Consequently, through purchase decisions, customers are in the position to influence the policies, and in particular the energy mix, of suppliers and therefore indirectly also stimulate sustainable electricity production: "The advent of electricity disclosure has the potential to bring in a new era of citizen involvement in determining the national electricity generation mix" [6].

To sum up, the objectives of electricity disclosure are related to an increased market transparency, the consumers' right for information for making informed choices, education of consumers and stimulation of sustainable electricity generation [6,11]. Most importantly regarding this article, it is emphasized that the disclosure information needs to be reliable in order for these objectives to be achieved.

2.2. Guarantees of Origin as the Instrument for Explicit Tracking in Building Reliable Electricity Disclosure

In order to provide consumers with electricity disclosure information, electricity suppliers must be able to track the energy origin of the electricity back to its production. Tracking generally means a methodology for the accounting of generation attributes (energy source and environmental impacts) in the electricity market and their allocation to final consumption of electricity [3]. This is much more difficult than it seems at first glance, because the electricity market itself is complex.

Theoretically, there are three options how tracking could be implemented:

- along the physical flows in the electricity grid;
- along the trading arrangements (contracts) in the electricity market; or
- in a separate accounting mechanism which is independent from the physical flows and from electricity contracts.

Typical criticism towards electricity tracking is driven by the argument that electrons cannot be traced, which is a well-known and acknowledged fact [11]. The electricity flows in the grid are a result of the demand patterns, the grid structure and the operation of power plants. A flow in a certain direction is usually caused by the balance of a large number of electricity contracts, and is influenced by technical constraints or incidents. The flows in the grid follow physical laws rather than market activities and thus can hardly be a good basis for the tracking of attributes [13].

Physical trading of electricity (which excludes purely financial instruments for price hedging) might be a more suitable basis for tracking of attributes, as it represents the market activity of generators, traders, suppliers and consumers of electricity. Such approach probably comes close to the intuitive expectations by end-consumers of how tracking should work. However, liberalized electricity markets usually encompass different types of bilateral trading, power exchanges with spot and futures markets as well as balancing power, and thus may be too complex to be used for tracking purposes [14]. Furthermore, if qualities of electricity, in terms of their generation attributes, would be added to the electricity market, creating differences in the wholesale market price of electricity from different energy sources, this would severely damage the liquidity of the electricity market [13].

Therefore, the tracking of attributes for purposes of disclosure should in principle be separated from physical flows and from electricity contracts [15]. This allows a tailor-made design of the tracking mechanism according to the needs of the disclosure scheme.

The most prominent instrument for tracking the origin of electricity is the guarantee of origin (GO), which was first introduced by the European Commission Directives for renewable energy (2001/77/EC) and cogeneration (2004/8/EC), and has been taken over recently in the new Directives 2009/28/EC [16] and 2012/27/EC [17], respectively. According to Article 15 of 2009/28/EC, the GO is an accurate, reliable and fraud-resistant electronic document, issued by nationally appointed competent body, representing the generation attributes of 1 MWh of electricity production from renewable energy sources. The directive goes on to define that a GO can be transferred independently from the trading arrangements and physical flows of the associated electricity and cancelled (*i.e.*, used) to verify the origin of 1 MWh of electricity consumption for the purpose of electricity disclosure. The GO is now a reasonably standardized instrument, at least for the tracking of electricity from renewable energy sources. It is important to note that a guarantee of origin is an objective accounting instrument which aims at providing objective information and can therefore be used to support consumer choice. It does not in itself guarantee to the consumer that the specific electricity is sustainable or complies with a specific definition of "greenness" or "environmental additionality".

An important part of today's GO system is the European Energy Certification System (EECS) maintained by the Association of Issuing Bodies (AIB) (The AIB is a not for profit umbrella organization of Issuing Bodies of different countries. The AIB aims to ensure "the reliable operation of international energy certificate systems" through a standardized system EECS—The European Energy Certificate System [18]). EECS enables technically harmonized procedures for issuing, transferring and cancelling

GOs [19] as well as international transfers between the AIB member countries through the AIB Hub [20]. There are currently 16 European countries in the AIB [21].

A reliable disclosure scheme has to be based on GOs, which can in some cases be complemented by other Reliable Tracking Systems (RTS) (such as: Homogenous disclosure mixes determined by a competent body for consumers in non-competitive electricity markets, as long as such segments still exist.

- Renewable energy support systems (e.g., German feed-in tariff) which require a defined allocation of the attributes of supported generation to consumers for disclosure purposes, which cannot be implemented reasonably based on GOs. In this case a pro-rata allocation of the attributes to all consumers which are paying for the support system can be the adequate solution;
- Under certain conditions, the contracts concluded by market participants in the physical electricity market (also known as contract-based tracking). The conditions include the central collection and supervision of such tracking data by an appointed competent body.

Such other Reliable Tracking Systems should only be introduced, if they provide added value to the tracking system (in addition to GOs and the residual mix). Furthermore they should be implemented in a reliable and transparent way, thus not endangering the reliability of disclosure information provided to consumers).

GOs feature the traceability of electricity back to its origin without disturbing the electricity market whilst avoiding double counting of electricity generation attributes in disclosure [13]. The E-Track project found out that having a reliable tracking system in place is the general expectation of all involved parties [4]. Reliability is also emphasized by the European Commission Directive: "Member States shall ensure that the same unit of energy from renewable sources is taken into account only once." [22].

When part of a reliable disclosure system, manifold possible effects of GOs have been identified:

- Support for renewable energy production

Producers can financially benefit from selling GOs. The prices on the international market are not public, but according to an estimate by the authors of this article, the price for large hydro power GOs currently ranges between 0.05 and 0.5 Euro/MWh [22]. However prices can be much higher especially for small hydro power and new renewable technologies. The extra revenue, albeit small compared to e.g., governmental renewable energy support schemes, might attract new investments [15].

- Interaction with governmental renewable support schemes

GOs can alleviate the budgetary burden of existing governmental renewable energy support schemes by allocating more of the burden to those willing to pay more. Besides relieving direct monetary load from governments, even more importantly, GOs relieve political burden. If consumers consider that renewable energy is worth a premium, they are encouraging the government to retain its current targets, and perhaps even to increase them, e.g., [15,23]. GOs give a voice to the people in an area traditionally dominated by the government: renewable energy support. It is especially important to involve the public into supporting renewable energy when considering the current problems faced by governments in maintaining their support schemes, e.g., [24–27].

- Supplier differentiation and consumer awareness

The customers choose their supplier based on the energy mix and the willingness to switch is higher when transparency on the market is high [28–30]. Further, the influence on energy behavior and the willingness to save energy are major drivers for having a functioning GO system in place [11,31].

- Harmonized tracking system in the EU

GOs are a harmonized mechanism within the EU and can cope with changes in the physical and market structures of electricity [13]. The international orientation of suppliers is becoming more relevant especially since the implementation of the new Internal Electricity Market Directive 2009/72/EC [2]. The fact that a growing number of countries fulfill the criteria on GOs and have adopted the EECS system opens an international demand and exchange of GOs. EECS-GOs alone currently cover more than 300 TWh of European renewable electricity production (Figure 1), which is impressive also due to the fact that large volumes of renewable electricity production are not eligible for GOs due to renewable energy support mechanisms and not all countries have yet adopted the EECS system.

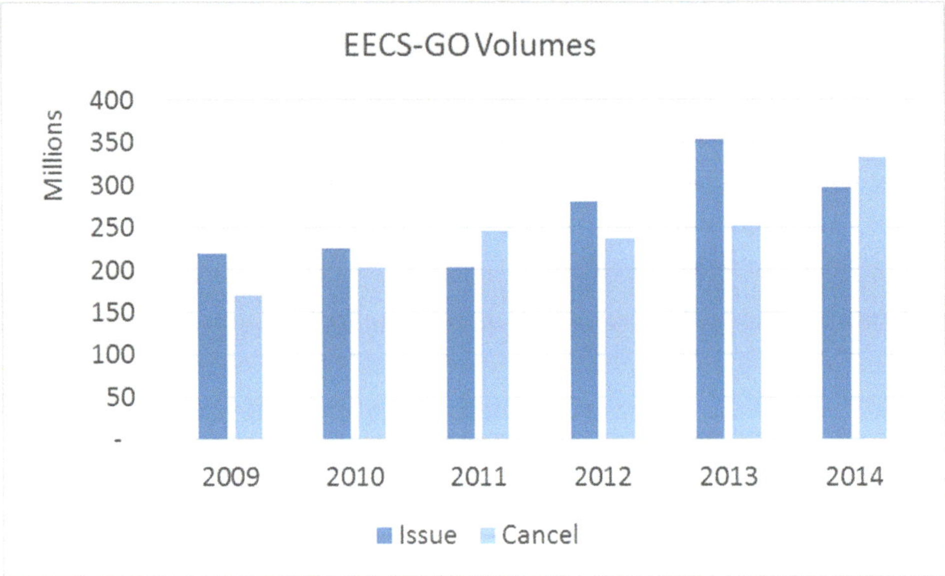

Figure 1. Millions of European Energy Certification System (EECS)-guarantees of origin (GOs) issued and cancelled (used) from 2009–2014 based on [32].

- Enabling operation of labels

According to [33], through the ability to track electricity, GOs enable the reliable operation of electricity labels. These labels use GOs to guarantee the origin of electricity, but add special criteria such as bird-, wildlife or nature conservation, CO_2 reductions, additionality (According to [15], in the context of renewable energy adoption, additionality indicates the additional renewable energy production compared to a Business As Usual case, due to a certain policy or action. For example GOs can deliver additionality, if they increase renewable energy production compared to the situation where no GOs would be issued) *etc.* into the electricity sold to a consumer. Labels can help customers in taking a decision in favor of a specific supplier, as labels are in general easy to understand and aim at a high recognition value.

To sum up, GOs are the mechanism of choice for explicit electricity tracking in the EU and have the potential to bring desired effects into the European electricity system. However, as outlined in Section 2.1, GOs will only be used if electricity disclosure is considered reliable by the consumers, *i.e.*, double counting of generation attributes is avoided [4]. The pronounced rules and guidelines for issuing, transferring and cancelling GOs, especially within the EECS system, efficiently hamper the possibilities for the origin to be tracked twice *explicitly*; but what about consumption not tracked with GOs?

The voluntary characteristic of using explicit tracking instruments is troublesome. Logically, if mostly renewable attributes are explicitly tracked, the remainder of consumption should be less rich in renewables than grid average. One of the most widespread reasons why double counting of renewable attributes still occurs is that competent bodies do not provide a *residual mix*, or do not require its usage for disclosure of consumption which is not explicitly tracked. This may lead to such consumption being disclosed with the production mix of the country where the renewable attributes presented by GOs (and RTSs) are included. The RE-DISS BPR recommend the explicit tracking of all electricity generation attributes, which supports reliable and more meaningful allocation of non-renewable attributes as well. If explicit tracking of electricity were extended to all types of generation attributes, the residual mix would not be required [34] (BPR number 11). (The RE-DISS project has recommended to expand the concept of GOs from electricity from renewable energy and high-efficient cogeneration to all energy sources, in order to support a reliable, explicit allocation of the attributes of any type of power generation [34]. This has already been done in e.g., Austria, Norway, Sweden and Switzerland [35])

The remainder of this article concentrates on the reliability of implicit disclosure, achievable through the correct implementation of the residual mix, because the aims of the GO system can only be achieved where the entire disclosure system is reliable. After all, who would be willing to pay for double counted renewable electricity?

2.3. Residual Mix as the Instrument of Reliable Implicit Electricity Disclosure

Apart from reports of E-TRACK [3,14], and later RE-DISS [36], electricity residual mixes are not well addressed in the academic literature. Raadal *et al.* [37] elaborate the need for a residual mix calculation in an electricity system with explicit tracking, with a focus on Nordic countries. [13] mention that "where default set of attributes is needed a residual mix should be used instead of uncorrected generation statistics in order to minimize multiple counting". This is also a key recommendation of the RE-DISS project [34]. Lastly, [15] has outlined problems in the residual mix calculation methodology, which will be further analyzed in Section 4.2.

Residual mix calculation prevents the double counting of explicitly tracked generation attributes by deducting them from the default mix, which is used for disclosure of non-explicitly tracked consumption. The residual mix is not to be used for any products which are differentiated with regard to the origin of electricity. Its use is restricted to bridging up the difference between the total electricity sales of a supplier and the attributes available from explicit tracking mechanisms where the supplier has not used explicit instruments for the disclosure of its entire sales [34].

The process of residual mix calculation might seem simple, but the international exchange of both electricity and GOs necessitates that also the calculation is coordinated among countries, which adds complexity. In a closed system where no connection to the external world exists, the consumption energy mix of a country equals, in volume and energy sources used, its production mix. But in real life, electricity as well as generation attributes (through GOs) are transferred across borders, which can significantly alter this equilibrium in a country.

The cross-border flow of explicit disclosure information (e.g., in the form of GOs) and of electricity can, at first sight, easily be described as a flow from one specific country to another specific country. One could therefore assume, that in an ideal case, each country should fill in the missing energy origin caused by net export of GOs with the residual mix of the country where the GOs were exported to (Also net physical import of electricity causes a deficit in available generation attributes of a country compared to its consumption, because part of the electricity is coming from abroad. In such case the missing attributes should ideally be derived from the residual mix of the exporting country. In practice however in the case of two European countries, the balancing is made with the European Attribute Mix (EAM), as with the balancing of attributes from GO transactions). In practice, such *bilateral balancing* would be highly complicated considering the international nature of the guarantee of origin markets (including also multiple cross-border transfers) and might likely not even be possible, because of the hen and egg problem: The residual mix of country A depends on that of B, which depends on that of C, which in turn depends on that of A. This might sound manageable with 3 countries, but the calculation needs to be Europe-wide.

One fundamental feature of the RE-DISS residual mix calculation methodology is the concept of a common attribute pool, generally known as the European Attribute Mix (EAM) (Which includes EU27, Norway, Iceland, Switzerland and Croatia but excludes Cyprus and Malta). Instead of different countries interacting with each other, they all interact with this common pool of attributes, which interconnects the domestic residual mixes the same way as the AIB Hub interconnects the explicit tracking of attributes (GOs). This approach does gain a "substitution mix" for the lack of attributes which does not to differentiate between the specific countries of net export of GOs. Although this obviously is a form of simplification of reality, it is consistent within the system borders of a group of countries which apply this approach, and corresponds to the concept of a pan-European internal market not only for electricity, but also for GOs.

The coordinated residual mix calculation for countries within the EAM area is divided into 4 phases (see Figure 2):

1. Data collection from each country:

    a. Net electricity production by tracking attribute including tracked externalities (CO_2 emissions and produced radioactive waste);
    b. Total electricity consumption;
    c. Data on explicitly tracked production attributes (imports, exports and cancellations of GOs and RTSs);
    d. Net electricity import and export *outside* the EAM area. In the case of net import, the residual mix (if exists; otherwise production mix) of the outside EAM country from where the electricity was imported.

2. Determination of the domestic residual mix and surplus/deficit of each country

    a. Determine *available attributes* by deducting exported and cancelled, and adding imported attributes to the generation mix (corrected with physical import or export to outside the EAM area). The mix of available attributes is the *domestic residual mix*;

b. Determine *untracked consumption* by deducting cancelled attributes from total electricity consumption;
c. Compare the volume of the available attributes with the volume of untracked consumption. If the amount of available attributes is greater/less than untracked consumption, the difference is surplus/deficit. The share of attributes in the surplus matches the domestic residual mix.

3. Determination of the European Attribute Mix including tracked externalities
   a. Combine the surpluses from all countries with a surplus.
4. Determination of the final residual mix of each country
   a. In case of surplus, the final residual mix of the country is equal to the domestic residual mix minus the surplus;
   b. Fill the domestic residual mix of deficit countries using the EAM until the volume of available attributes equals untracked consumption.

Figure 2. Residual mix calculation process based on [36].

A simplified numerical example of residual mix calculation is presented in Appendix section, Table A1.

Physically, electricity production and consumption in Europe equal each other in volume as long as electricity transfers to and from outside Europe are considered. International trading of GOs and electricity within Europe distort the equilibrium of available generation attributes and electricity consumption volumes on a national level, but on a European level the balance remains. The coordinated residual mix calculation, through EAM, returns this balance at the domestic level. Countries which have a surplus of generation attributes compared to their consumption (typically GO net importers and/or electricity net exporters), give attributes to the common pool and vice versa. Because of the physical balance, the total surplus equals in volume with the total deficit. The coordinated residual mix calculation is a simple yet powerful tool to allow international trading of generation attributes whilst avoiding double counting.

3. Analysis of Implicit Disclosure Problems before (2010) and after (2012) the RE-DISS Project

3.1. Implicit Disclosure Problems in Europe

Based on data collection of the disclosure practices of the 16 selected European countries, the RE-DISS project team found five different types of problems, which lead to double counting in implicit disclosure, due to deficiencies in the residual mix calculation:

- Uncorrected generation statistics used for implicit disclosure;
- Contract-based Tracking;
- Uncoordinated calculation within Europe;
- Overlapping Regions for Implicit Disclosure;
- Active GOs.

The following analysis of errors is based on qualitative and quantitative data collected for the 16 selected countries before (2010) and after (2012) the RE-DISS project.

3.1.1. Uncorrected Generation Statistics Used for Implicit Disclosure

Using uncorrected generation statistics for implicit disclosure appeared to be a straightforward approach followed by several countries, but it leads into the double counting of explicitly tracked attributes (often renewable), as explained in Sections 2.2 and 2.3. On the other hand, the relative shares of attributes which have not been explicitly tracked (particularly fossil and nuclear and their environmental impacts) are underestimated by such an uncorrected mix, because their share would be higher if the tracked attributes had been removed. The relevance of a proper residual mix calculation is best understood by comparing uncorrected generation statistics with residual mixes (Figure 3). The difference between the two is substantial.

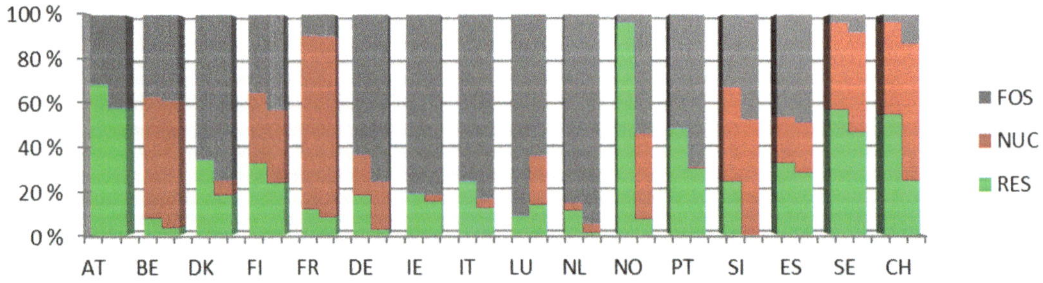

Figure 3. Generation (**left**) and residual (**right**) mixes of 2011 of selected European countries based on [38].

An alternative solution to the problem, without providing a residual mix, is utilized in Austria, Belgium, Germany, Luxembourg and The Netherlands, in which all renewable attributes are deducted from the generation statistics used for implicit disclosure [39]. This easy solution effectively excludes double counting of renewable attributes, but also leads to a disclosure error by neglecting possible remaining renewable shares in the residual mix and thus overestimating the fossil and nuclear shares including environmental impacts.

3.1.2. Contract-Based Tracking

Though in Section 2.2, it was explained that electricity can hardly be tracked along the trading arrangements in the electricity markets, contract-based tracking (CBT) is still allowed to be used for disclosure purposes in some countries. In CBT, the respective volumes which are physically delivered according to grid accounting rules to a supplier and his end-customers are, bilaterally between the buyer and seller, agreed to have been produced by a particular energy source or even a specific plant.

On top of the drawbacks already mentioned in Section 2.2, CBT is mostly not transparent for third parties and thus cannot be deducted from the residual mix of the country (Figure 4). This naturally results in double-counting of contract-based tracked attributes, which often include a higher than grid average share of renewable attributes. In order not to cause double counting, contract-based tracking has to fulfill the specifications for Reliable Tracking Systems as defined by the RE-DISS Best Practice Recommendations (BPR) by e.g., requiring central reporting of all tracked attributes to the national Competent Body, or that such contractual arrangements are accompanied with the transfer and use of GOs [34].

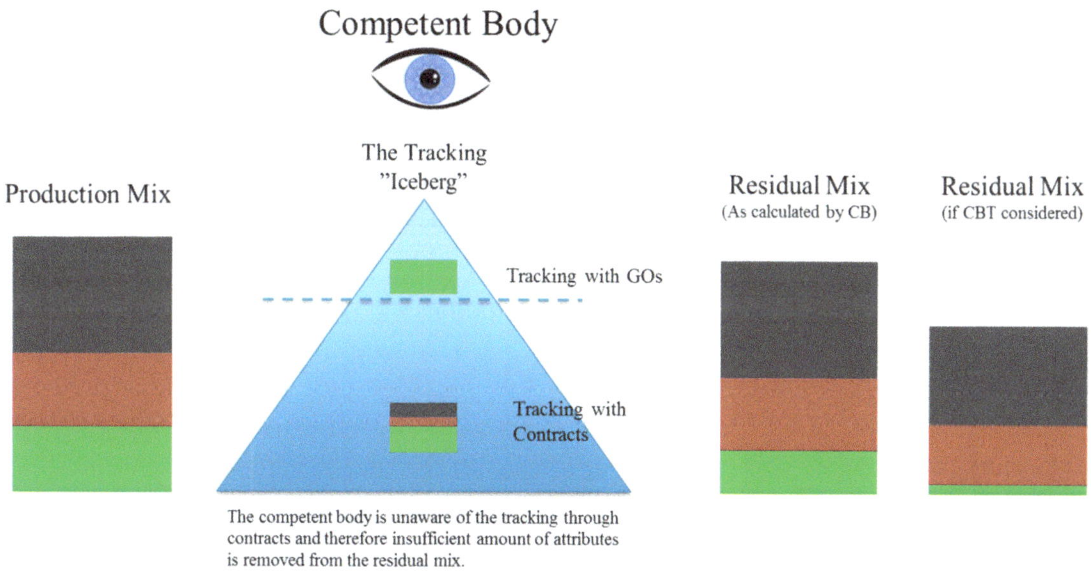

Figure 4. The problem of contract-based tracking as a reliable tracking system.

3.1.3. Uncoordinated Calculation of Residual Mixes within Europe

Uncoordinated calculation of residual mixes within Europe is a particularly relevant problem due to the international trade of both electricity and GOs. As explained in Section 2.3, such international trade means that a country might lack disclosure attributes. Without international coordination and the provision of the European Attribute Mix to balance surpluses and deficits, the deficit countries can either disclose the deficit as *unknown origin*, or expand the domestic residual mix to account for the entire untracked consumption.

For example, before adopting the EAM, Norway included in its residual mix a very high share of unknown origin, because most of the generation attributes were exported through GOs. Yet, given the production mix of Norway, it is likely that consumers presumed this unknown origin as renewable. Extrapolation of the domestic residual mix would have resulted in an even worse solution, because after accounting for the exported GOs the Norwegian domestic residual mix still contained a high share of renewable attributes, as practically all power production in Norway derives from renewable sources. Therefore the deficit would have been fulfilled by a much greener mix (expanding the Norwegian domestic residual mix) than the EAM, leading exported Norwegian GOs to be partially double counted. It should be highlighted that Norway has corrected the problem of unknown origin and currently uses the EAM for the filling up of its residual mix [39].

3.1.4. Overlapping Regions for Implicit Disclosure

The RE-DISS Best Practice Recommendations recommend that a residual mix should by default be calculated on a national level, but under a common decision, several countries can have a mutual residual mix if the electricity markets are closely integrated [34] (Recommendation 28). If some of the involved countries apply a common multi-country mix, and others only their national mix, the disclosure information of the latter countries will be overestimated. Although the problem has been acknowledged for several years, the Nordic countries (Denmark, Finland, Norway and Sweden) have yet to agree on a common solution. Sweden applies the common Nordic residual mix, while Norway and Denmark use their respective national mixes [39]. Finland has in the past unofficially used the common Nordic residual mix, but starting from 2013, the domestic residual mix will be officially applied [39]. This problem is especially relevant if the Nordic residual mix contains a higher renewable share than the Swedish mix, so that Sweden would benefit from such practice, while the other countries' renewable attributes would be double counted without them having a possibility to prevent this. But also in the opposite situation a disclosure error would occur due to loss of renewable attributes and double counting of others. A similar error also exists for countries which apply the overall ENTSO-E mix for all volumes of unknown origin (see Table 1).

3.1.5. Active GOs

Depending on how the residual mix is calculated, the problem of active GOs may occur if GOs issued for electricity production of year X are exported or cancelled after a defined disclosure deadline of the year X. The attributes presented by these GOs may already have been included in the residual mix of year X and thus the subsequent cancellation or export of the GOs would lead to double counting. This area was not well addressed in the residual mix calculation methodology developed by E-Track, but was taken on by RE-DISS and currently the methodology is robust against the active GO problem. The problem will be further discussed in Section 4.2.

3.2. Presence of Implicit Disclosure Problems in Europe before (2010) and after (2012) the RE-DISS Project

The RE-DISS project helped competent bodies of electricity disclosure to resolve problems described in Section 3.1 through consultation and dissemination actions, which led to the spreading of the RE-DISS Best Practice Recommendations for disclosure (BPR). During RE-DISS, seven workshops were carried out that counted with the participation of competent bodies from 16 countries. The first workshops focused on developing the BPR whereas towards the end of the project the emphasis was on supporting implementation. The BPR has been a living document throughout RE-DISS and the competent bodies have actively participated in its improvement.

The RE-DISS team also organized 16 (from Portugal, France, Greece, Latvia, Germany, Slovenia, Romania, Bulgaria, Czech Republic and Iceland) bilateral face-to-face meetings and a number of telephone conferences with competent bodies, performed consultancies of legal texts (for Germany, France, Poland and Iceland) and developed Country Profiles of 22 European countries that included an in-depth analysis of their electricity tracking system and provided tailor-made recommendations.

Through these actions, the RE-DISS project succeeded in triggering important enhancements in the field of electricity disclosure. Table 1 presents the status quo of the five problems in the 16 analyzed countries both before (1) and after (2) RE-DISS. For the sake of clarity it should be highlighted that the improvements listed in Table 1 are due to both efforts of RE-DISS and actions taken independently by the Competent Bodies.

In Table 1 marking X signifies that the problem is fully existing, (X) that it is partially existing and no marking that it is not existing in that country. Furthermore, if no residual mix is calculated in the country (problem 1), problems 3 and 5 are non-applicable. In such cases, the respective cell is marked grey.

4. Results

4.1. Effects from Corrections of Implicit Disclosure Problems during the RE-DISS Project

The effects of the improvements presented in Table 1 are quantified on an overall level in Figure 5 and by country in Figure 6. The results have been derived by simulating (By simulation it is meant that the residual mix of a specific country is calculated according to the national practices of that country. In other words practices, which are known to be unreliable in some cases are simulated and then compared to calculation results according to RE-DISS Best Practices) the implicit disclosure practices of the countries before 2010 (left column) and after 2012 RE-DISS (right column) according to the settings of Table 1. Columns in Figures 5 and 6 indicate volumes of erroneous disclosure compared to a situation where implicit disclosure is implemented according to RE-DISS BPR. Hence a positive column signifies double counting of the corresponding energy source, including externalities, and a negative column that the energy source is under-reflected.

Table 1. Improvements in implicit disclosure during RE-DISS [39].

<table>
<tr><th rowspan="2">Country</th><th colspan="5">Problem</th><th rowspan="2">Description</th></tr>
<tr><th>1</th><th>2</th><th>3</th><th>4</th><th>5</th></tr>
<tr><td>Austria (1)</td><td>X</td><td></td><td></td><td>X</td><td></td><td>Before RE-DISS:
No residual mix.
ENTSO-e mix used for implicit disclosure.</td></tr>
<tr><td>Austria (2)</td><td>(X)</td><td></td><td></td><td>X</td><td></td><td>Improvements:
All renewables filtered out of the ENTSO-mix before used for implicit disclosure.</td></tr>
<tr><td>Belgium (1)</td><td>(X)</td><td></td><td></td><td></td><td></td><td>Before RE-DISS:
No residual mix.
Production mix from which all RES filtered out used for implicit disclosure.</td></tr>
<tr><td>Belgium (2)</td><td>(X)</td><td></td><td></td><td></td><td></td><td>Improvements:
No improvements.</td></tr>
<tr><td>Denmark (1)</td><td rowspan="2">X</td><td rowspan="2">(X)</td><td rowspan="2">---</td><td rowspan="2"></td><td rowspan="2">---</td><td>Before RE-DISS:
No residual mix.
CBT for nuclear and fossil.</td></tr>
<tr><td>Denmark (2)</td><td>Improvements:
Reliable and coordinated residual mix calculation.
CBT of nuclear and fossil supervised.</td></tr>
<tr><td>Finland (1)</td><td rowspan="2">(X)</td><td>X</td><td rowspan="2"></td><td rowspan="2">X</td><td rowspan="2"></td><td>Before RE-DISS:
Residual mix of Finland based on the Nordic region.
No legal status for residual mix: given as a recommendation by the Association of Energy Industries.
Contract based tracking allowed.</td></tr>
<tr><td>Finland (2)</td><td>(X)</td><td>Improvements:
Reliable and coordinated residual mix calculation set by legislation.
CBT only for nuclear and fossil.</td></tr>
<tr><td>France (1)</td><td>X</td><td>X</td><td></td><td>X</td><td></td><td>Before RE-DISS:
No residual mix.
Mix of own production, contracts and ENTSO-e mix used for disclosure.</td></tr>
<tr><td>France (2)</td><td>X</td><td>X</td><td></td><td>X</td><td></td><td>Improvements:
No improvements.</td></tr>
<tr><td>Germany (1)</td><td>X</td><td>X</td><td></td><td>X</td><td></td><td>Before RE-DISS:
No residual mix.
ENTSO-e mix as default value for disclosure.
CBT, GOs, RECS and labels used for disclosure.</td></tr>
</table>

Table 1. *Cont.*

Country	Problem 1	Problem 2	Problem 3	Problem 4	Problem 5	Description
Germany (2)	(X)	(X)				Improvements National production mix, from which all renewables filtered out, used for implicit disclosure.
Ireland (1)			X			Before RE-DISS: Disclosure based on contracts and residual mix (residual mix accounts for contracts). Residual mix is not coordinated with other countries.
Ireland (2)						Improvements: Coordinated residual mix calculated.
Italy (1)	X	X	---		---	Before RE-DISS: No residual mix. Disclosure based on fuel mixes.
Italy (2)	X	X	X		(X)	Improvements: Residual mix calculated but not coordinated (deficit disclosed with Eurostat mix). It is not clear whether residual mix accounts for Active GOs.
Luxemburg (1)	---	---		---		Before RE-DISS: No disclosure.
Luxemburg (2)	(X)	X		X		Improvements: Disclosure system implemented. ENTSO-e mix from which all renewables filtered out used for implicit disclosure.
Netherlands (1)	(X)	X	X			Before RE-DISS: Residual mix calculated, but all renewables filtered out. Does not consider contracts and is not coordinated.
Netherland (2)	(X)	(X)	X			Improvements: Residual mix calculation considers contracts and is coordinated, but all renewables are filtered out.
Norway (1)			X		X	Before RE-DISS: Residual mix calculated, but not coordinated. Deficit attributes disclosed as unknown.
Norway (2)						Improvements: Deficit attributes replaced with the European Attribute Mix. Residual mix only accounts for year X certificates.
Portugal (1)	X	X				Before RE-DISS: No residual mix. Disclosure through contracts.
Portugal (2)	X	X				Improvements: Approach to a kind of residual mix.

Table 1. *Cont.*

Country	Problem					Description
	1	2	3	4	5	
Slovenia (1)	X	X		X		Before RE-DISS: No residual mix. Disclosure is based on contracts, GOs and ENTSO-e mix.
Slovenia (2)	X	X		X		Improvements: No improvements.
Spain (1)			X		(X)	Before RE-DISS: Residual mix is calculated, but not coordinated with other countries (domestic attributes expanded if needed). A problem with Active GOs might exist and GOs do not necessarily have to be cancelled in order to be used.
Spain (2)			X		(X)	Improvements: No improvements.
Sweden (1)	(X)	(X)		X		Before RE-DISS Residual mix based on the Nordic region. Contract based tracking allowed but accounted for. No legal status for residual mix: given as a recommendation by the Association of Energy Industries.
Sweden (2)				X		Improvements: Contract-based tracking not allowed (disclosure based on GOs or residual mix). Use of the residual mix obligated by law.
Switzerland (1)	X	X				Before RE-DISS No residual mix. Contract-based tracking allowed.
Switzerland (2)	---	(X)		---		Improvements: All electricity explicitly tracked with GOs (no residual mix needed).

The input data in all cases is that collected by RE-DISS for the 2011 residual mix calculation. This depicts that the volume to be disclosed remains constant throughout the analysis and hence that for each country the positive column equals in volume with the negative column, *i.e.*, if renewable attributes are double counted, nuclear and/or fossil attributes are automatically replaced by this amount in disclosure.

Other important settings and framework assumptions for the analysis are as follows:

- Effect of possible other problems besides the five listed have been neglected, e.g., problems relating to explicit tracking;
- in case CBT is allowed in the country, this has been assumed to cover:
  - 50% of untracked (not tracked with GOs or RTSs) domestic renewable production;
  - 20% of untracked domestic nuclear and fossil production;
  - for France (53% of RES and 56% of NUC and FOS) and Sweden (26% of RES, 17% NUC and 0% FOS), country specific estimates have been used.

- In case new legislation is currently being implemented and is scheduled to come into force in the near future, such progress has been taken into account for the after RE-DISS scenario. This is relevant for Germany, Sweden and Switzerland where the law or regulation is already ratified as well as for Finland and Italy where the ratification is in process.

Table 1 indicated that 12 (Austria, Denmark, Finland, Germany, Ireland, Italy, Luxembourg, The Netherlands, Norway, Portugal, Sweden, Switzerland) out of the 16 analysed countries made improvements regarding the 5 disclosure problems during the RE-DISS project. In 7 (Denmark, Finland, Germany, Italy, Luxembourg, Norway, Switzerland) of these countries, this significantly reduced the disclosure error as shown in Figure 6. Figure 5 illustrates that on a total level, double counting of renewable attributes decreased by 70 TWh (106−36) and the overall disclosure error by 168 TWh (243−75). Based on [40] these volumes represent 10% of electricity production from renewable energy sources and 7% of electricity consumption, respectively, of these 16 countries. Remaining double counting of renewable attributes after the project amounts to 36 TWh whereas the total disclosure error adds up to 75 TWh. However it should be noted that further errors may reside at individual supplier level as highlighted in Chapter 5.

No unknown origin was disclosed in the after RE-DISS case, compared to 102 TWh in the before case. This is also a significant improvement, because 95 TWh of the unknown origin was disclosed in Norway, where, given the production mix of the country, it is probable that consumers assume a green origin without better knowledge. The decreased amounts of renewable and unknown origin were correctly replaced by nuclear and fossil attributes, for which the negative disclosure error contracted by 58 TWh (76−8) and 110 TWh (134−24), respectively. The decreased disclosure error of fossil and nuclear attributes also increased the amount of disclosed externalities: CO_2 and radioactive waste, but these volumes were not covered by the analysis.

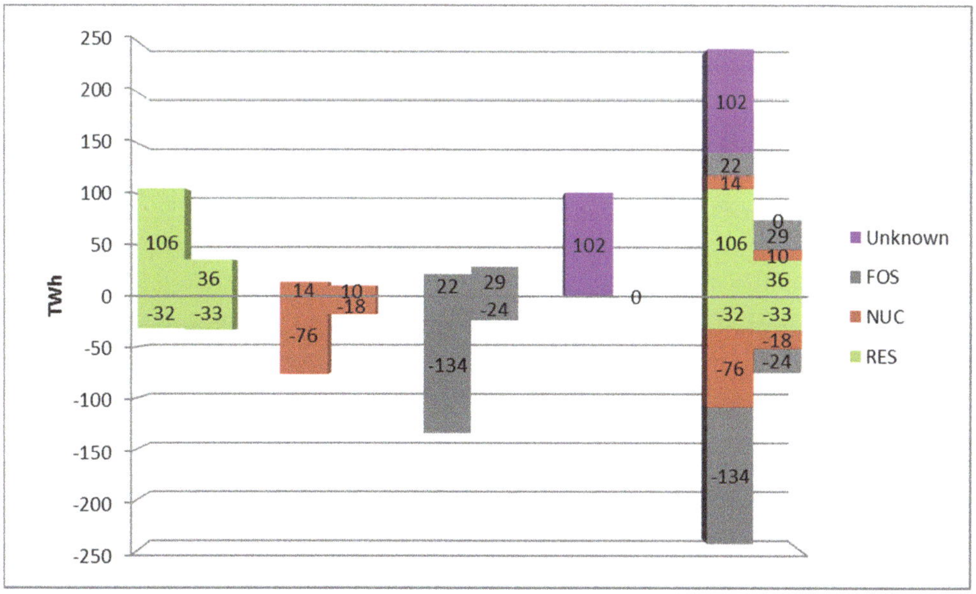

Figure 5. Total quantified implicit disclosure errors per energy source before 2010 (left-hand columns) and after 2012 (right-hand columns) RE-DISS.

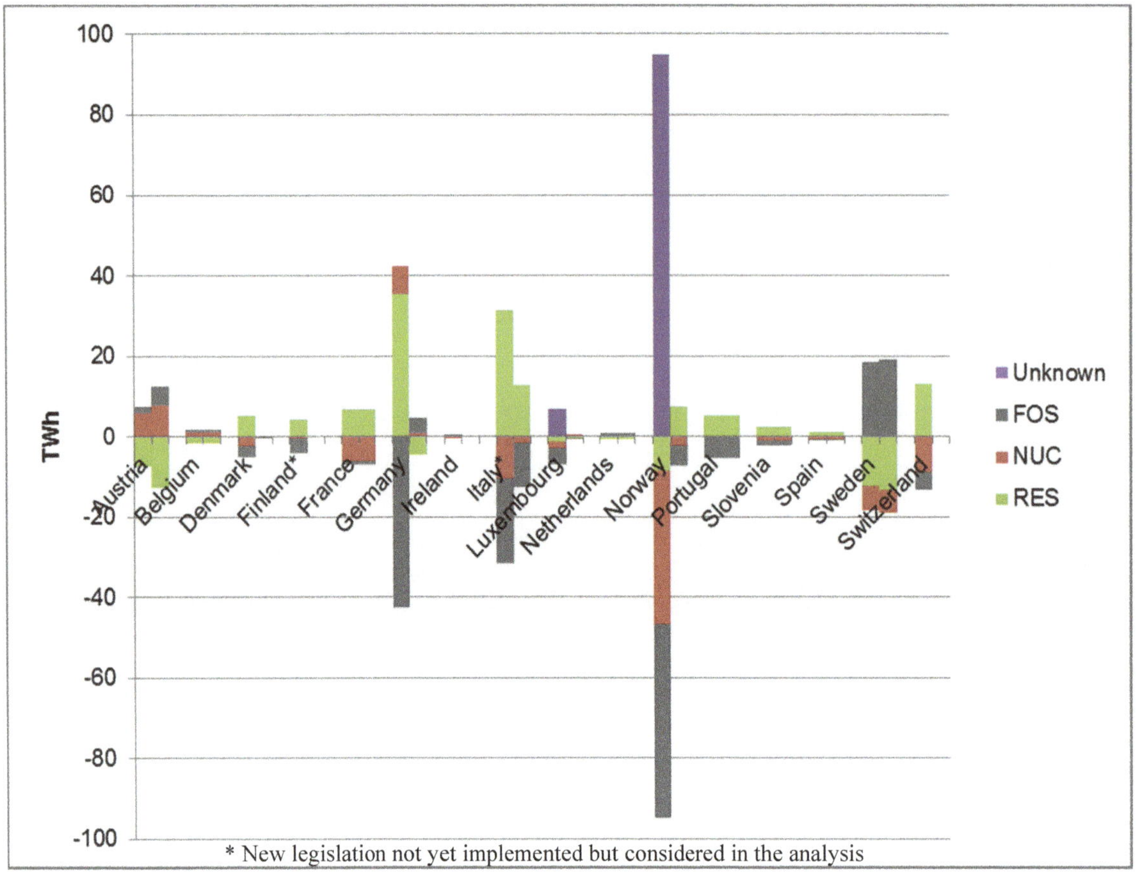

Figure 6. Quantified implicit disclosure errors per energy source and per country before 2010 (left-hand columns) and after 2012 (right-hand columns) RE-DISS.

Out of individual countries, Norway improved its disclosure the most, measured by volume of disclosure error avoided. This resulted from adopting the EAM for fulfilling deficit attributes (problem 3). Germany succeeded in making the largest reduction in double counting of renewable attributes, mostly by enhancements concerning problem 1. Instead of using uncorrected generation statistics, any renewable attributes are now eliminated from implicit disclosure, in Germany. Same amendment was made by Austria and Luxembourg, and was already used by Belgium and The Netherlands before RE-DISS. Although this approach does not eliminate the disclosure error altogether, it is effective against implicit double counting of renewable attributes. Also regarding problem 1, Italy made substantial progress by adopting a residual mix for implicit disclosure. The remainder of the implicit disclosure error in Italy would be avoided by using the EAM.

In Figure 6, the disclosure error is completely corrected in Denmark, Finland, Ireland and Switzerland. Switzerland overcame all problems by forcing all consumption to be explicitly tracked, which removes the need for implicit disclosure. Whereas the other three countries adopted a residual mix according to the RE-DISS BPR (although Finland still has a problem of unmonitored contract-based tracking of nuclear and fossil attributes).

Sweden was one of the top improvers on a qualitative level through the removal of contract-based tracking and adoption of mandatory use of the residual mix where GOs are not used. However in the quantitative analysis, these improvements were out shadowed by the persistence of problem 4: Sweden still uses the Nordic residual mix without a mutual decision between Nordic countries. In the residual mix of 2011, the Nordic residual mix contained a lower renewable share than the Swedish national mix, which meant that renewable attributes were not double counted according to the analysis, but rather under-reflected. This might change from year to year and hence the problem is important.

Throughout the analysis, improvements regarding problems 1 and 3 yielded highest results in the avoidance of disclosure errors and double counting of renewable attributes. However, the impact of developments regarding problem 2, contract-based tracking, was based on an estimated volume, and it is likely that these are in reality as important as those regarding problems 1 and 3. It should also be outlined that the analysis focused on measuring improvements per country, not per problem.

4.2. Improvements to the Residual Mix Calculation Methodology by RE-DISS

Based on data collection of the disclosure practices and quantitative tracking data of European countries, RE-DISS sought improvements to the residual mix calculation methodology. Though pioneering, the major deficiency of the E-TRACK's residual mix calculation model was the lack of detail regarding the time-frame and production year of GO transactions considered in the calculation [15]. Ideally, residual mix calculation of a certain year should reflect transactions of GOs issued for that year electricity production, which can be seen as the intention of the E-TRACK model. However, the 12 months' lifetime of GOs poses a problem for this *production year-based* approach, since not all GOs of production year X are cancelled or expired by the deadline for the disclosure of year X consumption (31st of March year X + 1 according to RE-DISS BPR number 5 [34]). Hence the respective attributes are included in the residual of year X. If in turn the calculation of year X + 1 only considers GOs of production year X + 1, the transactions of production year X GOs after the deadline fall out of the scope of both year calculations. As the case of 2010 depicts, it is crucial to account for the transactions of production year 2009 GOs also after the deadline:

When collecting residual mix calculation data for 2010, the RE-DISS project team observed that GO imports exceeded exports by 32 TWh. Comparison with the AIB statistics indicated that both exports and imports seemed to be missing, but exports in particular. Secondly, the total amount of GO cancellations was suspiciously low in the RM data (174 TWh) given the total cancellation volume during 2010 (211 TWh) [41]. Solving both of these issues was crucial, since according to the residual mix methodology described in Section 2.3, both lacking volume of exports and cancellations in the RM data, lead to production mixes not being corrected with a sufficient amount of explicitly tracked attributes, *i.e.*, implicit double counting.

The difference of imports and exports was a result of two different methods of collecting GO transaction data: Production year-based method (PYBM) and transaction-based method (TBM). The PYBM, used mainly for the Nordic countries, considered the transactions of only production year 2010 GOs that occurred before 31st of March 2011. This, as mentioned, was seen as the original intention of the calculation, but neglected the transactions of production year 2009 GOs which occurred after March 31, 2010. However, the registry systems of some countries were not capable of separating transaction statistics per production year of the GO. This signified that the transaction data for these countries consisted of all GO transactions that occurred during calendar year 2010 (TBM was used), without consideration of the production year of the GOs. The data was clearly inconsistent and neglected a significant volume of especially exports (Figure 7). This was due to the fact that PYBM was used for large exporters and TBM for significant importers and cancellers.

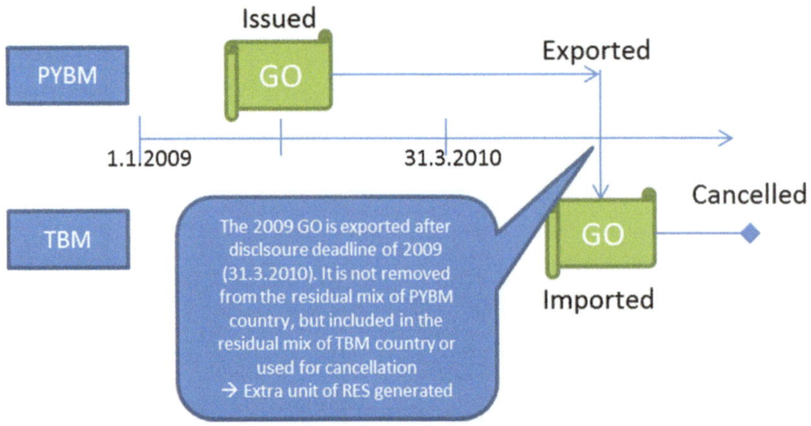

Figure 7. The problem of the production year-based method in residual mix calculation.

Based on detailed analysis of the data, the RE-DISS team was able to correct most of the problem for the 2010 calculation. The analysis also revealed that without corrections, explicit tracking of approximately 48 TWh of renewable and 21 TWh of nuclear attributes would have neglected in the residual mix calculation. For the 2011 calculation, the RE-DISS team set out to find a robust solution for the two major problems of the calculation:

- Transaction data being inconsistent between countries due to use of both PYBM and TBM;
- Active GOs from previous production years causing double counting.

As mentioned, the registry solutions of certain countries were, and still are, incapable of differentiating between production years in GO transaction data. Hence, the only applicable solution for the first problem was to apply TBM for all countries, and neglect the production year of GOs in the transaction data. Crucially, this also solved the problem of transactions of active GOs from previous production years, which were now considered automatically in the residual mix of 2011. In the 2011 calculation the double counting of Active GOs, prevented through TBM, was estimated by RE-DISS to be 40 TWh of renewable and 25 TWh of nuclear attributes.

As a consequence, the TBM, which was previously seen inferior to the PYBM, was lifted as the preferred method for residual mix calculation in RE-DISS. This was due to its ability to deliver a consistent methodology with the available data and to secure the avoidance of double counting in the existing legislative setting, which allows GOs to be usable for 12 months before expiry.

However the weakened accuracy of the residual mix in portraying explicit tracking of the correct production year attributes has stirred up discussions: "The residual mix is used to portray the generation attributes of a calendar year that are not used for explicit tracking… Thus, evidently, the theoretical argumentation to use PYB is better justified, because TB portrays the desired attribute flows poorer" [15]. The discussions have been further fueled by negative renewable volumes occurring in the domestic residual mixes of some countries. This may come about if the leftover of previous production year GOs is significant and if GOs are issued for a large share of the renewable production of the country. Negative renewable balances in final residual mixes of countries can be avoided through transferring the negativity to the European Attribute Mix, but this is not an optimal solution.

To enhance the accuracy of the calculation, the Shifted Transaction-Based Method (STBM) was introduced. STBM considers GO transactions between April 1st X and March 31st X + 1 for year X residual mix calculation, instead of the transactions of the calendar year X. This is logical since the time period to make cancellations for disclosing year X electricity consumption is accordingly shifted from the calendar year. Often most cancellation for year X disclosure occur in February and March of year X + 1, because it is the time suppliers close their electricity product portfolio for year X. The STBM considers this *cancellation peak* (see Figure 8) and the preceding transfer peak correctly in the residual mix of year X, whereas TBM would reflect it in the residual mix of year X + 1.

Figure 8. EECS-GO cancellations in 2012 [41].

A weakness of the Shifted Transaction-Based Method is that, though it decreases the chance for negative renewable balances, by increased accuracy, it does not eliminate the possibility entirely. Furthermore, a significant drawback of all methods that utilize import and export data, is their vulnerability to the GO market conditions. For example, GOs can be stocked in a country which has not yet implemented the expiry rule of GOs or which imposes little or no transaction costs. These patterns have been emphasized in the recent years, and can cause the renewable share in the residual mix to fluctuate significantly from year to year depending on whether the stock is built up or used that year. Even if the fluctuations are not powerful enough to turn the available renewable balance negative in a given year, they are hardly explainable and disturb confidence in the disclosure system. For example, the residual mix of Luxembourg contained a higher renewable share than its production mix in the 2011 calculation [38]. Robustness against market fluctuations is increasingly important due to the future emergence of exchanges in the GO market, e.g., [42].

The STBM was used for the residual mix calculation of 2012, though further solutions have also been considered to address the remaining weaknesses. To address these weaknesses, the Issuance-Based Method (IBM) might be a suitable approach, when certain preconditions are met. This method is presented in Section 5 in principle, but the detailed analysis is out of the scope of this paper.

5. Outlook and Discussion

As outlined, during the RE-DISS project (4/2010–10/2012), 16 selected European countries were able to reduce approximately 70 TWh of double counting of renewable attributes and 168 TWh of total implicit disclosure error. However, this only related to implicit disclosure errors on a national level and it is very likely that even greater errors reside in the individual supplier level. For example, if suppliers do not disclose product related mixes along with the total supplier mix as proposed in RE-DISS BPR number 39 [34], the resulting double counting can be even higher than that found in the analysis of Section 4.1. Therefore, in the future, emphasis should be put on the implementation and supervision of electricity disclosure at the supplier level.

But problems still persist on the national level as well. Out of the 16 countries in the analysis, only Denmark, Ireland and Switzerland had solved all five problems related to implicit disclosure (as well as Austria and Norway after the end of the project as explained in Table 2). Residual mix is still not calculated in France and Slovenia whereas Portugal has only taken initial steps towards a residual mix (problem 1). Likewise, Belgium, Germany and Luxembourg are not calculating a residual mix, but avoid double counting of renewable attributes (same applies partly for the Netherlands).

Contract-based tracking (problem 2) is still an issue in many countries as illustrated in Table 2, but it is better accounted for than in the beginning of the project. Spain and Italy have yet to coordinate their domestic calculation with other European countries (problem 3), whereas Sweden uses a regional approach for its residual mix calculation, which is not agreed upon with other countries in the Nordic region (problem 4). Table 2 presents the remaining steps to be made on a country level in correcting the five problems explained in Section 3.1.

Regarding improvements in the residual mix calculation methodology, RE-DISS solved the issue of double counting concerning Active GOs and introduced the Shifted Transaction-Based Method (STBM). The active GO problem was a result of emphasizing *accuracy* over *reliability* in the residual mix calculation. With RE-DISS coordinating efforts for Best Practices of electricity disclosure, such problems are unlikely in the future. However, it needs to be reminded that an explicit tracking system, requires a *reliable* implicit disclosure system as the backbone of operation. Hence reliability should always be emphasized over accuracy when the implicit system is developed. After all, the primary task of the residual mix is to secure full ownership of renewable attributes to consumers opting for green; not to perfectly portray attributes of correct production year to consumers buying regular electricity.

With this in mind, a future option for residual mix calculation, might be the Issuance-Based Method (IBM), which proposes a new perspective to the calculation. Instead of removing used production attributes (cancelled or net exported), the residual mix would be calculated by deducting attributes of issued guarantees of origin from the production mix and returning unused attributes (expired GOs) into it. The IBM is already correctly used by some competent bodies in the domestic residual mix calculation, but careful consideration is required in order to choose this as an approach for the determination of the EAM.

Table 2. Next steps in improving implicit disclosure on a national level based on [39].

Countries	Recommended Steps for Improving Implicit Disclosure
Austria	In Austria, explicit tracking of all electricity consumption will be required starting from 2013 electricity disclosure. Hence errors in implicit disclosure will be avoided in the future, but this was not covered by this analysis.
Belgium	To increase the accuracy of implicit disclosure, it is recommended that Belgium implement the RE-DISS residual mix, although the current method does not cause double counting of renewable attributes.
Denmark	Denmark has resolved the five problems related to implicit disclosure, presented in Section 3.1.
Finland	It is recommended that Finland ban unmonitored contract-based tracking of nuclear and fossil attributes.
Italy	It is recommended that Italy cooperate with other European countries in forming its residual mix to correctly account for international transfers of electricity and GOs in the residual mix and to prevent double counting of renewable attributes. Furthermore, it needs to be clarified whether the residual mix calculation accounts for previous production year GOs.
Netherlands	It is recommended that The Netherlands cease to filter out the possibly remaining renewable share from the residual mix. This would make the disclosure more accurate although the current form of implicit disclosure does not cause double counting of renewable attributes.
Sweden	It is recommended that Sweden use its domestic residual mix for implicit disclosure instead of the Nordic residual mix, unless Denmark, Finland and Norway also agree to use the Nordic residual mix.
Switzerland	Implicit disclosure is not used as all consumption is explicitly tracked. Hence errors in implicit disclosure are avoided.
Luxemburg	To obtain accurate implicit disclosure, it is recommended that Luxemburg implement the residual mix calculation according to the RE-DISS Best Practices, although the current method of Luxemburg does not cause double counting of renewable attributes.
Norway	Norway resolved the issue of active GOs after the end of the project (but this was not considered in the analysis) and has thus solved the five problems related to implicit disclosure, presented in Section 3.1.
Ireland	Ireland has resolved the five problems related to implicit disclosure, presented in Section 3.1.
France	It is recommended that France implement a residual mix calculation according to the RE-DISS BPR to eliminate implicit double counting of renewable attributes and to achieve more accurate disclosure.
Germany	To obtain accurate implicit disclosure, it is recommended that Germany implement the residual mix calculation according to the RE-DISS BPR, although the current method of Germany does not cause double counting of renewable attributes.
Portugal	It is recommended that Portugal implement a residual mix calculation according to RE-DISS BPR to eliminate implicit double counting of renewable attributes and to achieve more accurate disclosure.
Slovenia	It is recommended that Slovenia implement a residual mix calculation according to RE-DISS BPR to eliminate implicit double counting of renewable attributes and to achieve more accurate disclosure.
Spain	It is recommended that Spain cooperate with other European countries in forming its residual mix to correctly account for international transfer of electricity and GOs in the residual mix and to prevent double counting of renewable attributes. Furthermore, it needs to be clarified whether GOs need to be cancelled before they are used and whether the residual mix calculation accounts for previous production year GOs.

A clear benefit of IBM is that attributes presented by GOs are removed from the production mix of the year of production and country of issue. Hence, it is not affected by peculiarities in the GO market and cannot yield negative renewable balances, because, at maximum, renewable GOs can be issued for the volume of electricity production from renewable sources in the country. GOs of production year X, which are not cancelled by 31st March X + 1, are not without problems for the IBM either, because they disturb the balance between yearly production attributes and consumption to be disclosed. If a large part of GOs is unused before the disclosure deadline, the calculation removes more from the production than consumption side. This means that deficit exceeds surplus in the residual mix calculation and that the EAM should be expanded. Of course, if the volume of active GOs remains constant from year to year, the effect cancels itself out, but this can hardly be relied on.

Expanding of the EAM is not something that should be considered lightly, since the foundation of the coordinated calculation is on the balance of generation attributes and consumption. Therefore, from the perspective of calculating the EAM, the IBM might be a viable option only when implemented in connection with a so-called early expiry solution. The early expiry suggests that year X electricity consumption can only be disclosed with GOs of production year X and hence all remaining production year X GOs should expire at the disclosure deadline of year X. This should in theory not be contradictory to the lifetime of GOs as set by Article 15, paragraph 3 of Directive 2009/28/EC as it states that a GO should be used *within* 12 months of the associated electricity production. However, this is highly debated and countries interpreted the Directive differently in national legislation. Furthermore, to avoid potential arbitrage, all countries should use a harmonized disclosure deadline, which is not currently the case.

Linking the production year of the GO to the consumption year for which it can be used is a logical solution, since electricity consumed in a given year is effectively derived from the energy sources used for electricity production during that year. However, a change of this magnitude into existing procedures on a European level requires wide acceptance from national competent bodies of electricity disclosure and might not be possible without a revision of the EC Directive.

Acknowledgments

The authors of this article would like to thank all competent authorities who participated in the workshops and accepted to endorse the Best Practice Recommendations, which contain the RE-DISS methodology for the calculation of the Residual Mix. Without their involvement and commitment to change national regulations and practices, the RE-DISS work would not have led to the improvements described in this paper.

The authors would also like to acknowledge AIB's and RECS International's constructive and active support of the team in discussing and disseminating the findings of the project.

The authors are grateful to the European Commission for supporting the RE-DISS project through the IEE programme (Contract No. IEE/09/761/SI2.558253).

Appendix: Numerical Example of Simplified Residual Mix Calculation

In this chapter an example is given on how the residual mix is calculated based on the steps elaborated in Chapter 2.3. This example is to be used together with the process description in Chapter 2.3. For simplification, environmental indicators as well as exchange with countries outside of the European Attribute Mix are not included in the example.

Table A1. Simplified example of residual mix calculation process.

	Country A	Country B	Country C	Country D	SUM		Explanation
1. Data Collection							
1a. Production Mix							
RES	5	10	80	40	135	TWh	
NUC	30	0	40	0	70	TWh	
FOS	40	60	10	20	130	TWh	
1b. Consumption	100	40	140	55	335	TWh	
1c. Certificate Transactions (all RES)							
Imp.	60	30	5	0	95	TWh	
Exp	5	5	60	25	95	TWh	
Canc.	50	15	20	10	95	TWh	
2. Domestic RM							
2a. Available Attributes							
RES	10	20	5	5	40	TWh	1a.Prod+ 1c.Imp-1c.Exp-1c.Canc.
NUC	30	0	40	0	70	TWh	1a.Prod
FOS	40	60	10	20	130	TWh	1a.Prod
SUM	80	80	55	25	240	TWh	
2b. Untracked Consumption	50	25	120	45	240	TWh	1b.Cons-1c.Canc.
2c. Surplus/Deficit							
Surplus	30	55			85	TWh	IF 2a.SumAvailable Attributes >2b. Untracked Consumption—> Surplus = Sum 2a-2b
Deficit			65	20	85	TWh	IF 2a.SumAvailable Attributes <2b. Untracked Consumption—> Deficit = 2b-sum 2a
3. European Attribute Mix							
3a. Surplus to EAM							
RES	3.75	13.75	0	0	17.5	TWh	2c.Surplus*2a.Share of RES in Available Attributes
NUC	11,25	0	0	0	11.25	TWh	2c.Surplus*2a.Share of NUC in Available Attributes
FOS	15	41.25	0	0	56.25	TWh	2c.Surplus*2a.Share of FOS in Available Attributes
SUM	30	55	0	0	85	TWh	

Table A1. *Cont.*

	Country A	Country B	Country C	Country D	SUM		Explanation
4. Final Residual Mixes							
4a. RM of Surplus domain							
RES	6.25	6.25				TWh	2a.Av. Att RES—3a.Surplus RES
NUC	18.75	0				TWh	2a.Av. Att NUC—3a.Surplus NUC
FOS	25	18.75				TWh	2a.Av. Att FOS—3a.Surplus FOS
4b. RM of Deficit domain							
RES			18.38	9.12		TWh	2a.Av. Att. RES + 2c.Deficit*3a.ShareofRESinEAM
NUC			48.60	2.65		TWh	2a.Av. Att. NUC + 2c.Deficit*3a.ShareofNUCinEAM
FOS			53.01	33.24		TWh	2a.Av. Att. FOS + 2c.Deficit*3a.ShareofFOSinEAM
Final RM (%)							
RES	13%	25%	15%	20%			
NUC	38%	0%	41%	6%			
FOS	50%	75%	44%	74%			
Production Mix (comparison)							
RES	7%	14%	62%	67%			
NUC	40%	0%	31%	0%			
FOS	53%	86%	8%	33%			

Conflicts of Interest

The authors declare no conflict of interest.

References

1. Trevino, L. *Liberalization of the Electricity Market in Europe: An overview of the Electricity Technology and the Market Place. Chair Management of Network Industries—MIR*; College of Management of Technology, Ecole Polytechnique Federale de Lausanne: Lausanne, Switzerland, 2008.
2. 2009/72/EC. Directive 2012/27/EC of the European Parliament and of the Council of 25 October 2012 on Energy Efficiency, Amending Directives 2009/125/EC and 2010/30/EC and Repealing Directives 2004/8/EC and 2006/32/EC. Available online: http://eur-lex.europa.eu/LexUriServ/LexUriServ.do?uri=OJ:L:2009:211:0055:0093:EN:PDF (accessed on 15 April 2015).
3. Timpe, C.; Seebach, D. Best Practice for the Tracking of Electricity. Recommendations from the E-Track II Project. Deliverable 10 of the IEE Project "A European Tracking System for Electricity—Phase II (E-TRACK II)". Available online: http://www.e-track-project.org/docs/E-TRACK%20II_WP7_Recommendations.pdf (accessed on 15 April 2015).

4. Jansen, J.C.; Seebach, D. Requirements of Electricity End-Users on Tracking of Electricity Generation Attributes and Related Policies. D7 of WP 5 of the E-Track II Project (Final Report). A Report Prepared as Part of the EIE Project "A European Tracking System for Electricity—Phase II (E-Track II)". Available online: http://www.e-track-project.org/docs/final/WP5_D7_Consumer%20requirements%20report_V2.pdf (accessed on 15 April 2015).
5. The E-TRACK Project. Available online: http://www.e-track-project.org (accessed on 17 April 2015).
6. Boardman, B.; Palmer, J. Electricity disclosure: The troubled birth of a new policy. *Energy Policy* **2007**, *35*, 4947–4958.
7. Wilhite, H. An assessment of experiences in the U.S.A. with power and emission disclosure information for energy consumers. Available online: https://www.duo.uio.no/handle/10852/32691 (accessed on 15 April 2015).
8. Markard, J.; Holt, E. Disclosure of electricity products—Lessons learnt from consumer research as guidance for energy policy. *Energy Policy* **2003**, *31*, 1459–1474.
9. CEER 2012. Draft Work Programme 2013 of the Council of European Energy Regulators. Available online: http://www.energy-regulators.eu/portal/page/portal/EER_HOME/EER_CONSULT/CLOSED%20PUBLIC%20CONSULTATIONS/CROSSSECTORAL/2013%20Work%20Programme/CD/C12-WPDC-22–06_WP%202013-PC_18-Jun-12.pdf (accessed on 3 November 2012).
10. Miller, R.D.; Ford, J.M. *Shared Savings in the Residential Market: A Public/Private Partnership for Energy Conservation*; Energy Task Force, Urban Consortium for Technology Initiatives: Baltimore, MD, USA, 1985.
11. Aasen, M.; Westskog, H.; Wilhite, H.; Lindberg, M. The EU electricity disclosure from business perspective—A study from Norway. *Energy Policy* **2010**, *38*, 7921–7928.
12. EC 2013. Homepage of Citizens' Energy Forum of the European Council. Available online: http://ec.europa.eu/energy/gas_electricity/forum_citizen_energy_en.htm (accessed on 30 August 2012).
13. Lise, W.; Timpe, C.; Jansen, J.C.; ten Donkelaar, M. Tracking electricity generation attributes in Europe. *Energy Policy* **2007**, *35*, 5855–5864.
14. Timpe, C. A European Standard for the Tracking of Electricity. Available online: http://ec.europa.eu/energy/intelligent/projects/sites/iee-projects/files/projects/documents/e-track_summary_slides.pdf (accessed on 15 April 2015).
15. Klimscheffskij, M. Tracking of Electricity in the EU—From Directives to Practice. Master Thesis, School of Science, Aalto University, Espoo, Finland, 2011.
16. Directive 2009/28/EC of the European Parliament and of the Council of 23 April 2009 on the Promotion of the Use of Energy from Renewable Sources and Amending and Subsequently Repealing Directives 2001/77/EC and 2003/30/EC. Available online: http://eur-lex.europa.eu/legal-content/EN/TXT/PDF/?uri=CELEX:32009L0028&from=en (accessed on 15 April 2015).
17. 2012/27/EC. Directive 2009/72/EC of the European Parliament and of the Council of 13 July 2009 Concerning Common Rules for the Internal Market in Electricity and Repealing Directive 2003/54/EC. Available online: https://www.energy-community.org/pls/portal/docs/36275.PDF (accessed on 15 April 2015).
18. AIB 2011. Annual Report 2010. Association of Issuing Bodies. Available online: http://www.aib-net.org/portal/page/portal/AIB_HOME/NEWSEVENTS/Annual_reports (accessed on 15 April 2015).

19. AIB 2013a. EECS Rules. Association of Issuing Bodies. Available online: http://www.aib-net.org/portal/page/portal/AIB_HOME/EECS/EECS_Rules (accessed on 15 April 2015).
20. AIB 2013b. The AIB Hub. Association of Issuing Bodies. Available online: http://www.aib-net.org/portal/page/portal/AIB_HOME/FACTS/EECS%20Registries/AIB_Hub (accessed on 11 December 2012).
21. AIB 2013c. AIB Members. Association of Issuing Bodies. Available online: http://www.aib-net.org/portal/page/portal/AIB_HOME/FACTS/AIB%20Members/AIB%20Members (accessed on 1 August 2012).
22. EEX 2015. Guarantees of Origin—EEX Power Derivatives. Available online: https://www.eex.com/en/market-data/power/derivatives-market/guarantees-of-origin#!/2015/03/19 (accessed on 24 March 2015).
23. Raadal, H.; Dotzauer, E.; Hanssen, O.J.; Kildal, H.P. The interaction between electricity disclosure and tradable green certificates. *Energy Policy* **2012**, *42*, 419–428.
24. Iisd 2012. Trends in investor claims over feed-in tariffs for renewable energy. News article of International Institute for Sustainable Development. Available online: http://www.iisd.org/itn/2012/07/19/trends-in-investor-claims-over-feed-in-tariffs-for-renewable-energy/ (accessed on 1 December 2012).
25. Ref 2011. Turkey implements feed-in tariff, Spain cuts. News article of Renewable Energy Focus. Available online: http://www.renewableenergyfocus.com/view/14994/turkey-implements-feed-in-tariff-spain-cuts/ (accessed on 13 October 2012).
26. Reuters 2010. Czech government adopts a plan to rein in renewables. News article of Reuters. Available online: http://www.reuters.com/article/2010/08/25/us-czech-renewables-idUSTRE67O4DP20100825?utm_source=feedburner&utm_medium=feed&utm_campaign=Feed%3A+reuters%2FUSgreenbusinessNews+%28News+%2F+US+%2F+Green+Business%29&utm_content=Google+Reader (accessed on 20 September 2012).
27. Italy cuts FiTs in an effort to balance renewables growth. News article of Renewable Energy Focus. Available online: http://www.renewableenergyfocus.com/view/25145/italy-cuts-fits-in-an-effort-to-balance-renewables-growth/ (accessed on 13 October 2012).
28. Bird, L.; Wüstenhagen, R.; Aabakken, J. A review of international green power markets: Recent experience, trends, and market drivers. *Renew. Sustain. Energy Rev.* **2002**, *6*, 513–536.
29. Hansla, A.; Gamble, A.; Juliusson, A.; Gärling, T. Psychological determinants of attitude towards and willingness to pay for green electricity. *Energy Policy* **2007**, *36*, 768–774.
30. Statistics on switching customers in Austria 2013. Unpublished material.
31. Sogge, C.V.; Øen, J. Klima for Redusert Strøm for Bruk? En Focus Gruppe Studie I Spenningsfeltet Mellom Forbruk Og Miljø, Kunnskap Og Verdier. Master Thesis, University of Life Science, Ås, Norway, 2007.
32. AIB 2015. AIB statistics. Association of Issuing Bodies. Available online: http://www.aib-net.org/portal/page/portal/AIB_HOME/FACTS/Market%20Information/Statistics (accessed on 20 April 2015).
33. PWC 2009. Green Electricity Making a Difference: An International Survey of Renewable Electricity Labels. Available online: https://www.pwc.dk/da/publikationer/assets/green-electricity-making-a-difference.pdf (accessed on 11 April 2015).

34. RE-DISS 2012a. Best Practice Recommendations for the implementation of Guarantees of Origin and other tracking systems for disclosure in the electricity sector in Europe. Available online: http://reliable-disclosure.org/ (accessed on 15 April 2015).
35. RE-DISS 2012c. Country Profiles. Available online: http://www.reliable-disclosure.org/documents/ (accessed on 15 April 2015).
36. Klimscheffskij, M. Electricity Residual Mix Calculation According to the RE-DISS Project—Methodology of Residual Mix Calculation. Appendix to the RE-DISS Best Practice Recommendations; 2012. Available online: http://reliable-disclosure.org/ (accessed on 15 April 2015).
37. Raadal, H.L.; Nyland, C.A.; Hanssen, O.J. Calculation of residual mixes when accounting for the EECS (European Electricity Certificate System)—The need for a harmonised system. *Energies* **2009**, *2*, 477–489.
38. RE-DISS 2012b. European Residual Mixes 2011. Available online: http://www.reliable-disclosure.org/upload/147-RE-DISS_2011_Residual_Mix_Results_v1_1.pdf (accessed on 15 April 2015).
39. RE-DISS 2012d. *Country-specific Qualitative Data Sheets*; RE-DISS project; European Commission: Brussels, Belgium, 15 October 2012.
40. ENTSO-e 2012. Detailed Monthly Production Data for All Countries for a Specific Range of Time. Excel-Spreadsheet. Available online: https://www.entsoe.eu/data/data-portal/production/ (accessed on 1 May 2012).
41. AIB 2013d. EECS Market Statistics. Association of Issuing Bodies. Available online: http://www.aib-net.org/portal/page/portal/AIB_HOME/FACTS/Market%20Information/Statistics (accessed on 31 October 2012).
42. EEX 2013. Guarantees of origin—Overview. Available online: http://www.eex.com/en/goo (accessed on 17 November 2012).

7

A New Protection System for Islanding Detection in LV Distribution Systems

Anna Rita Di Fazio *, Mario Russo and Sara Valeri

Department of Electrical and Information Engineering, University of Cassino and Southern Lazio, via Di Biasio 43, I-03043 Cassino, Italy; E-Mails: russo@unicas.it (M.R.); s.valeri@unicas.it (S.V.)

* Author to whom correspondence should be addressed; E-Mail: a.difazio@unicas.it

Academic Editor: Antonella Battaglini

Abstract: The growth of penetration of Distributed Generators (DGs) is increasing the risk of unwanted islanded operation in Low Voltage (LV) distribution systems. In this scenario, the existing anti-islanding protection systems, installed at the DG premises and based on classical voltage and frequency relays, are no longer effective, especially in the cases of islands characterized by a close match between generation and load. In this paper, a new protection system for islanding detection in LV distribution systems is proposed. The classical voltage and frequency relays in the DG interface protections are enriched with an innovative Smart Islanding Detector, which adopts a new passive islanding detection method. The aim is to keep the advantages of the classical relays while overcoming the problem of their limited sensitivity in detecting balanced islands. In the paper, to define the requirements of the anti-islanding protection system, the events causing the islanded operation of the LV distribution systems are firstly identified and classified. Then, referring to proposed protection system, its architecture and operation are described and, eventually, its performance is analyzed and validated by experimental laboratory tests, carried out with a hardware-in-the-loop technique.

Keywords: distribution system; distributed generation; smart grid; islanding; anti-islanding relay; distribution system protection

---

1. Introduction

In distribution systems, events caused by reconfiguration, maintenance or faults can leave all or part of a LV distribution network operating without the connection to the Medium Voltage (MV) distribution system and remaining energized as an autonomous entity by the Distributed Generators (DGs) connected to the isolated grid. Such an operation is referred to as islanding or loss-of-mains of the LV distribution network [1,2]. Unintentional islanding creates hazards for the utility personnel, who may intervene on parts of the LV network which should be isolated but are actually still energized by the DGs. Furthermore, such an unsupervised system operation may entail adverse power quality and impact on the equipment and customers within the islanded area. Eventually, the electrical equipment can be damaged by out-of-phase closure occurring when, after a fault, the unwanted island is reconnected to the utility network by automatic reclosing of MV breakers without any synchronizing action [3]. For all these reasons, the adoption of an adequate protection system able to detect any islanding condition and disconnect the DGs from the isolated network is essential to ensure the safe and secure operation of LV distribution systems.

Until a few years ago, the adoption of low-and high-voltage and low-and high-frequency relays, installed in the interface protection system at the DG premises, ensured the detection of islanding conditions. In fact, the low level of DG penetration in distribution networks caused a significant unbalance between generation and load during islanded operation; the resulting voltage and frequency perturbations were large enough to be detected by the low-and high-voltage and low-and high-frequency relays that tripped the DGs [4]. Over the last decade, the wide spread of DGs in LV distribution networks, especially of photovoltaic systems, has made it possible that, following the loss-of-mains, generation may balance load, maintaining the voltage and frequency within the admissible limits and preventing the intervention of the classical anti-islanding relays adopted by DGs [5].

A further issue is related to the false tripping of the DGs which is caused by classical anti-islanding relays. In fact, although the low-and high-voltage and low-and high-frequency relays do not permit the detection of "balanced islands", their sensitivity is high enough to confuse voltage and frequency excursions due to islanded operation with the ones due to perturbations generated at the transmission or distribution level, such as faults or sudden load changes [5,6]. Nowadays, false islanding detections can cause a massive loss of generation through a domino effect in the distribution systems, with consequences on the stability of the transmission system [7]. Then, many national and international standards for the connection of DGs to the distribution networks have recently been revised and less restrictive thresholds for the classical DG relays have been adopted, with the adverse effect of increasing the inadequacy of the classical relays to detect balanced islanding operating conditions [1,8].

To avoid missing the detection of balanced islands as well as misclassifying other perturbations as loss-of-mains, a number of innovative solutions for the anti-islanding protection systems of the LV distribution networks have been proposed, including new relays based on communication systems [5,9] or classical relays equipped with additional new passive anti-islanding detection methods [5,10]. In the former solution, when an utility circuit breaker opens yielding an islanding condition, a communication-based relay detects the status of the breaker and sends signals to remotely disconnect the DGs. This solution could be very effective but may incur in prohibitive costs, especially for the LV distribution networks because it requires the realization of dedicated communication infrastructures between MV/LV substations and DGs. The second type of solution appears to be more adequate and economically-viable for LV distribution systems. The idea is to add new passive detection methods to the existing protection systems so as to correctly detect balanced islands without increasing DG false tripping in the case of other perturbations.

This paper proposes a new protection system for islanding detection in LV distribution systems which enriches the classical voltage and frequency relays installed at the DG premises with an innovative Smart Islanding Detector (SmartID), recently proposed from the authors to guarantee a reliable detection of the balanced islands when the classical relays fail. The SmartID adopts a new passive method based on a linear representation of the distribution system as seen from the SmartID installation point, which is named Thevenin-like model. Details about the method and its implementation in an actual device as well as about its validation in real grids can be found in [1,2]. The specific contribution of this paper is to analyze the features and the performance of the overall LV anti-islanding protection system, including both classical relays and the SmartID.

The paper is organized as follows: in Section 2 the events causing the islanded operation of the LV distribution system are identified and classified, so as to define the requirements of a reliable anti-islanding protection system. The architecture and operation of the proposed protection system is described in Section 3. Finally, experimental simulations results are reported in Section 4. The performance analysis is carried out by a hardware-in-the-loop test facility, whose core is the Real-Time Digital Simulator (RTDS) [11] that allows interfacing the real-time distribution system simulation with an actual SmartID device.

2. The Islanding Events in LV Distribution System

An islanding condition arises when a portion of the distribution system, including the DGs and loads, is inadvertently isolated from the upstream utility network and kept energized by the DGs [2]. The islanded operation of all/part of a LV distribution system can occur when there is the loss of the main supply delivered by the MV distribution system through the MV/LV substation. The islanded operation presents different features according to the cause that determines the loss of the main supply. In general, four cases can be recognized

1. loss of the MV supply in normal operation of the MV distribution system,
2. loss of the MV supply due to faults in the MV distribution system,
3. loss of the MV/LV substation in normal operation of the LV distribution system,
4. loss of the MV/LV substation due to faults in the LV distribution system.

To analyze in details these cases, let the distribution system in Figure 1 be considered. The MV distribution system is composed of two feeders, indicated as feeder_1 and feeder_2, supplied by two HV/MV transformers. Each MV feeder is protected by a circuit breaker at its head, named CB_1 and CB_2 respectively, and can be sectionalized by the switch disconnectors SD_1, ..., SD_{13}. In normal operation the two feeders are separated by SD_7 which is normally opened but can change its status to guarantee a flexible reconfiguration of the MV distribution system. Two LV networks including DGs are evidenced and indicated as LV network_1 and LV network_2: each one is supplied by a MV/LV substation, named Sub_1 and Sub_2 respectively, which are protected at their heads by two circuit breaker, CB_3 and CB_4; the two LV networks operate separately and the normally opened disconnector D_1 can be used only for load transfer [12].

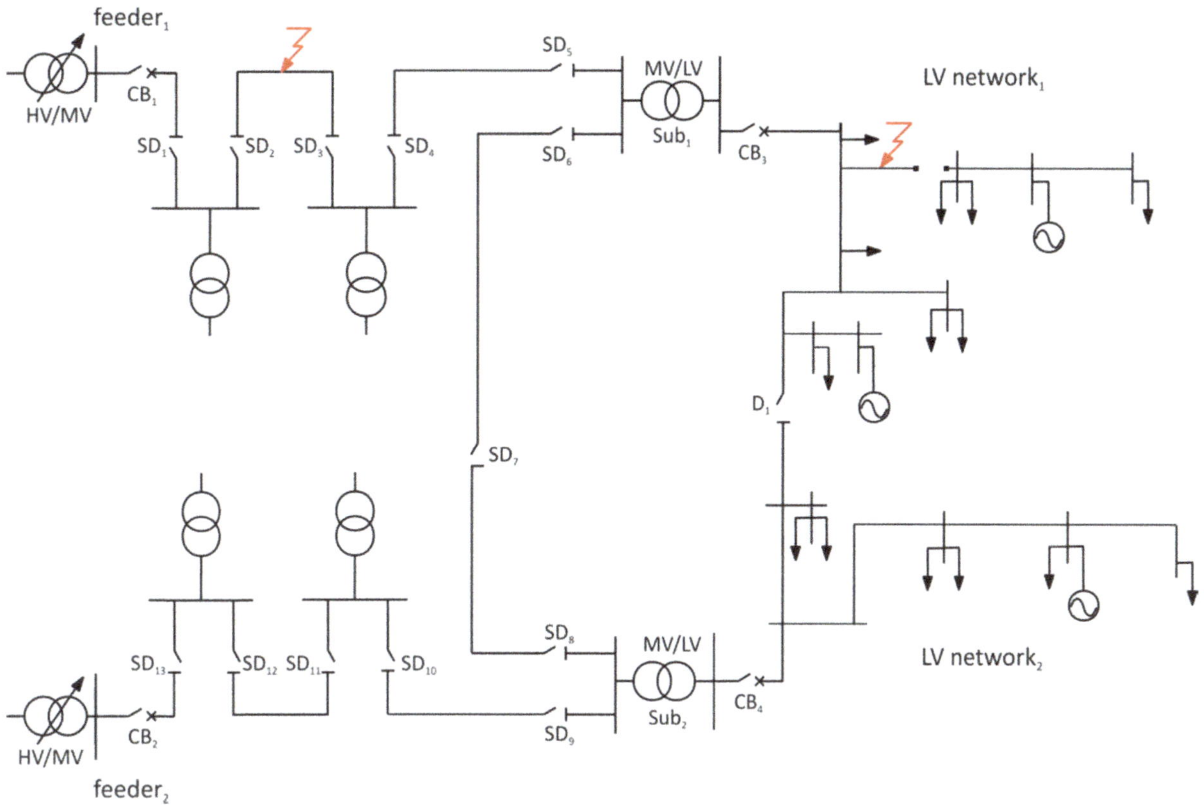

Figure 1. Simple distribution system with DGs.

2.1. Loss of the MV Supply in Normal Operation of the MV Distribution System

The loss of the MV supply can occur during the normal operation of the MV distribution system, as a consequence of switching operations for network reconfiguration or maintenance, or of false tripping in the MV networks. The MV network reconfiguration is used to reduce the distribution power losses and/or to relieve overloads; the maintenance is usually performed to improve the reliability of the MV distribution system; false tripping of the circuit breakers is caused by unexpected events [13–15]. Depending on the considered case, the loss of the MV supply can last for different time intervals. With reference to Figure 1 some examples are described in the following. A network reconfiguration can lead to supply Sub_1 from $feeder_2$ by automatically closing SD_7. In most cases this procedure does not cause the loss of the MV supply. However, if the network reconfiguration involves the simultaneous opening of the switch disconnectors in $feeder_1$ and $feeder_2$, it will cause the loss of the MV supply for several seconds. The maintenance of a component of $feeder_1$ can require the intervention of the utility personnel, that manually opens CB_1 to electrically isolate the component to be maintained; such an operation can last from minutes to hours. Finally, consequent to a false tripping of CB_1, a loss of the MV supply can last for around few hundreds of milliseconds (typically 200 ms), after which $feeder_1$ is re-energised by the automatic reclosure of the circuit breaker.

2.2. Loss of the MV Supply due to Faults in the MV Distribution System

The loss of the MV supply can occur as a consequence of switching operations due to single/multiple phases faults in the MV distribution system; the duration of the faults is minimized by automatic reclosing procedures of the MV protection system [16]. In Figure 1, for example, subsequent to the fault along $feeder_1$, the relays of the MV distribution system will detect the fault operating condition and will send an opening command to CB_1. A first fast reclosure of CB_1 is around 200 ms after the first CB_1 opening. If the fault is cleared before the CB_1 reclosure, the re-energization of the distribution system is successful; otherwise, CB_1 is opened again together with SD_1, ..., SD_6. In this latter case, a second slow reclosure of CB_1 is expected around 30 s after its first opening. After this reclosure, all the switch disconnectors are sequentially closed starting from SD_1. If the fault is not cleared yet, the MV protection system can foresee further steps of the reclosing procedure, otherwise the fault has to be repaired.

2.3. Loss of the MV/LV Substation in Normal Operation of the LV Distribution System

In comparison with the MV distribution system, the LV distribution system does not present an high level of automation [17]. Thus, network reconfiguration, maintenance and false tripping require the intervention of the utility personnel, which causes lack of supply of LV system for long time intervals (minutes-hours). In Figure 1, for example, the maintenance of Sub_1 requires to open both SD_5, SD_6 and CB_3. If a network reconfiguration is expected, the LV network can be re-energized by opening CB_4, changing the status of D_1 and closing CB_3 again. This manual procedure takes at least several minutes.

2.4. Loss of the MV/LV Substation due to Faults in the LV Distribution System

The loss of the MV/LV substation can occur as a consequence of single/multiple phases faults in the LV distribution system. For example, subsequent to the fault along the LV $network_1$ in Figure 1, the opening of CB_3 yields the loss of Sub_1. The reclosure of CB_3 is not automatic and requires the manual intervention of the utility personnel that takes a long time (minutes-hours). If the fault is cleared, the first reclosure attempt is successful; otherwise the failure component must be repaired and the loss of the MV/LV substation lasts for even more time. Moreover, also in the case of a successful reclosure, part of the LV network could still remain unfed by Sub_1, because the fault may cause a joint disconnection, evidenced in Figure 1 as a line interruption at the fault point in the LV $network_1$.

From the previous analysis, some general considerations about the islanded operation of the LV distribution system can be derived, that allows to define the requirements of the anti-islanding protection system. They can be summarized as follows

- the islanded operation of the whole LV distribution system can be caused by events occurring in the MV distribution system and can last for about 200 ms or for at least 30 s;
- the islanded operation of all/part of the LV distribution system can be caused by events occurring in the LV distribution system and can lasts for long time intervals (minutes-hours).

3. The Proposed Anti-Islanding Protection System

3.1. Architecture of the Anti-Islanding Protection System

The proposed anti-islanding protection system is composed of classical frequency and voltage relays, SmartIDs and voltage detectors. Classical relays are installed at each DG site in the interface protection system. They have different thresholds depending on the type of DG involved. They trip very fast so as to avoid islanding conditions characterized by a significant power mismatch and to permit a successful fast reclosure in MV distribution system. Their detection time increases as the power imbalance decreases.

The innovative SmartIDs are integrated in the interface protection system of some or all the DGs; they are able to detect the islanding events when the variations of the electrical quantities are so small that are undetectable by the classical relays; they trip slowly but fast enough to guarantee a successful slow reclosure in MV and LV distribution systems.

Eventually, to increase the safety of the personnel a voltage detector can be installed at the head of the feeder in the MV/LV substation, so as to give a visual information about the absence or presence of the voltage on the LV network when the circuit breaker is opened.

3.1.1. Voltage and Frequency Relays

The voltage relay is designed to detect voltage variations subsequent to power imbalances in the islanded system. If the voltage at the DG interface exceeds or falls below a threshold value, the relay ceases to energize the power island by sending a trip signal to the circuit breaker of the DG after a time delay. The purpose of the time delay is to ride through short-term disturbances to avoid excessive false tripping. Commonly, the voltage relay assumes different tripping thresholds with different tripping delays according to the standards adopted by different jurisdictions. Table 1 shows the minimum requirements for the voltage protection relays for the IEEE and the IEC standards together with the limits adopted by the Italian standard [18–20].

Table 1. Voltage requirements for LV distribution system.

	IEEE 1574		**IEC 61727**		**CEI 0-21**	
DG size	V (%) *	t_c ** (s)	V (%) *	t_c ** (s)	V (%) *	t_c ** (s)
≤30 kW	≥120	0.16	≥135	0.05	≥110	≤3.00
	$110 < V < 120$	1.00	$110 < V < 135$	2.00	≥115	0.20
	$50 \leq V < 88$	2.00	$50 \leq V < 85$	2.00	<85	0.40
	<50	0.16	<50	0.10	<40	0.20
>30 kW	≥120	≤0.16				
	$110 < V < 120$	≤1.00				
	$50 \leq V < 88$	≤2.00				
	<50	≤0.16				

* % of the nominal distribution system voltage; ** the clearing time t_c is the time between the occurrence of the island and the opening of the DG circuit breaker.

The frequency relay is designed to detect frequency variations subsequent to power imbalances in the islanded system. It incorporates two types of detector, namely the low- and high-frequency and the rate-of-change-of-frequency (ROCOF) relays, which operate in a combined manner. When the frequency in the island is lower/higher than a threshold value, a trip signal is immediately sent to the circuit breaker of the DG. The ROCOF detector is used to accelerate the detection and to solve the problem of the fast reclosure in MV distribution networks, as shown in the Appendix. In addition, a minimum voltage block function is implemented in such relay to block the trip signal during the generator start-up or the short-circuits. Again, the frequency relay presents different tripping thresholds with different trip delays, which vary according to the standards adopted by different jurisdictions. Table 2 shows the minimum requirements for the frequency protection relays for the IEEE and the IEC standards together with the limits adopted by the Italian standard [18–20].

Table 2. Frequency requirements for LV distribution system.

	IEEE 1574		**IEC 61727**			**CEI 0-21**	
DG size	f (Hz)	t_c (s)	f (Hz)	t_c (s)	DG type	f (Hz)	t_c (s)
≤30 kW	>60.5	0.16	>51.0	0.20	Rotating	>50.5	0.10
	<59.3	0.16	<49.0	0.20		<49.5	0.10
>30 kW	>60.5	0.16			Static	>51.5	0.10 or 4.00 **
	<(59.8–57.0) *	(0.16–300) *				<47.5	0.10 or 4.00 **
	<57.0	0.16					

* adjustable; ** the clearing time t_c is equal to 0.10 s or 4.00 s according to whether any integrative remote control protection system is absent or present.

3.1.2. SmartID

The SmartID is able to detect the islanding events even in the case in which the variations of the electrical quantities are so small that are undetectable by the classical relays. The passive method adopted by the SmartID is described by the block scheme in Figure 2. On the basis of local voltage $v(t)$ and current $i(t)$ measurements, the corresponding phasors $\bar{V}$ and $\bar{I}$ at the fundamental frequency are evaluated to estimate the parameters $\bar{a}$ and $\bar{b}$ of a linear Thevenin-like model, representing the LV distribution system as seen from the DG. The dimensions of the parameters $\bar{a}$ and $\bar{b}$ are, respectively, a voltage and an impedance. The detection method is based on the variations of the absolute values a and b assumed by $\bar{a}$ and $\bar{b}$ in grid-connected and islanded operation. Comparing a and b with their respective thresholds ref_a and ref_b, if:

$$a \geq ref_a \text{ and } b \leq ref_b$$

an output signal OS is generated which is representative of the grid-connected operation; on the other hand, if:

$$a < ref_a \text{ and } b > ref_b$$

a different output signal OS is generated which is representative of the islanded operation. Details about the islanding detection method and its implementation in the SmartID prototype can be found in [10,21].

Table 3 shows the thresholds ref_a and ref_b adopted by the SmartIDs, which are set according to the procedure described in [22]. Two interdiction zones are determined which clearly separate the values assumed by the estimated parameters a and b, respectively, in grid-connected and islanded operation; then, ref_a and ref_b can be fixed inside such interdiction zones. Indeed, for a given operating condition (grid-connected or islanded), the values of a and b are anyway subject to variations, due to changes of the loads and DG power injections in the distribution system and/or events occurring at higher voltage levels. Such variations of a and b are tracked by the estimator block in Figure 2; if the estimated values of a and b fall inside the interdiction zones and pass across the thresholds, a false detection occurs. To overcome this problem and improve SmartID stability with respect such variations of a and b, in this paper a hysteresis is introduced by using two different values for each threshold, ref_a and ref_b, in dependence of the type of the transition (from grid connected to islanded operation and viceversa) according to Table 3. To explain how the two threshold values are set and the benefits introduced by the hysteresis, let parameter a be considered for example. In grid-connected operation a generally lies above the interdiction zone, whereas in islanded operation it lies below the interdiction zone. The transition from grid-connected to islanded operation must be detected when the estimated value of a goes below the interdiction zone: by fixing ref_a in the lowest part of the interdiction zone, the SmartID correctly detects actual transitions from grid-connected to islanded operation and does not erroneously detect the variations that may cause the value of a falling in the interdiction zone. On the contrary, the inverse transition from islanded to grid-connected operation is detected when the estimated value of a goes above the interdiction zone: by fixing ref_a in the highest part of the interdiction zone, the SmartID correctly detects actual transitions from islanded to grid-connected operation while avoiding misdetections. Similar considerations can be made for the other parameter b.

Table 3. SmartID thresholds.

Transition operation	ref_a **(V)**	ref_b **(Ω)**
grid connected→ islanded	120	2.0
islanded→ grid connected	200	1.0

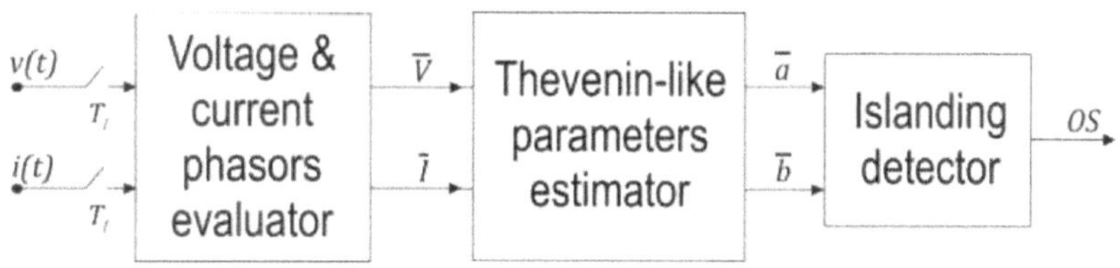

Figure 2. Block diagram of the SmartID.

The clearing time is not fixed but depends on many factors related to the detection method adopted by the SmartID. In general, the clearing time is of the order of seconds and has not to exceed 30 s to guarantee the successful MV slow reclosure.

3.2. Operation of the Protection System

The operation of the protection system is considered with reference to the events summarized at the end of Section 2.

3.2.1. Islanded Operation of the Whole LV Distribution System due to MV Network Events

As described in Section 2, subsequent to the loss of the MV supply due to MV network events, the whole LV distribution system can operate in islanding for either about 200 ms or at least 30 s. In most cases the imbalance between DGs and loads causes variations of frequency and voltage in the islanded system, such that the classical relays provide to disconnect all the DGs within 200 ms and the islanded operation ends before any possible recovery of the main supply. On the contrary, if the DGs match loads, the classical relays do not detect islanding within 200 ms and the fast reclosure occurs while the DGs is still connected to the distribution system. Then, two cases can occur. If the MV fault is extinguished, the reclosure is successful but can be out-of-phase. In the Appendix, it is shown how adequate thresholds adopted by classical relays avoid in this case a reclosure with more than 20° of voltage phase displacement [5]. If the MV fault is not extinguished, the reclosure fails and the islanded operation of the whole LV distribution system continues; the adoption of classical relays and SmartIDs assures that DGs are tripped before the slow reclosure of the circuit breaker, occurring 30 s after the failure of the first reclosure.

3.2.2. Islanded Operation of All/Part of the LV Distribution System due to LV Network Events

As described in Section 2, subsequent to the loss of the MV/LV substation due to LV network events, an islanding operation of all/part of the LV distribution system can last for long time (minutes-hours). The adoption of classical relays together with the SmartIDs guarantees the detection of balanced and unbalanced islands and the tripping of DGs within 30 s.

4. Numerical Simulations

The experimental studies have been performed with reference to the test distribution system in Figure 3. It is composed of a $20\ \text{kV} - 50\ \text{Hz}$ distribution system, represented by a Thevenin equivalent and the circuit breaker CB_1, which supplies through a $20/0.4\ \text{kV}$ - $0.25\ \text{MVA}$ transformer a $0.4\ \text{kV}$ three-phases feeder protected by the utility breaker CB_2. The electric parameters of the lines l_1,..., l_7 are reported in Table 4. The active and reactive powers absorbed by the loads $L_1, \dots, L_8$ supplied by the LV feeder are shown in Table 5; loads L_4, L_7 are subject to random variations within the range reported in Table 5. The LV distribution system includes two $0.25\ \text{MVA}$ DGs, named DG_1 and DG_2. DG_1 is a photovoltaic system coupled with the grid by a power converter; the active power is controlled and varies following the random irradiation, while the reactive power is controlled with a power factor equal to 0.9. DG_2 is a small hydro-turbine moving a three-phase synchronous generator, which is equipped with active and reactive power control systems, whose set-points are fixed so as to create and sustain balanced islands, as described in the following. Both the DG interfaces include the low-and high-voltage and low-and high-frequency relays, whose thresholds are fixed according to the CEI standard as reported, respectively, in Tables 1 and 2. The SmartID device is connected at the terminals of either DG_1 or DG_2 and its thresholds are reported in Table 3.

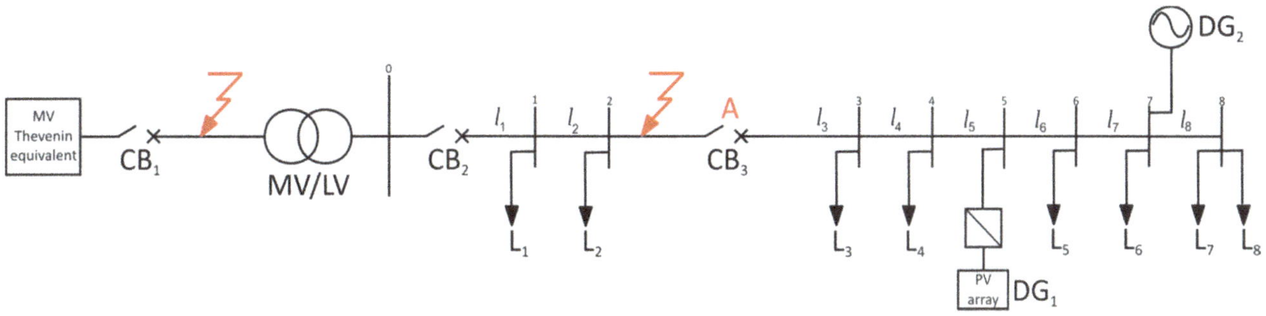

Figure 3. Test distribution system.

Table 4. Electric parameters of the lines.

l_1		l_2		l_3		l_4		l_5		l_6		l_7	
R (Ω)	X (Ω)	R (Ω)	X (Ω)	R (Ω)	X (Ω)	R (Ω)	X (Ω)	R (Ω)	X (Ω)	R (Ω)	X (Ω)	R (Ω)	X (Ω)
0.067	0.016	0.038	0.009	0.073	0.017	0.051	0.009	0.061	0.009	0.033	0.005	0.026	0.004

Table 5. Active and reactive powers absorbed by the loads.

Powers	L_1	L_2	L_3	L_4	L_5	L_6	L_7	L_8
P(kW)	1.5	0.58	21.0	0.0 ÷ 1.5	4.44	1.5	0.0 ÷ 1.6	3.0
Q(kVAr)	0.68	0.26	9.45	0.0 ÷ 0.68	2.0	0.68	0.0 ÷ 0.72	1.35

The analysis is carried out by a hardware-in-the-loop test facility. The core of the set-up is the RTDS [11] equipped with RSCAD/EMTDC, which is used to simulate in real-time the distribution network, the DGs and the classical interface protections. Conversely, the SmartID is an actual device, interfaced with the RTDS by D/A converters.

The aim of the case study is to compare the proposed anti-islanding protection system with the classical one. Two different analysis are carried out. The first one aims at comparing the performance of the two protection systems in detecting islands characterized by the exact balance between generation and load. The second one is a sensitivity analysis that aims at evaluating the impact of different imbalances of the active power ΔP and of the reactive power ΔQ on the performance of the two protection systems. In both the analysis two events are considered (see Section 2)

1. Event 1: islanded operation of the whole LV distribution system caused by a single-phase fault in the MV distribution system with the failure of the fast reclosure;
2. Event 2: islanded operation of part of the LV distribution system caused by a three-phase fault in the LV distribution system yielding joint-disconnection.

To take into account the stochastic nature of loads and irradiation, the analysis are carried out by adopting a statistical approach. Repeated simulations of the same distribution system changing from grid-connected to islanded operation are performed by imposing random variations of the two variable loads and the irradiation of the photovoltaic system.

The transition from grid-connected to islanded operation is simulated by opening a circuit breaker in the distribution system when the power imbalance in the island is equal to the desired value, achievable by acting on the DG_2 control system.

In the case of Event 1, the simulation starts with both the DGs in connection with the MV supplying system; acting on the DG_2 control system, the powers flowing through CB_2 are reduced to the desired power imbalances ΔP and ΔQ. Then, a single-phase fault occurs at the MV busbar, as shown in Figure 3. After 150 ms, CB_1 is opened thus simulating the action of the MV protection system against fault; after further 200 ms, the fast reclosure of CB_1 takes place but fails and CB_1 is opened again after 150 ms. In this way, the whole LV distribution system operates in island while the single-phase fault in the MV network is not extinguished.

In the case of Event 2, the joint is located in the point A of the line l_3 in Figure 3 and its disconnection is simulated by the opening of the breaker CB_3. The simulation starts with both the DGs in connection with the MV/LV substation; acting on the DG_2 control system, the powers flowing through CB_3 are reduced to the desired power imbalances ΔP and ΔQ. Then, a three-phase fault occurs along the line l_3, as shown in Figure 3. After 50 ms, CB_3 is opened thus simulating the joint disconnection and a part of the LV distribution system operates in island.

The performance of the anti-islanding protection systems is measured on the basis of the successful detections and the related detection times (the detection time represents the time interval elapsing between the occurrence of the island and the change of the relay output) for the voltage and frequency relays and the SmartID. The use of detection times rather than the clearing times allows the use of the same simulation to compare the performance of the two protection systems. The detection times are expressed in terms of Gaussian statistical distributions. For the sake of comparison, some performance indices are derived from the distributions, namely the percentage of missed detections p_{md}, the mean value μ_{sd} and the standard deviation σ_{sd} of the detection times in the cases of successful detections.

4.1. Performance Analysis in Balanced Islands

The performances of the two protection systems in detecting islands characterized by $\Delta P = 0$ and $\Delta Q = 0$ are compared. Four cases are analyzed, combining the two events causing the islanded operation and the SmartID installation points:

(a) Case A: Event 1 with the SmartID connected to DG_1;
(b) Case B: Event 1 with the SmartID connected to DG_2;
(c) Case C: Event 2 with the SmartID connected to DG_1;
(d) Case D: Event 2 with the SmartID connected to DG_2.

For each case twenty simulations have been performed.

4.1.1. Case A

Table 6 reports the detection times for the classical voltage and frequency relays installed at both the DG_1 and DG_2 interfaces, indicated respectively as V/f_{DG_1} and V/f_{DG_2}, and for the SmartID connected to the DG_1 terminals, indicated as SID_{DG_1} for each simulation. Detection times overcoming 120 s are not reported and considered as a missed detection. It is evident from Table 6 that V/f_{DG_1} and

V/f_{DG_2} often fail detection due to the balanced island and, in general, V/f_{DG_1} takes more time to intervene with respect to V/f_{DG_2} because its thresholds setting is less restrictive. On the contrary, the SmartID always identifies islanding with limited detection times. It is worth to notice that in trial 14 V/f_{DG_1} and V/f_{DG_2} are faster than SID_{DG_1}; this is due to the random variations of PV generation and loads which can sometimes cause large voltage and frequency excursions and the subsequent fast intervention of the classical relays.

Table 6. Detection times in the Case A.

Trial	Detection time (s)		
	V/f_{DG_1}	SID_{DG_1}	V/f_{DG_2}
1	-	6.96	87.84
2	6.51	1.99	6.30
3	-	4.73	100.70
4	72.88	1.87	50.98
5	-	4.62	-
6	7.13	2.98	6.92
7	72.85	1.75	50.63
8	7.61	6.33	7.40
9	-	3.44	-
10	-	3.77	-
11	-	5.71	53.22
12	77.83	1.78	54.64
13	-	3.06	55.37
14	6.44	7.27	6.24
15	7.34	3.35	7.56
16	72.85	2.56	50.59
17	-	3.30	87.98
18	72.85	3.35	50.59
19	-	6.01	-
20	-	5.58	48.64

Table 7 reports the performance indices derived from Table 6. They are evaluated not only for each relay but also for both the overall classical and proposed protection systems. Comparing the single relays, SID_{DG_1} successfully detects the condition of islanding while V/f_{DG_2} fails for the 20% of the simulations and V/f_{DG_1} fails for the 50% of the simulations. Referring to the successful detections, the fastest relays is SID_{DG_1} which averagely takes 4.02 s whereas V/f_{DG_1} and V/f_{DG_2} take more than 30 s, which is the requirement of a reliable anti-islanding protection system for LV networks. Also the standard deviation of SID_{DG_1} is very limited, differently from the other ones. Comparing the classical to the proposed protection systems, the latter one presents a more reliable and effective performance, because it always detects islanded operation and it averagely takes 3.97 s rather than 45.34 s to detect island. It is worth to notice that average detection time of the overall proposed protection system is smaller than the one of the single SmartID, because, as previously explained, there is a case in which the classical relays are faster than the SmartID.

Table 7. Comparison between the classical and the proposed protection systems in the Case A.

Performance indices	Single relay			Classical protection system	Proposed protection system
	V/f_{DG_1}	SID_{DG_1}	V/f_{DG_2}		
p_{md} (%)	50	0	20	20	0
μ_{sd} (s)	40.43	4.02	45.35	45.34	3.97
σ_{sd} (s)	35.26	1.77	31.05	31.07	1.68

4.1.2. Case B

Moving the SmartID from DG1 to DG2, the results in Table 8 confirm the reliability and the effectiveness of the proposed protection system in detecting island when generation and load closely match. The results related to SID_{DG_2} are even better than the ones of SID_{DG_1} in the previous case. In both Tables 7 and 8, the values of p_{md} and μ_{sd} are always worse for V/f_{DG_1} with respect to V/f_{DG_2}, due to the less restrictive threshold setting of V/f_{DG_1}. For this reason, in the following Section 4.2 the sensitivity analysis is performed assuming the SmartID installed at the DG1 terminals to empower its interface protection system.

Table 8. Comparison between the classical and the proposed protection systems in the Case B.

Performance indices	Single relay			Classical protection system	Proposed protection system
	V/f_{DG_1}	V/f_{DG_2}	SID_{DG_2}		
p_{md} (%)	30	15	0	15	0
μ_{sd} (s)	70.17	51.31	2.97	51.31	2.97
σ_{sd} (s)	10.58	19.72	1.47	19.72	1.47

4.1.3. Case C

Table 9 reports the performance indices for the single relay V/f_{DG_1}, V/f_{DG_2} and SID_{DG_1} as well as for both the overall classical and proposed protection systems. Comparing the single relays, V/f_{DG_1} fails for the 20% of the simulations, SID_{DG_1} always detects the condition of islanding and V/f_{DG_2} fails for 5% of the simulations. Referring to the successful detections, the fastest relay is SID_{DG_1} which on average takes 7.85 s and also presents a limited standard deviation with respect to the other relays. Comparing the classical to the proposed protection systems, the latter one presents a more reliable and effective performance.

Table 9. Comparison between the classical and the proposed protection systems in the Case C.

Performance indices	Single relay			Classical protection system	Proposed protection system
	V/f_{DG_1}	SID_{DG_1}	V/f_{DG_2}		
p_{md} (%)	20	0	5	5	0
μ_{sd} (s)	59.02	7.85	38.31	36.84	7.67
σ_{sd} (s)	24.06	2.52	21.57	18.13	2.46

4.1.4. Case D

Moving the SmartID from DG_1 to DG_2, the results in Table 10 confirm that the proposed protection system, thanks to presence of the SmartID, is always able to identify balanced islands in a limited interval of time. Comparing Tables 9 and 10, it is evident that, also in this case, the SmartID located at the DG_2 terminals averagely takes smaller time to detect island than the one located at the DG_1 terminals.

Table 10. Comparison between the classical and the proposed protection systems in the Case D.

Performance indices	Single relay			Classical protection system	Proposed protection system
	V/f_{DG_1}	V/f_{DG_2}	SID_{DG_2}		
p_{md} (%)	55	10	0	10	0
μ_{sd} (s)	35.13	34.78	6.05	34.65	5.65
σ_{sd} (s)	32.19	19.04	2.52	19.22	1.35

4.2. Sensitivity Analysis with Respect to Power Imbalances

The impact of different ΔP and ΔQ on the performance of the two protection systems is analyzed. Four cases are considered, combining the two events causing the islanded operation and different types and amplitudes of the power imbalance:

(a) Case A: Event 1 with $\Delta P = (-10, -5, +5, +10)\%$ and $\Delta Q = 0$;
(b) Case B: Event 1 with $\Delta P = 0$ and $\Delta Q = (-5, +5)\%$;
(c) Case C: Event 2 with $\Delta P = (-10, -5, +5, +10)\%$ and $\Delta Q = 0$;
(d) Case D: Event 2 with $\Delta P = 0$ and $\Delta Q = (-5, +5)\%$.

In all the cases the SmartID has been installed at the DG_1 premises whose voltage and frequency relays present less restrictive thresholds. As in the previous analysis, twenty simulations are carried out for each considered case.

4.2.1. Case A

The results of a sensitivity analysis of the performance indices with respect to an active power imbalance ΔP ranging from −10% to +10% are reported in Table 11. Also in the case of significant power imbalances, V/f_{DG_1} fails by either missing detection or presenting an unacceptable detection time. On the contrary, V/f_{DG_2} and SID_{DG_1} successfully detect the unwanted conditions with adequate detection times. Comparing the performance indices of the classical and of the proposed protection systems, the correct operation within 30 s of both the solutions is evident, although the proposed system always presents smaller μ_{sd} and σ_{sd}. This result could be misleading: the successful intervention of the classical protection system is due to the presence of V/f_{DG_2} characterized by strict thresholds, that are going out of use. It is worth noticing that further increases of ΔP are not analyzed because the larger the frequency and voltage excursions are, the faster the intervention of the classical relays.

Table 11. Comparison between the classical and the proposed protection systems in the Case A.

Performance indices		Single relay			Classical protection system	Proposed protection system
		V/f_{DG_1}	SID_{DG_1}	V/f_{DG_2}		
p_{md} (%)	$\Delta P = +5\%\ \Delta Q = 0$	30	0	0	0	0
	$\Delta P = +10\%\ \Delta Q = 0$	60	0	0	0	0
	$\Delta P = -5\%\ \Delta Q = 0$	30	0	0	0	0
	$\Delta P = -10\%\ \Delta Q = 0$	25	0	0	0	0
μ_{sd} (s)	$\Delta P = +5\%\ \Delta Q = 0$	63.38	6.05	11.54	11.53	5.98
	$\Delta P = +10\%\ \Delta Q = 0$	26.14	7.97	3.63	3.55	2.64
	$\Delta P = -5\%\ \Delta Q = 0$	31.13	7.08	11.14	11.12	5.75
	$\Delta P = -10\%\ \Delta Q = 0$	73.69	7.44	1.82	1.76	1.62
σ_{sd} (s)	$\Delta P = +5\%\ \Delta Q = 0$	18.05	2.19	3.38	3.40	2.06
	$\Delta P = +10\%\ \Delta Q = 0$	23.77	3.00	3.86	3.75	2.59
	$\Delta P = -5\%\ \Delta Q = 0$	31.65	3.36	4.91	4.94	2.05
	$\Delta P = -10\%\ \Delta Q = 0$	31.79	2.06	2.35	2.30	1.97

4.2.2. Case B

In Table 12, the results of a sensitivity analysis of the performance indices with respect to a reactive power imbalance ΔQ ranging from −5% to +5% are reported. Concerning V/f_{DG_1}, it presents an opposite behavior with respect to the sign of ΔQ: for a positive variation it always detects islanded operation whereas for a negative one it always misses detection. This is due to the asymmetrical setting of the voltage thresholds reported in Table 2. Concerning V/f_{DG_2}, it always detects the unwanted conditions in a small time thanks to the frequency relay with strict threshold settings. Concerning SID_{DG_1} , it always identifies islands with a performance similar to the one observed for imbalances of ΔP; it can be stated that the SmartID is quite insensitive to the type, the sign and the amplitude of the power imbalance. Concerning the classical and the proposed protection systems, considerations similar to the ones of the previous case can be made.

Table 12. Comparison between the classical and the proposed protection systems in the Case B.

Performance indices		Single relay			Classical protection system	Proposed protection system
		V/f_{DG_1}	SID_{DG_1}	V/f_{DG_2}		
p_{md} (%)	$\Delta P = 0\%\ \Delta Q = +5\%$	0	0	0	0	0
	$\Delta P = 0\%\ \Delta Q = -5\%$	100	0	0	0	0
μ_{sd} (s)	$\Delta P = 0\%\ \Delta Q = +5\%$	23.62	7.87	1.06	1.04	0.95
	$\Delta P = 0\%\ \Delta Q = -5\%$	-	7.40	2.52	2.43	2.21
σ_{sd} (s)	$\Delta P = 0\%\ \Delta Q = +5\%$	10.27	2.83	1.55	1.51	1.13
	$\Delta P = 0\%\ \Delta Q = -5\%$	-	1.67	3.72	3.64	3.22

4.2.3. Case C

Table 13 reports the performance indices of the single relays and of the two protection systems which are obtained by varying the active power imbalance ΔP from −10% to +10%. Even if significant power imbalances arise, V/f_{DG_1} misses detection several times. Even when V/f_{DG_1} detects island, it takes minutes and the order of magnitude of its μ_{sd} and σ_{sd} are too large to guarantee an adequate level of power quality so as to avoid electric equipment damages. On the contrary, V/f_{DG_2} and SID_{DG_1} and both the classical and the proposed protection systems successfully detect the unwanted conditions with adequate detection times. As already explained, the successful intervention of the classical protection system is due to the presence of V/f_{DG_2} characterized by strict thresholds, that are going out of use.

Table 13. Comparison between the classical and the proposed protection systems in the Case C.

Performance indices		Single relay			Classical protection system	Proposed protection system
		V/f_{DG_1}	SID_{DG_1}	V/f_{DG_2}		
p_{md} (%)	$\Delta P = +5\%\ \Delta Q = 0$	15	0	0	0	0
	$\Delta P = +10\%\ \Delta Q = 0$	35	0	0	0	0
	$\Delta P = -5\%\ \Delta Q = 0$	15	0	0	0	0
	$\Delta P = -10\%\ \Delta Q = 0$	30	0	0	0	0
μ_{sd} (s)	$\Delta P = +5\%\ \Delta Q = 0$	67.80	7.35	17.13	17.10	6.90
	$\Delta P = +10\%\ \Delta Q = 0$	53.54	6.95	0.77	0.77	0.77
	$\Delta P = -5\%\ \Delta Q = 0$	46.26	8.44	10.20	10.20	5.42
	$\Delta P = -10\%\ \Delta Q = 0$	35.67	7.96	0.66	0.66	0.66
σ_{sd} (s)	$\Delta P = +5\%\ \Delta Q = 0$	19.41	2.06	5.88	5.90	2.29
	$\Delta P = +10\%\ \Delta Q = 0$	28.35	1.54	0.01	0.01	0.01
	$\Delta P = -5\%\ \Delta Q = 0$	33.10	2.24	8.37	8.37	3.82
	$\Delta P = -10\%\ \Delta Q = 0$	31.14	1.68	0.07	0.07	0.07

4.2.4. Case D

Table 14 reports the performance indices of the single relays and of the two protection systems which are obtained by varying the reactive power imbalance ΔQ from −5% to +5%. For a positive variation of ΔQ, V/f_{DG_1} misses detection for 5% of the simulations whereas for a negative one it always misses detection. Concerning V/f_{DG_2}, SID_{DG_1} and both the classical and the proposed protection systems, they always identify islands with a small detection time.

Table 14. Comparison between the classical and the proposed protection systems in the Case D.

Performance indices		Single relay			Classical protection system	Proposed protection system
		V/f_{DG_1}	SID_{DG_1}	V/f_{DG_2}		
p_{md} (%)	$\Delta P = 0\%\ \Delta Q = +5\%$	5	0	0	0	0
	$\Delta P = 0\%\ \Delta Q = -5\%$	100	0	0	0	0
μ_{sd} (s)	$\Delta P = 0\%\ \Delta Q = +5\%$	13.67	11.75	0.64	0.64	0.64
	$\Delta P = 0\%\ \Delta Q = -5\%$	-	7.02	2.79	2.79	2.52
σ_{sd} (s)	$\Delta P = 0\%\ \Delta Q = +5\%$	9.72	5.24	0.07	0.07	0.07
	$\Delta P = 0\%\ \Delta Q = -5\%$	-	1.57	3.40	3.40	3.06

5. Conclusions

A new anti-islanding protection system for LV distribution systems has been proposed by enriching the classical voltage and frequency relays installed at the DG premises with an innovative Smart

Islanding Detector, which adopts a new passive islanding detection method. The proposed protection system retains the advantages of the classical voltage and frequency relays while overcoming the problem of their limited sensitivity in detecting balanced islands. Moreover, it guarantees that the first fast reclosure in MV networks is not out-of-phase and the second slow reclosure occurs after the detection of the islanded operation. The proposed protection system is a valid alternative for LV distribution systems to solutions based on communication methods, thanks to its technical effectiveness and economic viability. Its performance has been analyzed and validated by using a hardware-in-the-loop test facility, based on Real-Time Digital Simulator interfaced with an actual Smart Islanding Detector device.

Author Contributions

The islanding events in LV distribution systems have been identified and studied by A.R. Di Fazio and S. Valeri. The main contribution to defining the architecture and operation of the overall anti-islanding protection system has been given by A.R. Di Fazio. The numerical simulations and the considerations in the appendix have been developed principally by S. Valeri. The work coordination was up to M. Russo.

Appendix

In the following, it is shown how adequate thresholds for frequency relays avoid an out-of-phase fast reclosure between the MV supplying system and the islanded LV distribution system. Let's consider the frequency relays protecting the rotating generators, which can be damaged by a reclosure with more than 20° of phase displacement (static generators are less sensitive to this problem). Figure A1 shows the thresholds of the high-frequency relay, equal to 50.5 Hz, and of the positive ROCOF relay, equal to 2.5 Hz/s. The behaviors of the low-frequency and the negative ROCOF relays are symmetrical to the ones in Figure A1.

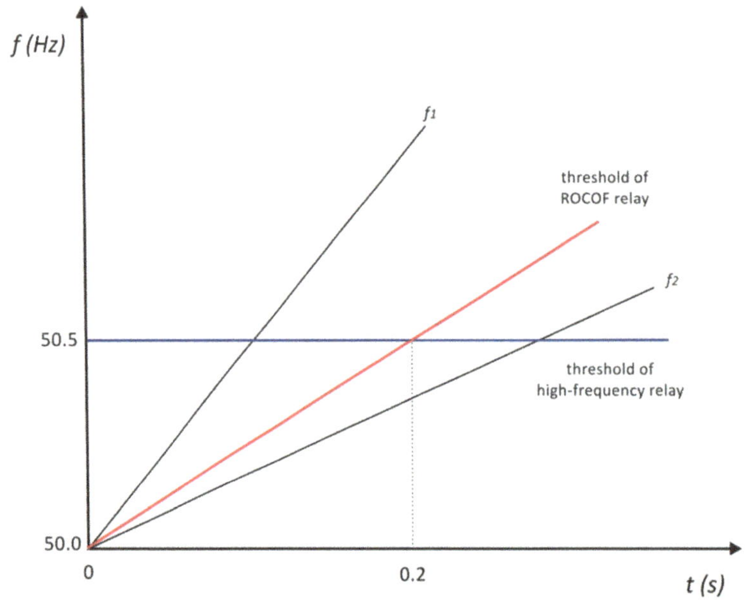

Figure A1. Threshold setting of the high-frequency and ROCOF relays for rotating generators.

Subsequent to a fault in the MV distribution system, the circuit breaker at the head of the MV feeder opens at t = 0 s, yielding the islanded operation of the LV distribution system. A first fast reclosure of the breaker occurs at t = 0.2 s and it is assumed to be successful, reconnecting the MV supply to the islanded LV distribution system. During the 200 ms interval, a linear increase of the frequency f in the island is assumed whereas the MV supply remains at 50 Hz. In the case of the time evolution f_1 in Figure A1, the ROCOF relay immediately detects that the rate-of-change of the frequency is higher than its threshold and trips the DG before reclosure. In the case of the time evolution f_2, both the ROCOF and the high-frequency relays do not intervene before the reclosure, which will occur with a phase displacement. From Figure A1 it is apparent that the maximum displacement between the MV supply and the LV island voltages arises when the time evolution of the frequency f superimposes the ROCOF threshold. In such a case the phase displacement θ can be evaluated as:

$$\theta = 2\pi \int_0^{0.2} (f - 50)dt \cong 0.314 \text{ rad} = 18°$$

and, then, it is smaller than the required 20°.

Conflicts of Interest

The authors declare no conflict of interest.

References

1. Caldon, R.; Coppo, M.; Sgarbossa, R.; Sgarbossa, L.; Turri, R. Risk of unintentional islanding in LV distribution networks with inverter-based DGs. In Proceedings of the 48th IEEE International Universities Power Engineering Conference, Dublin, Ireland, 2–5 September 2013; pp. 1–6.
2. Dyśko, A.; Burt, G.M.; Galloway, S.; Booth, C.; McDonald, J.R. UK distribution system protection issues. *IET Gener. Transm. Distrib.* **2007**, *1*, 679–687.
3. Timbus, A.; Oudalov, A.; Ho, C.N.M. Islanding detection in smart grid. In Proceedings of the IEEE Energy Conversion Congress and Exposition, Atlanta, GA, USA, 12–16 September 2010; pp. 3631–3637.
4. Redfern, M.A.; Usta, O.; Fielding, G. Protection against loss of utility grid supply for a dispersed storage and generation unit. *IEEE Trans. Power Deliv.* **1993**, *8*, 948–954.
5. Dyśko, A.; Booth, C.; Anaya-Lara, O.; Burt, G.M. Reducing unnecessary disconnection of renewable generation from the power system. *IET Renew. Power Gener.* **2007**, *1*, 41–48.
6. Freitas, W.; Xu, W. False operation of vector surge relays. *IEEE Trans. Power Deliv.* **2004**, *19*, 436–438.
7. Delfanti, M.; Merlo, M.; Monfredini, G.; Olivieri, V. Coordination of interface protection systems for DG applications in MV distribution networks. In Proceedings of the 22th IET International Conference on Electricity Distribution, Stockholm, Sweden, 10–13 June 2013; pp. 1–4.
8. Crăciun, B.; Kerekes, T.; Sera, D.; Teodorescu, R. Overview of recent grid codes for PV power integration. In Proceedings of the 13th IEEE International Conference on Optimization of Electrical and Electronic Equipment, Brasov, Romania, 24–26 May 2012; pp. 959–965.

9. Bufano, V.; D'Adamo, C.; D'Orazio, L.; D'Orinzi, C. Innovative solutions to control unintentional islanding on LV network with high penetration of distributed generation. In Proceedings of the 22th IET International Conference on Electricity Distribution, Stockholm, Sweden, 10–13 June 2013; pp. 1–4.
10. Di Fazio, A.R.; Fusco, G.; Russo, M.; Valeri, S.; Amura, G.; Noce, C. A smart device for islanding detection in distribution system operation. *Electr. Power Syst. Res.* **2015**, *120*, 87–95.
11. RTDS Technologies. Available online: http://www.rtds.com (accessed on 20 February 2015).
12. Nicolae, D.V.; Siti, M.W.; Jimoh, A.A. LV self balancing distribution network reconfiguration for minimum losses. In Proceedings of the IEEE PowerTech, Bucharest, Romania, 28 June–2 July 2009; pp. 1–6.
13. Siti, M.W.; Nicolae, D.V.; Jimoh, A.A.; Ukil, A. Reconfiguration and load balancing in the LV and MV distribution networks for optimal performance. *IEEE Trans. Power Deliv.* **2007**, *22*, 2534–2540.
14. Mohammadnezhad-Shourkaei, H.; Abiri-Jahromi, A.; Fotuhi-Firuzabad, M. Incorporating service quality regulation in distribution system maintenance strategy. *IEEE Trans. Power Deliv.* **2011**, *26*, 2495–2504.
15. Lee, J.; Chang, S.; Myung, S.; Cho, Y. Transient false tripping characteristic analysis of ground fault circuit interrupter. In Proceedings of the International Conference on Power System Technology, Singapore, 21–24 November 2004; pp. 522–526.
16. Njozela, M.; Chowdhury, S.; Chowdhury, S.P. Impacts of DG on the operation of auto-reclosing devices in a power network. In Proceedings of the IEEE Power and Energy Society General Meeting, San Diego, CA, USA, 24–29 July 2011; pp. 1–8.
17. Kadurek, P.; Cobben, J.F.G.; Kling, W.L. Future LV distribution network design and current practices in the Netherlands. In Proceedings of the 2nd IEEE PES International Conference and Exhibition on Innovative Smart Grid Technologies, Manchester, UK, 5–7 December 2011; pp. 1–6.
18. *IEEE Standard for Interconnecting Distributed Resources with Electric Power Systems*; IEEE Standard 1547; IEEE Standards Association: Piscataway, NJ, USA, 2008.
19. *Photovoltaic (PV) System—Characteristics of Utility Interface*; IEC 61727; IEEE Standards Association: Piscataway, NJ, USA, 2004.
20. *Reference Technical Rules for the Connection of Active and Passive Users to the LV Electrical Utilities*; Standard CEI 0-21; IEEE Standards Association: Piscataway, NJ, USA, 2011.
21. Di Fazio, A.R.; Fusco, G.; Russo, M. New islanding detection method based on a Thevenin-like model. *IET Gener. Transm. Distrib.* **2015**, accepted.
22. Di Fazio, A.R.; Valeri, S. Threshold setting of an innovative anti-islanding relay for LV distribution systems by real time simulations. In Proceedings of the 3rd IET Renewable Power Generation Conference, Naples, Italy, 24–25 September 2014; pp. 1–6.

8

Enhanced Load Power Sharing Accuracy in Droop-Controlled DC Microgrids with Both Mesh and Radial Configurations

Yiqi Liu *, Jianze Wang †, Ningning Li †, Yu Fu † and Yanchao Ji †

School of Electrical Engineering and Automation, Harbin Institute of Technology, Harbin 150001, China; E-Mails: 18686792201@163.com (J.W.); lnn4147@163.com (N.L.); 13163663651@163.com (Y.F.); gdjyc163@163.com (Y.J.)

† These authors contributed equally to this work.

* Author to whom correspondence should be addressed; E-Mail: hitliuyq0925@gmail.com

Academic Editor: Josep M. Guerrero

Abstract: The rational power sharing among different interface converters should be determined by the converter capacity. In order to guarantee that each converter operates at the ideal condition, considering the radial and mesh configuration, a modified strategy for load power sharing accuracy enhancement in droop-controlled DC microgrid is proposed in this paper. Two compensating terms which include averaging output power control and averaging DC voltage control of neighboring converters are employed. Since only the information of the neighboring converter is used, the complexity of the communication network can be reduced. The rational distribution of load power for different line resistance conditions is realized by using modified droop control that can be regarded as a distributed approach. Low bandwidth communication is used for exchanging sampled information between different converters. The feasibility and effectiveness of the proposed method for different network configurations and line resistances under different communication delay is analyzed in detail. Simulation results derived from a DC microgrid with three converters is implemented in MATLAB/Simulink to verify the proposed approach. Experimental results from a 3 × 10 kW prototype also show the performance of the proposed modified droop control scheme.

Keywords: DC microgrid; communication delay; droop control; load power sharing; mesh configuration; radial configuration

---

1. Introduction

Nowadays, development and utilization of green energy has become an important concern [1,2]. In order to achieve coordinated energy management of different types of green energy, the microgrid is presented which is currently attracting increasing attention [3]. Most microgrids adopt AC distribution the same as in conventional power systems. However, some renewable energy sources have DC output, such as the photovoltaic (PV) system, the fuel cell, and energy storages. Therefore, the DC-type microgrid can be more beneficial in efficiency enhancement and power quality improvement.

Since the power is delivered to load from different sources through the transmission lines, the power transmission efficiency and the rationality power sharing accuracy of distribution are particularly important [4,5]. At the same time, power electronic converters are indispensable in distributed generation (DG) due to the requirement of renewable energy regulations. In order to avoid circulating currents among the converters without using any dedicated communication between them [6], and to reduce the line loss and improve the overall efficiency of the DC power system [7–9], an improved droop control method was proposed in [10]. This method can be used simultaneously to compensate the voltage drop and enhance the load distribution accuracy. However, it also has the drawback of heavy communication traffic since sampled data from all the converters are needed. Meanwhile, only radial configuration is taken into account.

To overcome these drawbacks, a modified droop control method is proposed in this paper. The average voltage and average output power of the adjacent converters are selected as the control variables. Since only the data of the adjacent converters are used, the communication traffic is not so high. The information of voltage and power for each converter is transferred through a low bandwidth communication network. The influence of line impedance and power sharing accuracy are discussed in this method. At the same time, the stability of the method could be improved because the control system does not require a centralized controller. Both mesh and radial system architectures are studied. Simulation and experimental results demonstrate the feasibility of the proposed method with case studies based on both system configurations. The outline of the paper is as follows. In Section 2 the drawbacks of traditional droop control are discussed. This is followed by analysis of droop control issues in Section 3. Improved droop control is discussed in Section 4. Finally, the simulation results are compared with the experimental results in Section 5.

2. Traditional Droop Control

2.1. State-of-the-Art of DC Microgrids

In recent years, with the increasing penetration of distributed sources with DC output such as photovoltaic (PV), battery storage, and the fuel cell, DC microgrids have been intensively studied. Their advantages are shown in the following aspects, including the technical considerations of control, economical operation and efficiency:

- Control [11,12]: DC microgrids have simpler models and control since there is no phase angle, frequency or reactive power, while synchronization, reactive power flow and harmonics have to be considered for AC systems which leads to more complicated control system. Moreover,

DC is chosen over AC because it facilitates integrating most modern electronic loads, energy storage devices, and DG technologies—all of them inherently DC.

- Economical operation [13–15]: Economical operation in DC microgrids can be achieved without complex and computation-intensive optimization algorithms and as pointed out in the existing literature, the total cost of ownership, infrastructure, equipment, maintenance and operation are lower in DC microgrids.
- Efficiency [16–18]: The system efficiency becomes higher due to the reduction of conversion losses of inverters between DC output sources and loads. DC microgrids already have a fault-ride-through capability of their own due to the stored energy of the DC capacitor and the voltage control of the AC/DC converter.

2.2. Configuration of DC Microgrids

In DC microgrids, due to the distributed characteristics of renewable energy sources, the interface converters are connected parallel to each other [19]. The configuration of DG is generally divided into two groups, namely mesh configuration and radial configuration [20–22], as shown in Figure 1. By using Thevenin equivalent circuits, the simplified models of a DC microgrid are shown in Figure 2. Traditional droop control [23,24] is widely used in the DC microgrids, which can be expressed as

$$v_{dci} = v_{dc}^{*} - m_0 \cdot P_{dci} \tag{1}$$

where v_{dc}^{*} is the reference value of the DC output voltage of each converter, and v_{dci} is the actual output voltage of each converter. P_{dci} is the output power of each converter, m_0 is the droop coefficient, and $i = 1, 2, 3\cdots$.

Meanwhile, the reference of the DC output voltage is selected as 380 V [25–27], since this level is globally accepted for standardized components with the best balance of economics and safety. Power distribution at 380 V has some inherent advantages as follows:

- Equipment costs become lower.
- Distribution capabilities are enhanced.
- Sustainability is improved due to reduced copper use.

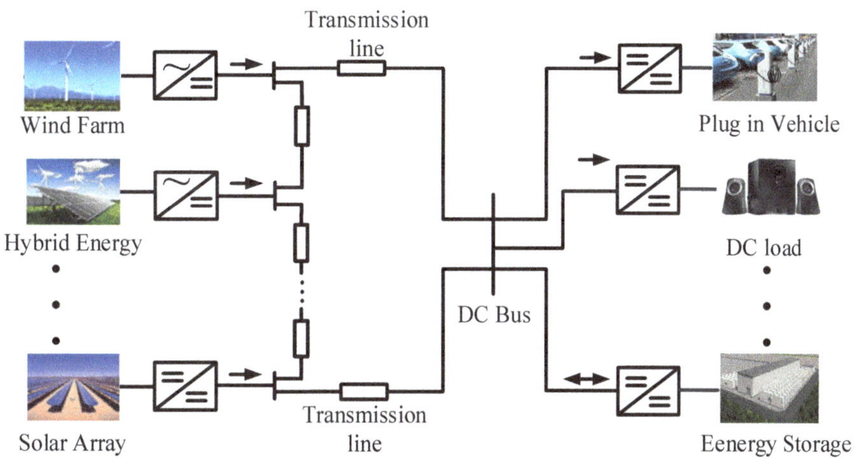

(a)

Figure 1. *Cont.*

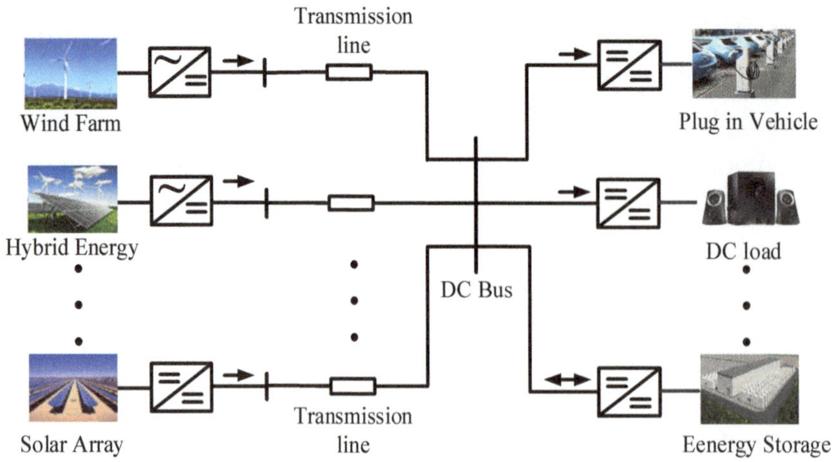

(b)

Figure 1. Distributed generation (DG) sources parallel operation diagrams. (**a**) Mesh configuration. (**b**) Radial configuration.

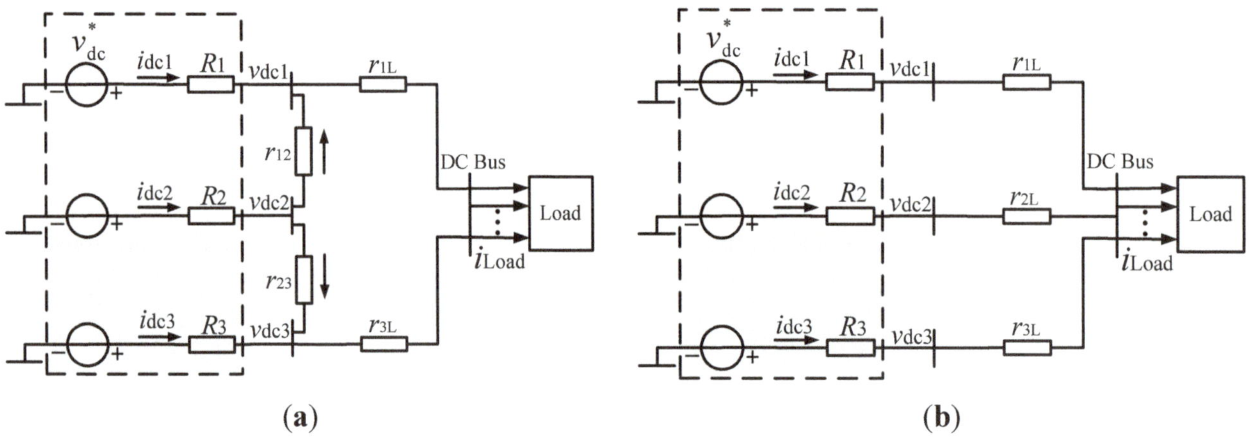

Figure 2. Simplified model of DC microgrid. (**a**) Mesh configuration (**b**) Radial configuration.

2.3. Drawbacks of Traditional Droop Control

The diagram of traditional droop control is depicted in Figure 3. According to Figure 3 and Equation (1), the principle of droop control is shown as follows: The value of the actual DC output voltage linearly decreases with increasing DC output power for each converter.

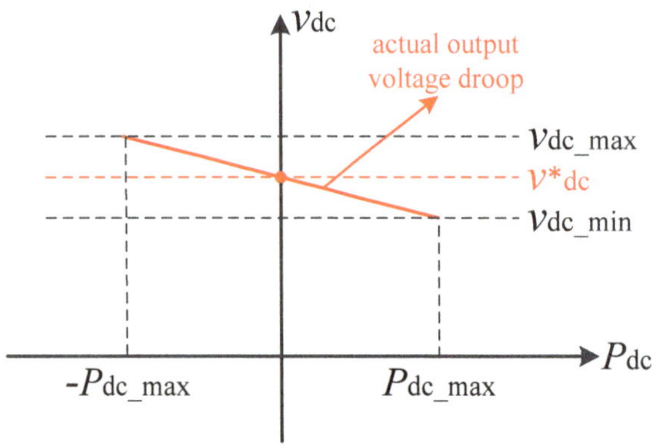

Figure 3. Droop control diagram at the DC side.

Some drawbacks in traditional droop control in DC microgrids are as follows [28–30]:

- Deviation of bus voltage is inevitable.
- There is trade-off between power-sharing accuracy and voltage deviation.

In general, if the scale of DC microgrids is small, the line impedance can be ignored. The bus voltages are approximately equal and load power distribution can be properly achieved. However, if the scale of DC microgrids increases, the impact of line impedance should be considered. In this case, sharing the accuracy of load power should be enhanced.

3. Analysis of Droop Control Issues

The aforementioned drawbacks of the traditional droop control method in both mesh and radial configurations are analyzed in detail as follows.

3.1. Mesh Configuration

By using the listed voltage and current equations based on the multiple nodes of the simplified system models, the converter output current for mesh and radial configuration in DC microgrids can be derived. According to Kirchhoff's law, the following circuit equations can be derived as in Figure 2a:

$$\begin{cases} v_{\mathrm{L}} = R_{\mathrm{L}} \cdot i_{\mathrm{L}} \\ i_{\mathrm{L}} = i_{\mathrm{dc1}} + \dfrac{v_{\mathrm{dc2}} - v_{\mathrm{dc1}}}{r_{\mathrm{1L}}} + i_{\mathrm{dc3}} + \dfrac{v_{\mathrm{dc2}} - v_{\mathrm{dc3}}}{r_{\mathrm{3L}}} \end{cases} \tag{2}$$

$$\begin{cases} i_{\mathrm{dc1}} = \dfrac{v_{\mathrm{dc1}} - v_{\mathrm{L}}}{r_{\mathrm{1L}}} - \dfrac{v_{\mathrm{dc2}} - v_{\mathrm{dc1}}}{r_{12}} \\ i_{\mathrm{dc2}} = \dfrac{v_{\mathrm{dc2}} - v_{\mathrm{dc1}}}{r_{12}} + \dfrac{v_{\mathrm{dc2}} - v_{\mathrm{dc3}}}{r_{23}} \\ i_{\mathrm{dc3}} = \dfrac{v_{\mathrm{dc3}} - v_{\mathrm{L}}}{r_{\mathrm{3L}}} - \dfrac{v_{\mathrm{dc2}} - v_{\mathrm{dc3}}}{r_{23}} \end{cases} \tag{3}$$

where v_{dci} is DC side output voltage of converter # i ($i = 1, 2, 3$), i_{dci} is DC side output current, v_{L} is load voltage, i_{L} is load side current, R_{L} is load side resistance, r_{ij} ($i = 1, 2, 3, j = 2, 3$) is line resistance between different converters; R_{i} is the virtual resistance.

Based on above circuit analysis and combining Equations (2) and (3), DC side output current can be obtained:

$$\begin{cases} i_{\mathrm{dc1}} = \dfrac{1}{\lambda}\begin{pmatrix} [-(R_3 + 2R_L + r_{3L})r_{23}^2 - (R_2R_3 + 2R_2R_{\mathrm{L}} + 2R_3R_{\mathrm{L}} - R_2r_{\mathrm{3L}} - R_3r_{\mathrm{3L}})r_{23} - 2R_2R_3R_{\mathrm{L}}]r_{12}^2\ v_{\mathrm{dc}}^* \\ -[R_2r_{23}^2(R_3 + r_{\mathrm{3L}}) + 2R_2R_{\mathrm{L}}r_{23}(R_3 + r_{23}) + R_2R_3r_{23}(r_{\mathrm{3L}} + r_{\mathrm{1L}})]r_{12}v_{\mathrm{dc}}^* \end{pmatrix} \\ i_{\mathrm{dc2}} = \dfrac{1}{\lambda}\begin{pmatrix} (-R_1R_3r_{23} - 2R_3R_{\mathrm{L}}r_{23} - R_3r_{23}r_{\mathrm{1L}})r_{12}^2v_{\mathrm{dc}}^* \\ -[R_1R_3r_{23}(r_{23} + r_{\mathrm{3L}} + r_{\mathrm{1L}}) + R_1r_{23}^2(2R_{\mathrm{L}} + r_{\mathrm{3L}}) + 4R_1R_3R_{\mathrm{L}}r_{23}]r_{12}v_{\mathrm{dc}}^* \end{pmatrix} \\ i_{\mathrm{dc3}} = \dfrac{1}{\lambda}\begin{pmatrix} [-(R_1 + r_{\mathrm{1L}})r_{23}^2 - (R_1R_2 + 2R_2R_{\mathrm{L}} + R_2r_{\mathrm{1L}})r_{23}]r_{12}^2v_{\mathrm{dc}}^* \\ -[(2R_{\mathrm{L}} + r_{\mathrm{1L}})(R_1 + R_2)r_{23}^2 + R_1R_2r_{23}(r_{\mathrm{3L}} + r_{23} + r_{\mathrm{1L}} + 2R_{\mathrm{L}})]r_{12}v_{\mathrm{dc}}^* - 2R_1R_2R_{\mathrm{L}}r_{23}^2v_{\mathrm{dc}}^* \end{pmatrix} \end{cases} \tag{4}$$

where

$$\lambda = \begin{pmatrix} 2R_2R_3R_L^2 - (r_{1L}r_{3L} + R_1R_3 + R_1R_L + R_1r_{3L} + R_3R_L + R_3r_{1L} + R_Lr_{1L} + R_Lr_{3L})r_{23}^2 - \\ [(r_{1L}r_{3L} + 2R_1r_{3L})(R_2 + R_3) + (R_3 + R_L)R_1R_2 + (R_1 + R_2)R_3R_L + (R_2R_3 + R_2R_L + R_3R_L)r_{1L}]r_{23} \end{pmatrix} r_{12}^2$$
$$- \begin{pmatrix} [(R_3 + R_L)R_1R_2 + (R_1 + R_2)(R_3R_L + R_3r_{1L} + R_Lr_{1L} + R_Lr_{3L} + r_{1L}r_{3L}) + R_1R_2r_{3L}]r_{23}^2 - \\ [2R_2R_L^2(R_3 + R_1) - [(R_1 + R_2)R_3R_L + (R_3 + R_L)R_1R_2](r_{1L} + r_{3L}) - R_1r_{1L}r_{3L}(R_2 + R_3) - R_2R_3r_{1L}r_{3L}]r_{23} \\ -2R_1R_2R_3R_L(2R_L + r_{3L}) \end{pmatrix} r_{12}$$
$$+ 2R_1R_2R_Lr_{23}(R_Lr_{23} + 2R_3R_L + R_3r_{1L})$$

In DC microgrids, by using traditional droop control method, accurate load power sharing accuracy can be obtained when the converter DC output power is set to be inversely proportional to the corresponding droop coefficient, the following expression can be obtained:

$$\frac{m_0}{k_1}\underbrace{[i_{dc1}(v_{dc}^* - \frac{m_0}{k_1}i_{dc1}v_{dc1})]}_{P_{dc1}} = \frac{m_0}{k_2}\underbrace{[i_{dc2}(v_{dc}^* - \frac{m_0}{k_2}i_{dc2}v_{dc2})]}_{P_{dc2}} = \frac{m_0}{k_3}\underbrace{[i_{dc3}(v_{dc}^* - \frac{m_0}{k_3}i_{dc3}v_{dc3})]}_{P_{dc3}} \tag{5}$$

Based on the above analysis, it can be seen that the power sharing error can be eliminated if and only if the droop coefficient and line impedance satisfy the relationship in Equation (5). However, this assumption is only suitable for an ideal system and the practical system is not satisfied. This is the limitation of the traditional droop control method in the mesh configuration of DC microgrids.

3.2. Radial Configuration

From the simplified model of a multi-bus radial configuration in Figure 2b, the load voltage and current equations can be obtained:

$$\begin{cases} v_L = v_{dc}^* - i_{dc1}R_1 - i_{dc1}r_{1L} \\ v_L = v_{dc}^* - i_{dc2}R_2 - i_{dc2}r_{2L} \\ v_L = v_{dc}^* - i_{dc3}R_3 - i_{dc3}r_{3L} \\ v_L = i_LR_L \\ i_L = i_{dc1} + i_{dc2} + i_{dc3} \end{cases} \tag{6}$$

Based on Equation (6) the relationship of i_{dci} and v_{dc}^* is shown as

$$\begin{bmatrix} R_1 + r_{1L} + R_L & R_L & R_L \\ R_L & R_2 + r_{2L} + R_L & R_L \\ R_L & R_L & R_3 + r_{3L} + R_L \end{bmatrix} \begin{bmatrix} i_{dc1} \\ i_{dc2} \\ i_{dc3} \end{bmatrix} = \begin{bmatrix} v_{dc}^* \\ v_{dc}^* \\ v_{dc}^* \end{bmatrix} \tag{7}$$

The DC side output current can be expressed as:

$$\begin{cases} i_{dc1} = \dfrac{[-R_2R_3 - 2R_2R_L - 2R_3R_L - (R_3 + 2R_L)r_{2L} - (R_2 + 2R_L)r_{3L} - r_{2L}r_{3L} - 4R_L^2]v_{dc}^*}{\lambda} \\ i_{dc2} = \dfrac{[-R_1R_3 - 2R_1R_L - 2R_3R_L - (R_3 + 2R_L)r_{1L} - (R_1 + 2R_L)r_{3L} - r_{1L}r_{3L} - 4R_L^2]v_{dc}^*}{\lambda} \\ i_{dc3} = \dfrac{[-R_1R_2 - 2R_1R_L - 2R_2R_L - (R_2 + 2R_L)r_{1L} - (R_1 + 2R_L)r_{2L} - r_{1L}r_{2L} - 4R_L^2]v_{dc}^*}{\lambda} \end{cases} \tag{8}$$

where

$$\lambda = -(R_2R_3 + R_2R_L + R_3R_L)r_{1L} - (R_1R_3 + R_1R_L + R_3R_L)r_{2L} - (R_1R_L + R_2R_L + R_1R_2)r_{3L}$$
$$-(R_1 + R_L)r_{2L}r_{3L} - (R_2 + R_L)r_{1L}r_{3L} - (R_3 + R_L)r_{1L}r_{2L} + 4R_L^3 - (R_1R_2 + R_1R_3 + R_2R_3)R_L - R_1R_2R_3 - r_{1L}r_{2L}r_{3L}$$

By combining Equations (5) and (8), the expected power sharing accuracy can be obtained. According to the aforementioned theoretical derivation and analysis, it was shown that droop control can induce the inevitable load power sharing error when considering the value of line impedance, and irrational power sharing can thereby cause voltage degradation. The serious imbalance between each line resistance may lead to the condition that the output power of some converters exceeds the maximum power rating, and consequently causes the failure of the converter.

4. Proposed Approach

In order to solve the two problems of induced traditional droop control, this paper proposes a method of controlling the average values of the DC-link voltage of adjacent converters as well as the output power to compensate the voltage deviation induced by droop control, and simultaneously improve the load power sharing accuracy. The whole control diagram of the system is given in Figure 4. Compared to the existing method of averaging the voltage and power based on global information, the improved control method alleviates the communication traffic and further reduces the dependence on the communication system. The DC-link voltage references of the two adjacent converters are as follows:

$$v_{dci}^* = v_{dc}^* - \frac{m_0}{k_i} \cdot P_{dci} \cdot G_{LPF} + \underbrace{(v_{dc}^* - \frac{G_d \cdot v_{dc(i-1)} + G_d \cdot v_{dc(i+1)}}{2}) \cdot G_{piv}}_{\text{Compensating Controller I}} - \underbrace{(P_{dci}/k_i - \frac{G_d \cdot P_{dc(i-1)} + G_d \cdot P_{dc(i+1)}}{2}) \cdot G_{pip}}_{\text{Compensating Controller II}} \quad (9)$$

where v_{dci}^* is output DC-link voltage reference of the ith converter, v_{dc}^* is line DC-link voltage reference, $v_{dc(i-1)}$ and $v_{dc(i+1)}$ are the output voltages of the ith converter's two adjacent converters, P_{dci} is the ith converter's output power, $P_{dc(i-1)}$ and $P_{dc(i+1)}$ are the output power of the ith converter's two adjacent converters, k_i is the proportional sharing accuracy of output power, m_0 is the coefficient of traditional droop control, G_{LPF} is low pass filter introduced in droop control, ω_s is the cut-off frequency of the filter, G_{piv} and G_{pip} are two compensating terms of improved droop control: The averaging voltage controller and averaging power controller, are both traditional PI controllers. The communication delay of the transfer variable is G_d, which can be expressed as the following:

$$G_{pip} = k_{pp} + \frac{k_{ip}}{s} \quad (10a)$$

$$G_{piv} = k_{pv} + \frac{k_{iv}}{s} \quad (10b)$$

$$G_d = \frac{1}{1 + \tau s} \quad (11)$$

In the control system, low bandwidth communication is employed to transfer the sampling data of the DC-link voltage and power values between different converter units. Voltage deviations induced by droop control can be eliminated by PI controllers I for the average values of DC-link voltage and power respectively. However, there is a variety of renewable energy distributed into microgrids by multiple converters, which are separate and have no need of high frequency telecommunication lines. Hence, the sampling for all the transmission voltage and power between the converters will cause unnecessary errors and increase communication complexity. In this paper, there is improved droop control method sampling of the average value of the DC-link voltages and power between the adjacent converters to eliminate voltage deviations or power sharing error which are induced by droop control. Meanwhile, load power can be proportionally shared by the power compensating controller II. Finally the goal of accurate proportional sharing of the DC output power is achieved without a centralized controller.

In the mesh configuration, the communication delay is a very sensitive parameter for system stability which is shown in Figure 5a. If the communication delay is set to a large value, poles III, IV and VI move towards the imaginary axis, and the system becomes unstable.

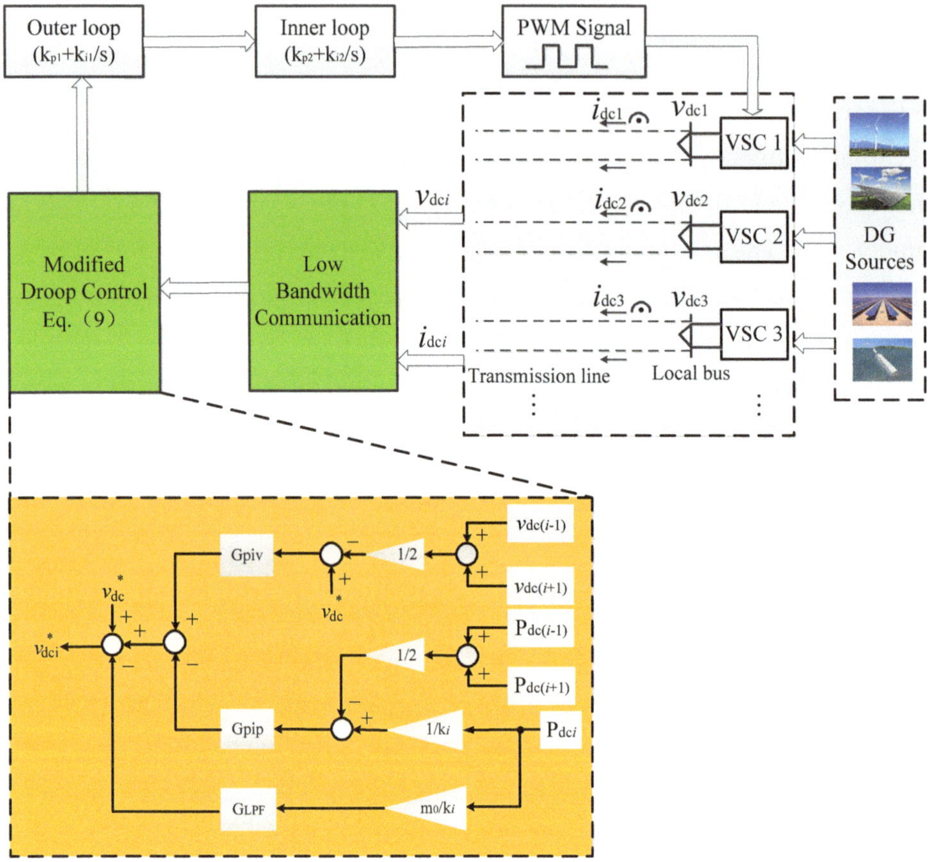

Figure 4. Detailed diagram of the proposed droop control system.

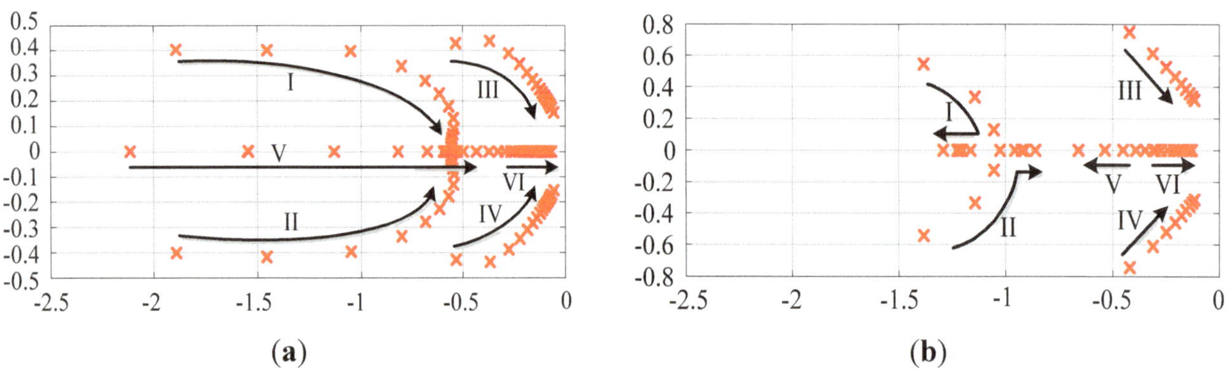

Figure 5. Closed-loop dominant poles for varying communication delay. (**a**) Mesh configuration (**b**) Radial configuration.

Figure 5b gives the location of the dominant closed-loop poles in the radial configuration while varying the communication delays from 1 to 10 s. Poles I, II, and V keep to the left side of the s plane, which means they do not harm the system stability. However, poles III, IV and VI move towards an imaginary axis which leads to an unstable status when increasing communication delay or encountering communication loss.

5. Simulations and Experimental Results

Simulations based on MATLAB/Simulink were performed to evaluate the performance of the proposed method of improved droop control, and the feasibility of the control method was verified by considering different line resistance r_{ij}, and communication delay τ, based on two kinds of configurations. The parameters for the DC microgrids system are listed in Table 1. The differences between the maximum and the minimum value of the DC side power of different converters as well as the DC side voltage are defined respectively as follows:

$$\varepsilon_j = \max(P_{dci}) - \min(P_{dci}) \tag{12a}$$

$$\varepsilon_j \leq 0.1 \times P_{dci} \tag{12b}$$

$$\delta_j = \max(v_{dci}) - \min(v_{dci}) \tag{13a}$$

$$\delta_j \leq 0.05 \times v_{dci} \tag{13b}$$

where i = 1, 2, 3 denotes the converter's sequence number, j = 1, 2, 3, 4 denotes the type of specific case.

Table 1. DC microgrids system parameters.

Item	Symbol	Value
Reference of DC output voltage	v_{dc}	380 V
Load resistance	R_L	19 Ω
Power sharing proportion (Converter #i=1,2,3)	k_i	1
LPF cutting frequency	f_c	20 Hz
Communication delay	τ	0.1 s or 1 s
Droop coefficient	m_0	0.0015
Averaging voltage controller proportion coefficient	k_{pv}	2×10^{-3}
Averaging voltage controller integral coefficient	k_{iv}	4.5×10^{-2}
Averaging power controller proportion coefficient	k_{pp}	2.3×10^{-3}
Averaging power controller integral coefficient	k_{ip}	5.5×10^{3}

Mesh configuration						**Radial configuration**				
Case No.	r_{1L}	r_{12}	r_{23}	r_{3L}	τ	Case No.	r_{1L}	r_{2L}	r_{3L}	τ
1	0.8 Ω	1.0 Ω	1.0 Ω	1.2 Ω	0.1 s	4	0.8 Ω	1.0 Ω	1.2 Ω	0.1 s
2	0.8 Ω	1.0 Ω	1.0 Ω	1.2 Ω	1 s	5	0.8 Ω	1.0 Ω	1.2 Ω	1 s
3	0.6 Ω	1.2 Ω	1.2 Ω	1.8 Ω	1 s	6	0.6 Ω	1.2 Ω	1.8 Ω	1 s

5.1. Simulation Results for Mesh Configuration

First, the same line resistance and different communication delay are used to test the performance of the proposed control system in Case 1 and Case 2. The responses for power sharing accuracy and voltage restoration enhancement are shown in Figure 6a–d. Before t = 1 s the compensating controllers are not activated. The traditional droop control sharing error ε is 2530 W among different converters. The DC output average voltage is increased 4.4 V. When $t > 1$ s, the compensating controllers are activated to eliminate the error of output power sharing and restore the DC voltage. At the same time, the communication delays of 0.1 s and 1 s are tested, respectively. The DC-side output power of each converter gradually becomes equal, and the DC-side voltage deviation δ of each converter is less than 5% of the voltage rating.

Second, the same communication delay and different transmission line resistance are used in Case 3. When the proportion of the line resistance increases, the power sharing error is changed from 2530 W to 4176 W compared to the results in Case 1 and Case 2. The DC output voltage deviation is increased from 11–15 V. The changes for DC output power and DC voltage are shown in Figure 6e,f.

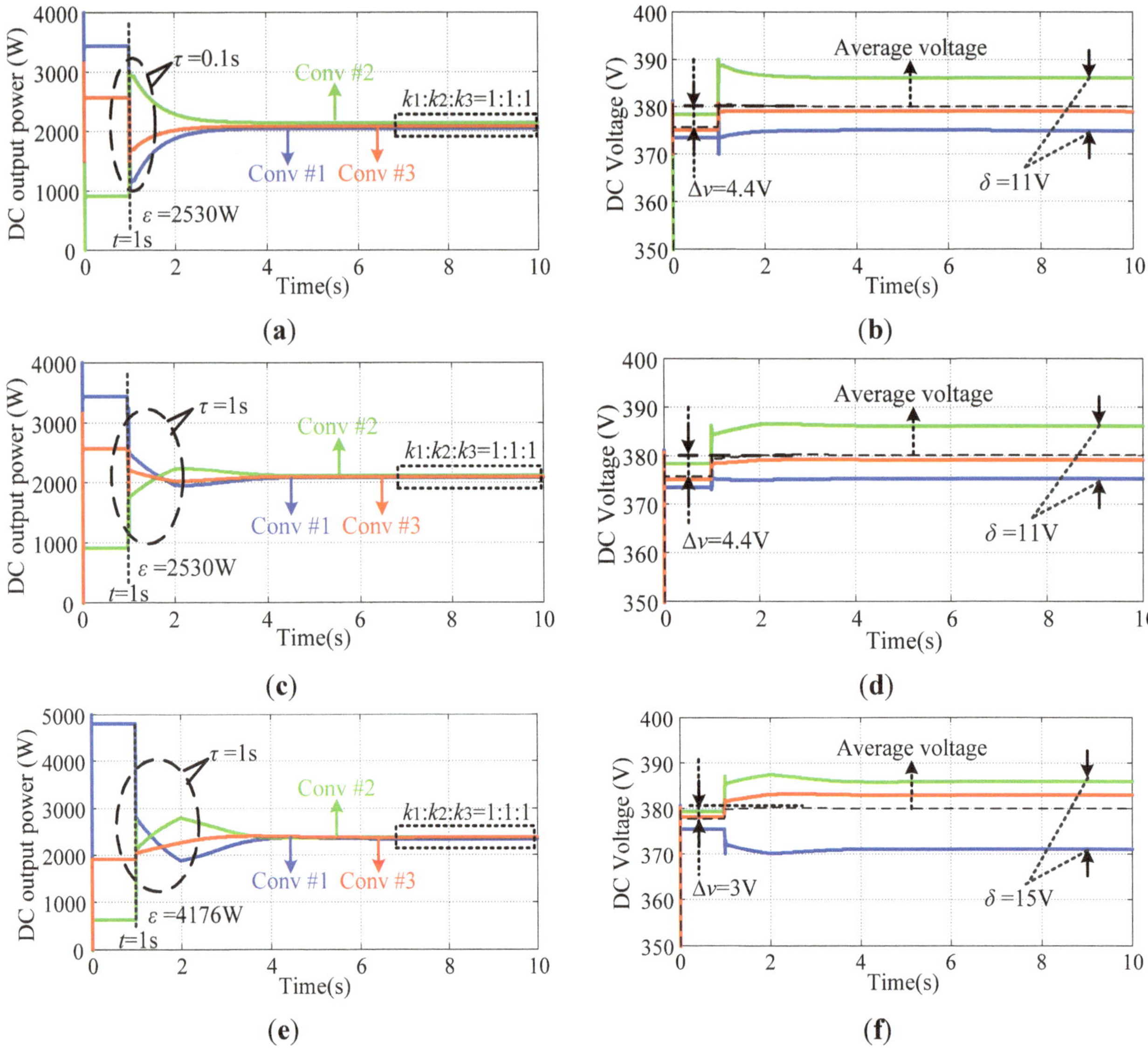

Figure 6. Transient responses for mesh configuration in the DC microgrid. (**a**) Case 1, Power sharing accuracy; (**b**) Case 1, Voltage restoration; (**c**) Case 2, Power sharing accuracy; (**d**) Case 2, Voltage restoration; (**e**) Case 3, Power sharing accuracy; (**f**) Case 3, Voltage restoration.

5.2. Simulation Results for Radial Configuration

In Case 4 and Case 5, the transient response for radial configuration is shown in Figure 7a–d. It should be noted that the simulation conditions in these two cases are the same as in Case 1 and Case 2. The maximum power error ε is 793 W among the different converters. The maximum deviation of DC-side voltage among each converter is 5.4 V which is less than 5% of the rated value. In Case 6, the same communication delay and different transmission line resistance are also used. With the proportion of the line resistance increasing, the power sharing error is increased from 793 W to 2156 W. The DC output voltage deviation is increased from 5.4 V to 15.6 V. The changes for output power and DC voltage are shown in Figure 7e,f.

In this paper, we chose the mesh configuration and the middle of the converter is not connected to load directly, the start of DC power is determined by some factors in the system e.g., the DC output voltage of two adjacent converters and transmission line resistance *etc.* It is more complicated than radial configuration. By using the modified droop control method, the expected results can be implemented.

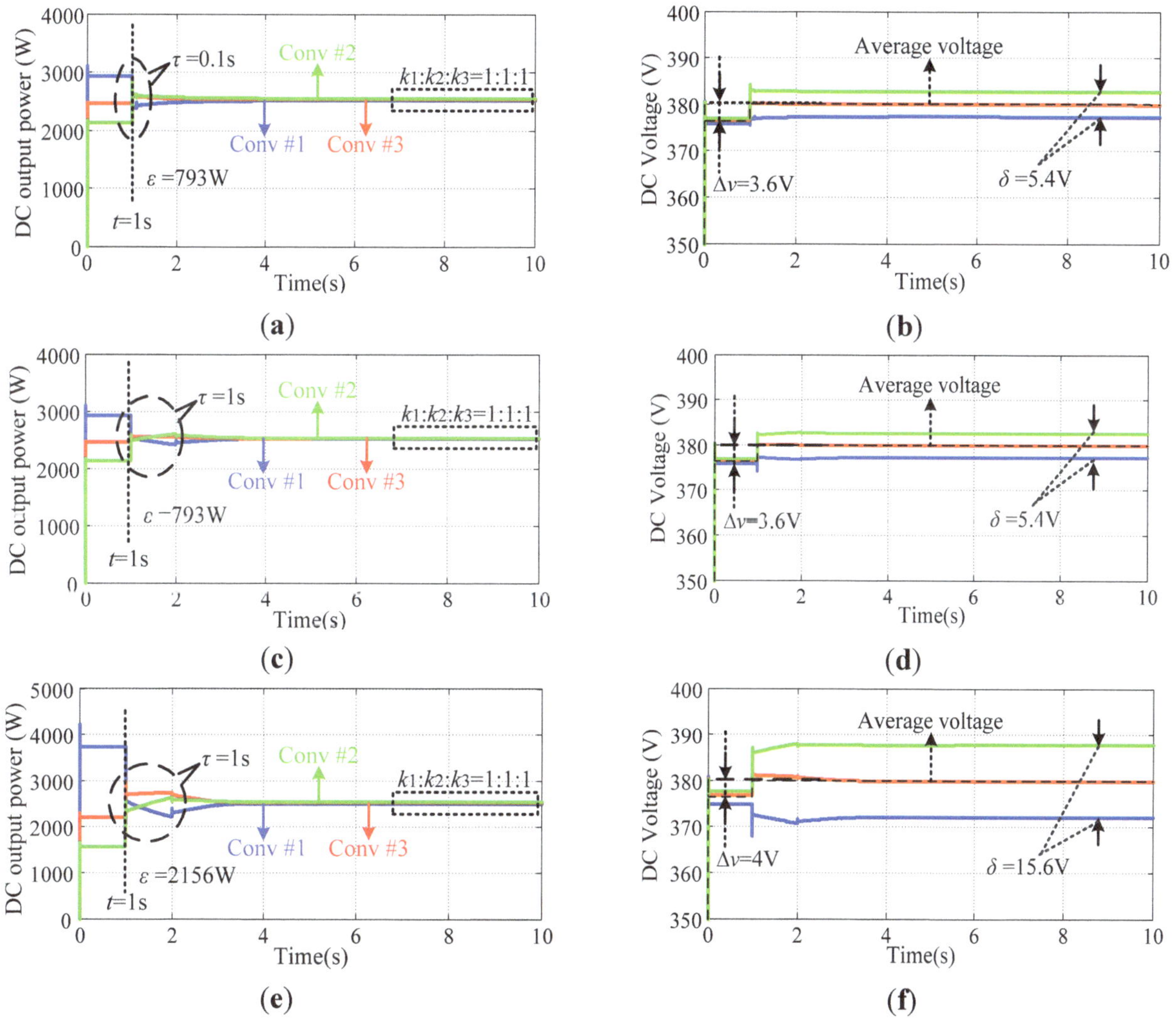

Figure 7. Transient responses for radial configuration in the DC microgrid. (**a**) Case 4, Power sharing accuracy; (**b**) Case 4, Voltage restoration; (**c**) Case 5, Power sharing accuracy; (**d**) Case 5, Voltage restoration; (**e**) Case 6, Power sharing accuracy; (**f**) Case 6, Voltage restoration.

5.3. Experimental Verification

A prototype with 3 × 10 kW was used to verify the theoretical results from the previous section. The system parameters are the same as in Table 1. The performance of the modified control system can be verified by the test results of the DC output power and voltage.

The experimental results for Case 3 in the mesh configuration are shown in Figure 8a,b. The modified control was started when $t = 0.5$ s. Before $t = 0.5$ s, compensating terms are not activated. Hence, the power sharing error can be found. The differences between the maximum and minimum DC output power among converters is 4149 W. After activating the modified droop control method, the power and voltage waveforms are modified. During this period, the output power gradually becomes equal which satisfies the objective of sharing the proportion ($k_i = 1$), and the system is also stable when the communication delay $\tau = 1$ s. The average value of the DC output voltage is increased by 3.1 V and the maximum error of the DC output voltage is kept at 16.2 V which is lower than 5% of the reference voltage.

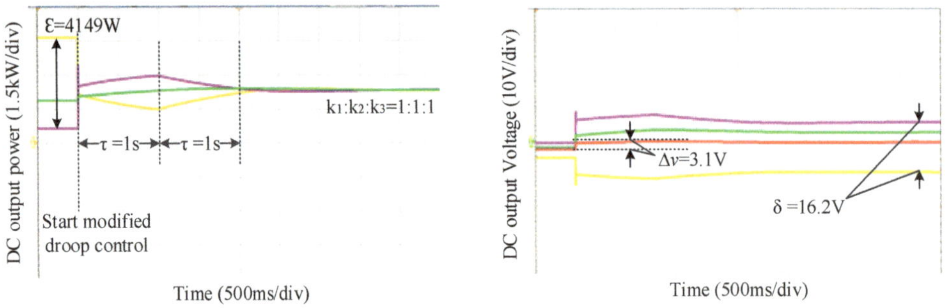

(a) (b)

Figure 8. Modified control method test result for Case 3 in the mesh configuration. (**a**) Power sharing accuracy; (**b**) Voltage restoration.

The experimental results for Case 6 in the radial configuration are shown in Figure 9a,b. The power sharing error is smaller than the mesh configuration before $t = 0.5$ s. The differences between the maximum and minimum DC power among converters is 2137 W. When the compensating items are activated at 2 s, the output power gradually becomes equal. The average value of the DC output voltage is increased by 3 V by using the modified control method. The corresponding maximum output voltage error is 16 V which is also lower than 5% of the reference voltage.

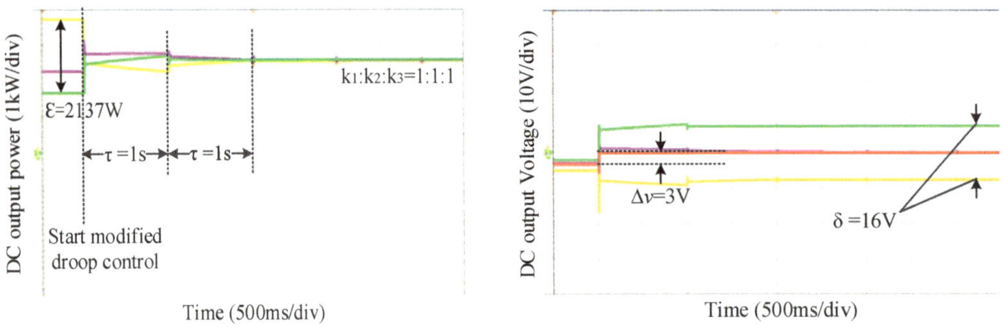

(a) (b)

Figure 9. Modified control method test result for Case 6 in radial configuration. (**a**) Power sharing accuracy; (**b**) Voltage restoration.

6. Conclusions

In order to improve the overall performance of a droop-controlled DC microgrid, a modified distributed control method for mesh and radial configuration was proposed and studied. The key points included in this paper are summarized as follows:

(1) It was demonstrated that the feasibility and stability of the control system can be guaranteed with both mesh and radial configurations.
(2) Local DC output voltage deviation can be eliminated and the load power sharing accuracy can be enhanced with the modified droop control method.
(3) The proposed control method only relies on the data of voltage and current sampled from two adjacent converters, which reduces the stress on the communication system.
(4) The control system is also stable considering the longer communication delay time and larger mismatch of line resistance.

Author Contributions

All authors contributed to this work by collaboration. Yiqi Liu was the main author of this manuscript. Jianze Wang, Ningning Li, Yu Fu, and Yanchao Ji provided some useful suggestions in the construction of the paper. All authors revised and approved the publication.

Conflicts of Interest

The authors declare no conflict of interest.

References

1. Dasgupta, S.; Mohan, S.N.; Sahoo, S.K.; Panda, S.K. Application of four-switch-based three-phase grid-connected inverter to connect renewable energy source to a generalized unbalanced microgrid system. *IEEE Trans. Ind. Electron.* **2013**, *60*, 1204–1215.
2. Hakimi, S.M.; Moghaddas-Tafreshi, S.M. Optimal planning of a smart microgrid including demand response and intermittent renewable energy resources. *IEEE Trans. Smart Grid* **2014**, *5*, 2889–2900.
3. Eghtedarpour, N.; Farjah, E. Distributed charge/discharge control of energy storages in a renewable-energy-based DC micro-grid. *IET Renew. Power Gener.* **2014**, *8*, 45–57.
4. Vandoorn, T.L.; de Kooning, J.D.M.; Meersman, B.; Guerrero, J.M.; Vandevelde, L. Automatic power-sharing modification of P/V droop controllers in low-voltage resistive microgrids. *IEEE Trans. Power Deliv.* **2012**, *27*, 2318–2325.
5. Augustine, S.; Mishra, M.; Lakshminarasamma, N. Adaptive droop control strategy for load sharing and circulating current minimization in low-voltage standalone DC microgrid. *IEEE Trans. Sustain. Energy* **2015**, *6*, 132–141.
6. Guerrero, J.M.; Vasquez, J.C.; Matas, J.; de Vicuña, L.G.; Castilla, M. Hierarchical control of droop-controlled AC and DC microgrids—A general approach toward standardization. *IEEE Trans. Ind. Electron.* **2011**, *58*, 158–172.

7. Rouzbehi, K.; Miranian, A.; Candela, J.I.; Luna, A.; Rodriguez, P. A generalized voltage droop strategy for control of multiterminal DC grids. *IEEE Trans. Ind. Appl.* **2015**, *51*, 607–618.
8. Bevrani, H.; Shokoohi, S. An intelligent droop control for simultaneous voltage and frequency regulation in islanded microgrids. *IEEE Trans. Smart Grid* **2013**, *4*, 1505–1513.
9. Dou, X.; Quan, X.; Wu, Z. Improved control strategy for microgrid ultracapacitor energy storage systems. *Energies* **2014**, *7*, 8095–8115.
10. Lu, X.; Guerrero, J.M.; Sun, K.; Vasquez, J.C. An improved droop control method for DC microgrids based on low bandwidth communication with DC bus voltage restoration and enhanced current sharing accuracy. *IEEE Trans. Power Electron.* **2014**, *29*, 1800–1812.
11. Maknouninejad, A.; Qu, Z.; Lewis, F.L.; Davoudi, A. Optimal, nonlinear, and distributed designs of droop controls for dc microgrids. *IEEE Trans. Smart Grid* **2014**, *5*, 2508–2516.
12. Wang, B.; Sechilariu, M.; Locment, F. Intelligent DC microgrid with smart grid communications: Control strategy consideration and design. *IEEE Trans. Smart Grid* **2012**, *3*, 2148–2156.
13. Liang, C.; Shahidehpour, M. DC Microgrids: Economic operation and enhancement of resilience by hierarchical control. *IEEE Trans. Smart Grid* **2014**, *5*, 2517–2526.
14. Kanchev, H.; Colas, F.; Lazarov, V.; Francois, B. Emission reduction and economical optimization of an urban microgrid operation including dispatched pv-based active generators. *IEEE Trans. Sustain. Energy* **2014**, *5*, 1397–1405.
15. Nutkani, I.U.; Wang, P.; Poh, C.L.; Blaabjerg, F. Autonomous economic operation of grid connected DC microgrid. In Proceedings of the 2014 IEEE 5th International Symposium on Power Electronics for Distributed Generation Systems (PEDG), Galway, Ireland, 24–27 June 2014; pp. 1–5.
16. Kwasinski, A. Quantitative evaluation of DC microgrids availability: Effects of system architecture and converter topology design choices. *IEEE Trans. Power Electron.* **2011**, *26*, 835–851.
17. Yukita, K.; Shimizu, Y.; Goto, Y.; Yoda, M.; Ueda, A.; Ichiyanagi, K.; Hirose, K.; Takeda, T.; Ota, T.; Okui, Y.; *et al.* Study of AC/DC power supply system with DGs using parallel processing method. In Proceedings of the International Power Electronics Conference (IPEC)-ECCE, Sapporo, Japan, 21–24 June 2010; pp. 722–725.
18. Kakigano, H.; Miura, Y.; Ise, T. Low-voltage bipolar-type DC microgrid for super high quality distribution. *IEEE Trans. Power Electron* **2010**, *25*, 3066–3075.
19. Shadmand, M.B.; Balog, R.S. Multi-objective optimization and design of photovoltaic-wind hybrid system for community smart DC microgrid. *IEEE Trans. Smart Grid* **2014**, *5*, 2635–2643.
20. Mehrizi-Sani, A.; Iravani, R. Potential-function based control of a microgrid in islanded and grid-connected modes. *IEEE Trans. Power Syst.* **2010**, *25*, 1883–1891
21. Prasai, A.; Du, Y.; Paquette, A.; Buck, E.; Harley, R.G.; Divan, D. Protection of meshed microgrids with communication overlay. In Proceedings of the 2010 IEEE Energy Conversion Congress and Exposition (ECCE), Atlanta, GA, USA, 12–16 September 2010; pp. 64–71.
22. Shukla, S.; Deng, Y.; Shukla, S.; Mili, L. Construction of a microgrid communication network. In Proceedings of the 2014 IEEE PES Innovative Smart Grid Technologies Conference (ISGT), Washington, DC, USA, 19–22 February 2014; pp. 1–5.
23. Luo, F.; Lai, Y.M.; Loo, K.H.; Tse, C.K.; Ruan, X. A generalized droop-control scheme for decentralized control of inverter-interfaced microgrids. In Proceedings of the 2013 IEEE International Symposium on Circuits and Systems (ISCAS), Beijing, China, 19–23 May 2013; pp. 1320–1323.

24. Wang, W.; Zeng, X.; Tang, X.; Tang, C. Analysis of microgrid inverter droop controller with virtual output impedance under non-linear load condition. *IET Power Electron.* **2014**, *7*, 1547–1556.
25. Becker, D.J.; Sonnenberg, B.J. DC microgrids in buildings and data centers. In Proceedings of the 2011 IEEE 33rd International Telecommunications Energy Conference (INTELEC), Amsterdam, The Netherlands, 9–13 October 2011; pp. 1–7.
26. Emerge Alliance. Available online: http://www.emergealliance.org/Standards/DataTelecom/Overview.aspx (accessed on 20 March 2015).
27. Wu, T.-F.; Chang, C.-H.; Lin, L.-C.; Yu, G.-R.; Chang, Y.-R. DC-bus voltage control with a three-phase bidirectional inverter for DC distribution systems. *IEEE Trans. Power Electron.* **2013**, *28*, 1890–1899.
28. Olivares, D.E.; Mehrizi-Sani, A.; Etemadi, A.H.; Canizares, C.A.; Iravani, R.; Kazerani, M.; Hajimiragha, A.H.; Gomis-Bellmunt, O.; Saeedifard, M.; Palma-Behnke, R.; *et al.* Trends in microgrid control. *IEEE Trans. Smart Grid* **2014**, *5*, 1905–1919.
29. Choi, J.; Jeong, H.Y.; Choi, J.-Y. Voltage control scheme with distributed generation and grid connected converter in a DC microgrid. *Energies* **2014**, *7*, 6477–6491.
30. Kim, J.; Guerrero, J.M.; Rodriguez, P.; Teodorescu, R.; Nam, K. Mode adaptive droop control with virtual output impedances for an inverter-based flexible AC microgrid. *IEEE Trans. Power Electron.* **2011**, *26*, 689–701.

Improved Performance of Electrical Transmission Tower Structure Using Connected Foundation in Soft Ground

Doohyun Kyung [1], Youngho Choi [2], Sangseom Jeong [1] and Junhwan Lee [1,*]

[1] School of Civil and Environmental Engineering, Yonsei University, 134 Shinchon-dong, Seodaemun-gu, Seoul 120-749, Korea; E-Mail: iamdh0630@yonsei.ac.kr (D.K.)
[2] KTP Consultants Pte Ltd., E-Centre@Redhill, 3791 Jalan Bukit Merah, Singapore 159471, Singapore; E-Mail: choiyh@ktp.com.sg

* Author to whom correspondence should be addressed; E-Mail: junlee@yonsei.ac.kr

Academic Editor: Stefan Gößling-Reisemann

Abstract: A connected foundation is an effective foundation type that can improve the structural performance of electrical transmission towers in soft ground as a resilient energy supply system with improved stability. In the present study, the performance of a connected foundation for transmission towers was investigated, focusing on the effect of connection beam properties and soil conditions. For this purpose, a finite element analysis was performed for various foundation and soil conditions. In order to validate the finite element analysis, the calculated results were compared with measured results obtained from field load tests. The use of connection beams was more effective for uplift foundations that usually control the design of transmission tower foundations. For the effect of soil condition, the use of connected foundation is more effective in soft clays with lower undrained shear strength (s_u). Smaller amounts of differential settlement were observed in all soil conditions for both unconnected and connected foundations when a bearing rock layer was present. When the foundation was not reinforced by connection beams, the values of lateral load capacity of tower structure (H_u) were similar for both with- and without-rock layers. It was confirmed that introducing haunch-shaped connection beams is effective for increasing connection beam stability.

Keywords: resilient energy infrastructure system; electrical transmission tower structures; connected foundation; soft grounds; field load test; finite element analysis

1. Introduction

For the electrical transmission system, transmission towers are often installed with certain foundations that support the upper tower structure and associated overhead power lines. The transmission tower structures are subjected to various unexpected damages from a wide range of extreme weather events, including hurricanes, tornadoes, snow, and ice storms, and human disasters such as terrorism [1–3]. In particular, as climate change has been an issue in various social and engineering fields, it has become important to prepare a resilient infrastructure system that can guarantee or improve the stability of the energy supply system [4–6]. For the subsurface soil zone where the foundations of transmission towers are embedded, the issue of climate change also needs to be addressed for both design and construction of the structures. Increasing freezing-thawing cycles within the soil zone cause changes in various soil properties such as permeability, volume change behavior, strength, and compressibility [7–9]. Unusual fluctuation of groundwater level due to changes in annual precipitation characteristics causes additional settlements and unexpected reduction in the bearing capability of foundations [10–12]. All these threaten the stability and sustainability of the electrical transmission tower system, highlighting the need for a more robust and resilient structural system with certain reinforcements.

The types of tower foundation often used are pile, pier, inverted T-type, and mat foundation [13–16]. The inverted T-type foundation is widely used in transmission towers and can be used for small load conditions in good quality soils composed of sand. The pier foundation can be used in steep grade or deep bearing strata. This foundation is effective to support large loads of transmission towers and is frequently used for Ultra High Voltage (UHV) transmission towers. Pile foundation is generally used in weak soils, such as clay and reclaim soil, and often suffers structural damage and geotechnical instability due to insufficient foundation resistance and large differential settlements [14,15,17].

The size of transmission towers increases with electricity demand; the foundation size also increases to efficiently support the larger transmission towers. A foundation reinforced with additional structures is often used to improve foundation performance. For example, various researchers proposed setting rock bolt on a pier foundation to reduce settlement and to increase the resistance of the foundation; and setting a protective slab under the tower foundation to reduce the differential settlement of the foundation [18–20].

TEPCO [21] and IEEE [13] proposed the use of connection beams placed between the individual tower foundation components. According to TEPCO [21], the increase in load capacity for a foundation reinforced with a connection beam can be estimated based on the mobilized shear stresses and bending moments due to the weight of the connection beam. In the IEEE [13], a general description of the use of connection beams is presented with emphasis on the reduction effect of differential settlements. The significant effects of connection beams were investigated by Kyung *et al.* [22]. They performed small-scale model tests and finite element analyses in a clay condition and found that a 25% relative stiffness of the connection beams to that of the mat foundation is most effective to improve foundation performance.

All these results were obtained for certain assumed soil and foundation conditions. There were no changes in soil conditions such as strength and compressibility, which were indicating limited condition of the application. As a connected foundation can be more effectively used in soft, clayey soil, it is important to check any possible effect of soil and foundation conditions on variation in mechanical performance.

In the present study, the effects of a connected foundation on the performance of transmission tower structures were investigated with consideration of different types of connection beam and various soil conditions. Soft clay conditions with and without a bearing rock layer were assessed as well. For this purpose, a series of finite element analyses were performed and used to analyze the effects of a connected foundation. A large-scale field load test using a prototype model structure was performed and compared with the results from the finite element analyses. Improved performance of transmission tower structures was analyzed in detail for various connection beam stiffness and soil conditions.

2. Transmission Tower Foundation

2.1. Foundation Types and Design Procedure

A transmission tower structural system consists of overhead power lines, steel lattice tower structures, and foundations. The foundations are generally installed at the four corners of the tower structure. The foundations of transmission towers can be classified as axial load foundations and moment load foundations. Axial load foundations indicate the cases where lateral load acting on a tower would be transferred as uplift and compressive loads on the individual foundations at each corner. Pile foundations, pier foundations, and inverted T foundations, shown in Figure 1a to c, are considered as axial load foundations, which are effective in resisting lateral tower loads but vulnerable to differential settlements. A mat foundation, shown in Figure 1d, is a moment-load foundation, which is effective in preventing structural damage from differential settlements, while the lateral resistance tends to be lower than that of axial load foundations.

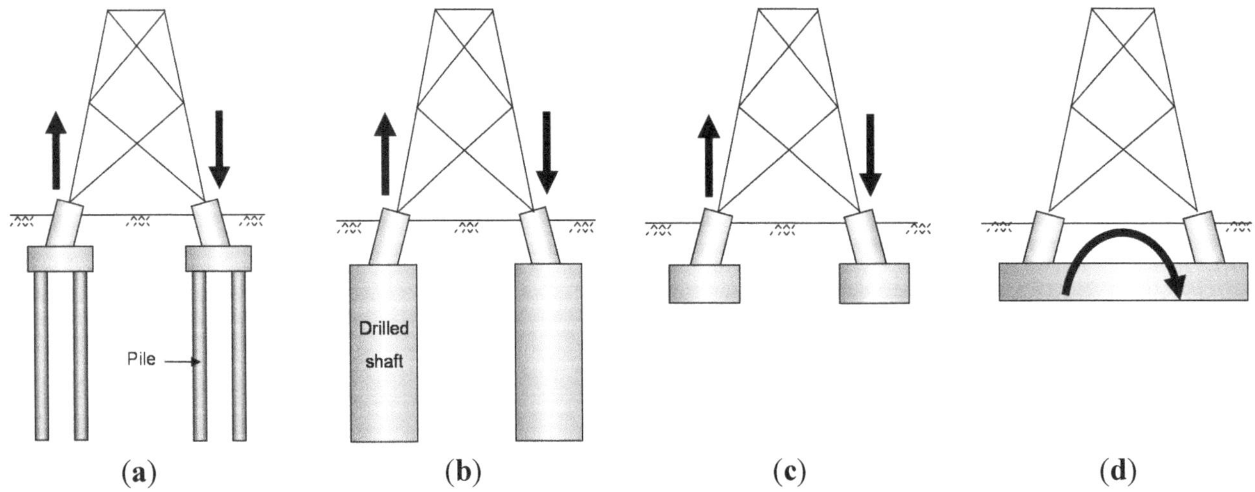

Figure 1. Types of transmission tower foundations: (**a**) pile foundation; (**b**) pier foundation; (**c**) inverted-T foundation; and (**d**) mat foundation.

The type of transmission tower foundation used is determined based on soil conditions. A pile foundation is generally used in soft soil conditions. For stability analysis, the load capacity of an individual foundation is evaluated for a given soil condition, ensuring that the load capacity is greater than the design load. Figure 2 shows the typical configuration of loads and resistances for a transmission tower structure where an applied lateral load on a tower (*H*), transferred loads to lower foundations (*Q*), and mobilized foundation resistances (*R*) are indicated. The lateral load (*H*) acting on a tower is transferred to the lower foundations of the front compressive and rear uplift sides. The transferred loads are composed of the vertical (Q_{vc} and Q_{vt}) and horizontal (Q_{hc} and Q_{ht}) load components, as indicated in Figure 2. The magnitude of each load component can be calculated from the geometric characteristics of the transmission tower. The stability of a transmission tower foundation is then determined as follows [21,23]:

$$Q_{vc} \leq R_{vc,m} = \frac{R_{vc}}{FS} \text{ and } Q_{vt} \leq R_{vt,m} = \frac{R_{vt}}{FS} \tag{1}$$

$$Q_{hc} \leq R_{hc,m} = \frac{R_{hc}}{FS} \text{ and } Q_{ht} \leq R_{ht,m} = \frac{R_{ht}}{FS}, \tag{2}$$

where Q_{vc} and Q_{vt} = transferred compressive and uplift tensile loads on the front and rear sides; Q_{hc} and Q_{ht} = transferred horizontal loads on the front and rear sides; $R_{vc,m}$ and $R_{vt,m}$ = allowable compressive and uplift resistances; $R_{hc,m}$ and $R_{ht,m}$ = allowable horizontal front and rear resistances; R_{vc}, R_{vt}, R_{hc}, and R_{ht} = ultimate compressive, uplift, and horizontal front and rear resistances; and FS = factor of safety. While stabilities for both vertical and horizontal loads must be guaranteed, the vertical stability against uplift load (Q_{vt}) frequently controls the design, as uplift resistance is usually smaller than vertical compressive resistance.

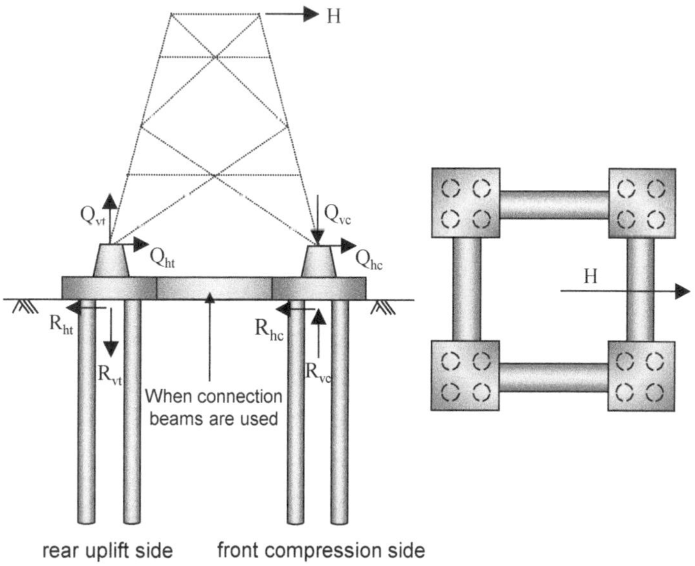

Figure 2. Configurations of loads and resistances for transmission tower structure.

2.2. Connected Foundation

When lateral loads act on the upper transmission tower, the lower foundations at the corners are subject to uplift and compressive loads. This indicates that a larger amount of differential settlement could occur due to the opposite directions of induced settlements at the uplift and compressive sides, which may cause structural damage to the entire transmission tower structure system. Modified foundations reinforced using additional structural components are an option to reduce differential settlement and improve the performance of a foundation. For example, Yang *et al.* [24] and Wang *et al.* [25] presented connected H-shaped girders to prevent instability in transmission towers. Yuan *et al.* [20] also presented tower foundations with protective slabs in order to reduce differential settlement.

A connected foundation is a type of reinforced foundation using connection beams placed between foundations, as illustrated in Figure 2. The design guidelines and performance analysis of connected foundations can be found in TEPCO [21], IEEE [13], and Kyung *et al.* [22]. According to TEPCO [21], connection beams are regarded as rigid components; their mechanical properties are not considered in the design. IEEE [13] also referred to the use of connection beams for the same purpose of increasing foundation resistance and reducing differential settlements. Kyung *et al.* [22] analyzed the performance of connected foundations for different structural and load conditions. It should be noted that all these investigations were performed on the limited soil and foundation conditions available in the experimental testing program. Further investigation is warranted to examine the detailed performance of connected foundations in different foundation and soil conditions.

3. Finite Element Analysis

3.1. Description of Analysis

Full-scale model tests of foundations for transmission towers are difficult due to the great expense and effort required in constructing test models and conducting field load tests. For this reason, numerical modeling and analysis are often introduced as an alternative. In the present study, the finite element (FE) analysis of foundations for full-scale transmission tower structures was performed to analyze the behavior and improved performance of foundations reinforced with connection beams. The PLAXIS 3D Foundation [26] was used in the study, which is a widely used finite element analysis program for geotechnical engineering.

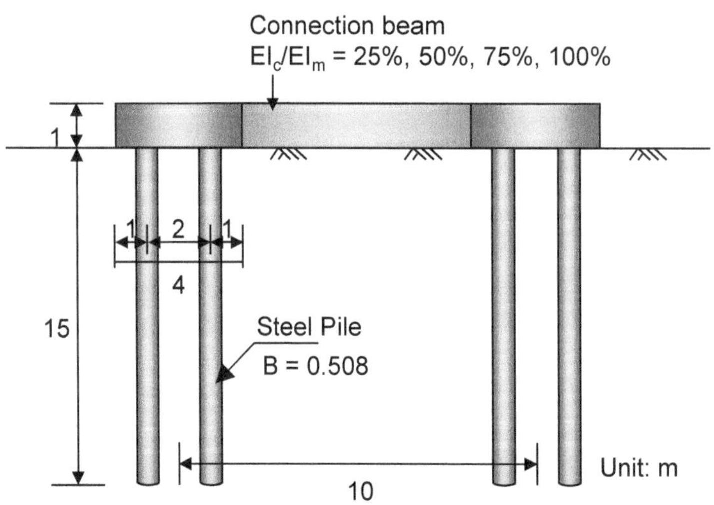

Figure 3. Configuration of foundation parts for finite element analysis.

The finite element analyses were planned and prepared based on the typical size and configuration of 345 kV transmission tower structures. Figure 3 shows the configuration of transmission tower foundation considered in the finite element analyses. As shown in Figure 3, the foundation parts are placed at the four sides of the transmission tower, and the foundation at each side consists of a mat and four piles. The width and height of the mat are 4 and 1 m, respectively, and the distance between the foundation sides is 10 m. The diameter and length of the piles are 0.508 and 15 m, respectively, and the pile-to-pile spacing equals 2.7 m.

Four different sizes of connection beams were considered in the analyses, as the rigidity of connection beam varies with its size and affects the effectiveness of connection beams. While the height of the connection beams was always 1 m, different widths of connection beams equal to 25%, 50%, 75%, and 100% of the mat width were considered. The material properties of the connection beams were the same as those of the mat with an elastic modulus (E) and Poisson's ratio (υ) of 40 GPa and 0.2, respectively. This means that the flexural stiffness of the connection beams (EI) was equal to 25%, 50%, 75%, and 100% of the mat EI. The height of the tower structure was 27 m, and the transferred loads on the foundations were determined from the geometric condition of the transmission tower and foundations when the lateral load (H) was applied on the top of the tower structure.

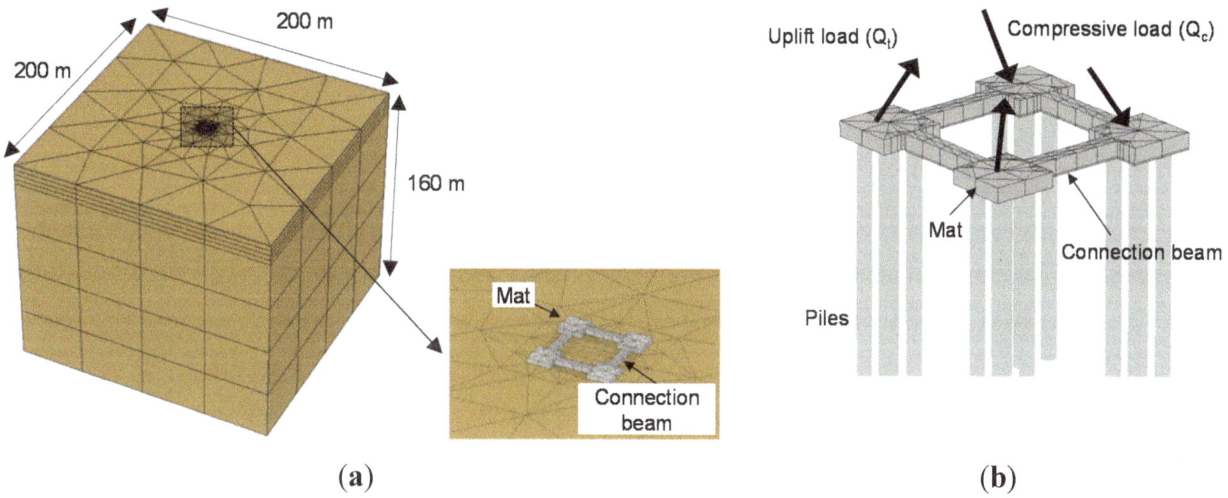

Figure 4. Finite element model for transmission tower foundation: (**a**) configuration of connected foundation model and (**b**) configuration of applied loads.

Figure 4 shows the finite element model prepared in this study. The width and depth of the finite element model in this study were 200 and 160 m, respectively, corresponding to a size larger than 10 times the pile length or 23 times the mat width. According to Liang *et al.* [27], this represents a sufficiently larger model size to avoid the boundary effect. Finer mesh was used in the zone near the model structure where higher stress concentration was expected. Fixed-end boundary conditions were set along the lateral and bottom sides of the mesh. Mat foundations were assumed to be linear-elastic concrete material, as described previously. Piles were regarded as linear-elastic steel piles with an elastic modulus (E) and Poisson's ratio (υ) equal to 205 GPa and 0.3, respectively. The interface elements were used between the soil and the pile surfaces, which allowed elastic small displacements and plastic slip behaviors. The shear strength of the interface element was defined based on the strength reduction factor R_{int} [26], which is given by the following relationships:

$$c_{int} = R_{int} c_{soil} \quad (3)$$

$$\tan\phi_{int} = R_{int} \tan\phi_{soil}. \quad (4)$$

where c_{int} and ϕ_{int} = cohesion and friction angle of interface; and c_{soil} and ϕ_{soil} = cohesion and friction angle of soil materials; and R_{int} = reduction factor. The value of R_{int} was set to 0.4, as obtained from the model load test results that will be further described in Section 3.3. Figure 4b shows the detailed configuration of the applied loads on the foundations at corners. With lateral load acting on the tower, the two front-side foundations are subjected to compressive loads and the other two rear-side foundations are subjected to uplift loads.

3.2. Soil Conditions

Various clay soil conditions were considered in the finite element analyses. Figure 5 shows the soil profiles considered in the finite element analyses. The soil conditions considered were a soft clay deposit with a lightly over-consolidated (OC) layer near the surface, as is commonly encountered in practice. Below the surface layer, soils were all assumed to be normally consolidated (NC) clay. Other cases with a bearing rock layer at the pile base level, as indicated in Figure 5a, were also considered in the finite element analyses. The unit weight (γ_{sat}) of clay was equal to 16 kN/m^3 with a Poisson's ratio (υ) of 0.495 assuming undrained condition. For the NC clay condition, the undrained shear strength (s_u) varied linearly with depth given as follows:

$$s_u = \alpha \cdot \gamma' \cdot z \quad (5)$$

where α = strength increase ratio; z = depth; and γ' = effective unit weight. The values of s_u within the OC zone near the surface were assumed to be 10 kPa. α is the ratio of s_u to the vertical effective stress (*i.e.*, $\gamma' \cdot z$). The value of α for most clays ranges from 0.2 to 0.4 [28–32]. In the present study, the value of α was set to 0.2, assuming a soft clay condition.

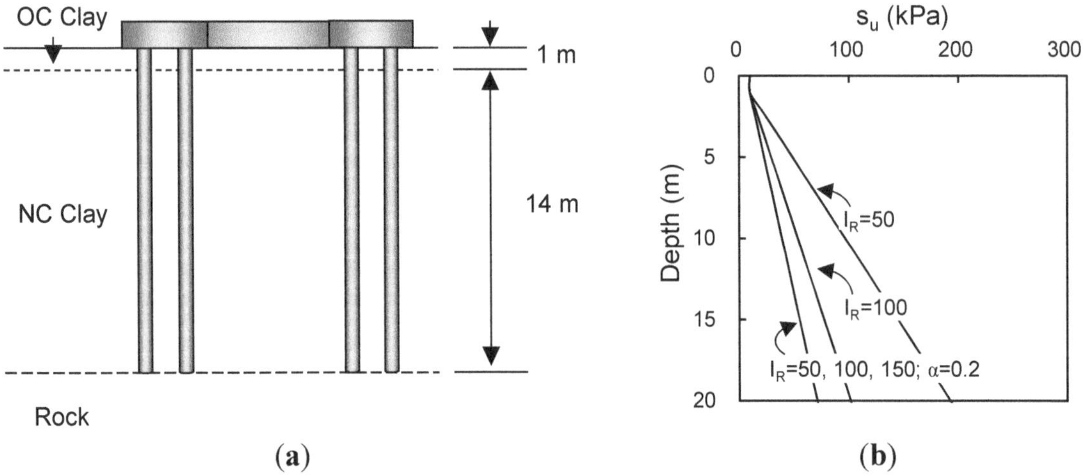

Figure 5. Soil conditions for finite element analysis: (**a**) soil layer condition and (**b**) depth profile of considered soil condition.

The rigidity index is often used to characterize the mechanical state of clay and is defined as the ratio of shear modulus (G) to s_u given by:

$$I_R = \frac{G}{s_u} = \frac{E_s}{2\cdot(1+\nu_u)\cdot s_u} = \frac{E_s}{3\cdot s_u} \tag{6}$$

where G = shear modulus; E_s = elastic modulus; and υ_u = Poisson's ratio for undrained condition = 0.5. Three different values of I_R equal to 50, 100, and 150 were considered in the analyses. Figure 5b shows the profiles of soil conditions considered in the finite element analyses. For the cases with a bearing rock layer, the values of E and υ for rock were 50 MPa and 0.15, respectively, which represent the conditions of weathered rock [33,34].

3.3. Comparison of Finite Element Analyses with Field Load Tests

To assess the validity of the finite element analysis, results from field load tests for the transmission tower structures were obtained and compared with those from the finite element analyses. Two field test cases were used, the small-scale model load tests reported in Kyung *et al.* [22] and larger-scale prototype model load tests conducted in the present study. For both cases, the transmission tower structures were prepared using connected foundations.

The small-scale model transmission structures consisted of an upper tower structure and lower foundations, as shown in Figure 6a. The foundation at each corner included a mat and piles that were embedded into the ground to a depth of 0.8 m. The width and height of the mat were 0.1 and 0.05 m and the diameter and length of the piles were 0.05 and 0.8 m, respectively. Two different types of connection beams were used in the tests: (1) low-stiffness beams with EI = 6.135 N·m^2 and (2) high-stiffness beams with EI = 1571 N·m^2.

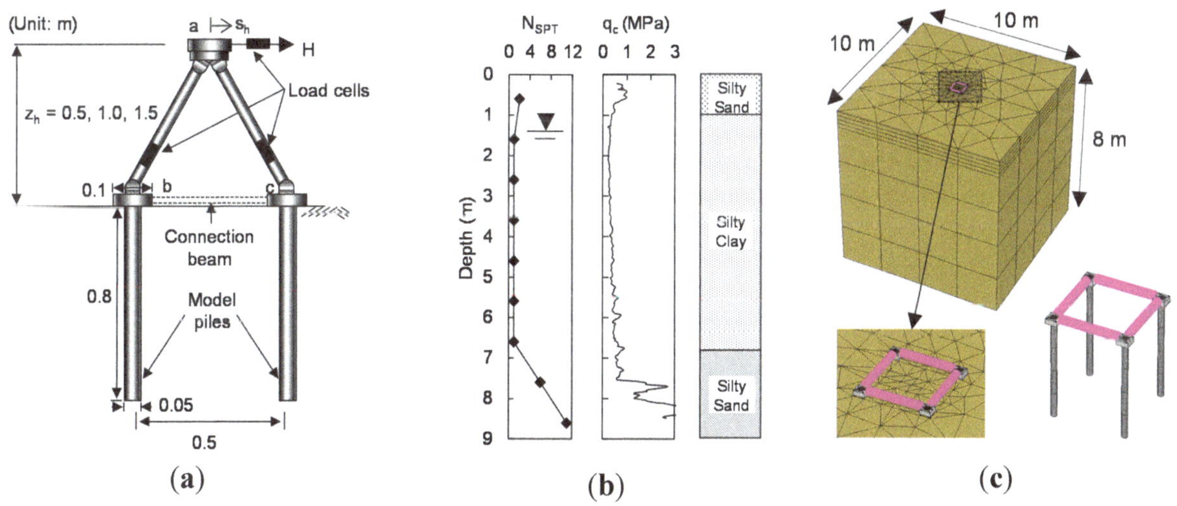

Figure 6. Small-scale model load tests with connected foundations [22]: (**a**) model structures; (**b**) depth profiles of soil layer, SPT, and CPT results; and (**c**) FE model configuration.

The soils at the test site were clays classified into CL according to the unified soil classification system (USCS). The unit weight (γ_{sat}), specific gravity (G_s), water content (w), and plasticity index (PI) were 16.59 kN/m^3, 2.69, 43.3%, and 23.3%, respectively. Unconfined compression and unconsolidated undrained (UU) triaxial tests were conducted, and s_u ranged from 8.4 to 11.1 kPa. Figure 6b shows the detailed soil profiles of the standard penetration (SPT) and cone penetration (CPT) test results. The detailed configuration of the finite element model for the small-scale model load tests is shown in

Figure 6c. As the clay was quite homogeneous and pile length was only 0.8 m, a single layer was assumed in the finite element analysis with material properties given in Table 1. Mat and pile were assumed as linear-elastic materials with the elastic modulus (E) and Poisson's ratio (υ) equal to 205 GPa and 0.3, respectively. The Mohr–Coulomb model was adopted with the undrained shear strength given in Table 1. Figure 7 shows the compared uplift load-displacement curves from the small-scale model tests and finite element analyses for two connection beam cases of EI = 6.135 and 1571 N·m². As shown in Figure 7, reasonably close agreements are observed between measured and estimated results.

Table 1. Material properties for small-scale model load test.

Material	Model	Depth (m)	γ_t^1 (kN/m³)	E^2 (kN/m²)	s_u^3 (kN/m²)	υ^4
Clay	Mohr–Coulomb	0–8	16.59	3,000	11.08	0.5
Mat	Linear elastic	-	75	205,000,000	-	0.3
Pile	Linear elastic	-	75	205,000,000	-	0.3

γ_t^1 = total unit weight, E^2 = elastic modulus, s_u^3 = undrained shear strength, υ^4 = Poisson's ratio.

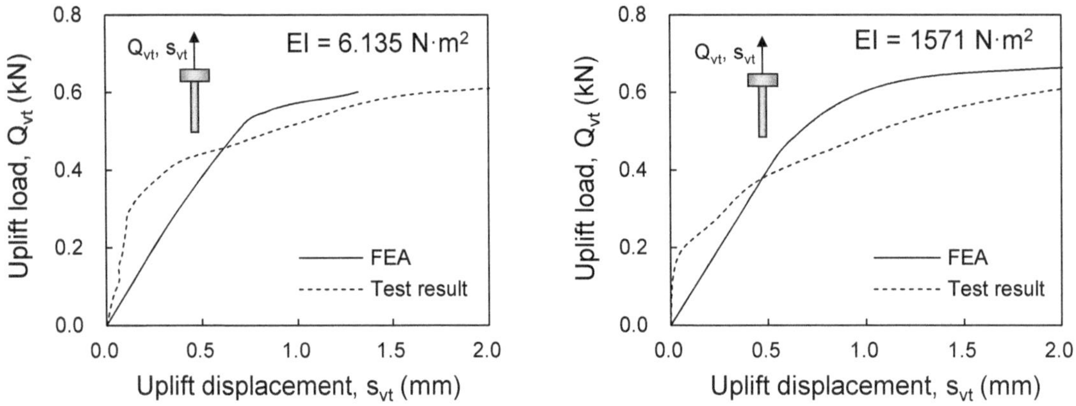

(a) (b)

Figure 7. Compared load-displacement curves of small-scale model tests and finite element analyses for connection beam cases of (**a**) EI = 6.135 N·m² and (**b**) EI = 1571 N·m².

The other field load test using a larger-scale prototype model was conducted in this study at Hwaseong, Korea. Figure 8a shows the configuration of the prototype model structure. The load height (z_h) and contiguous length (*w*) of the transmission tower structures were 15 and 4.84 m, respectively. The mat foundations were made of concrete with width and height equal to 2.4 and 0.8 m, respectively. Four piles with a diameter of 0.318 m were used at the corners and were embedded to the depth of 17 m. The connection beams with 25% mat stiffness were used for this prototype structure. The width, height, and length of the connection beams were 0.6, 0.8, and 2.44 m, respectively.

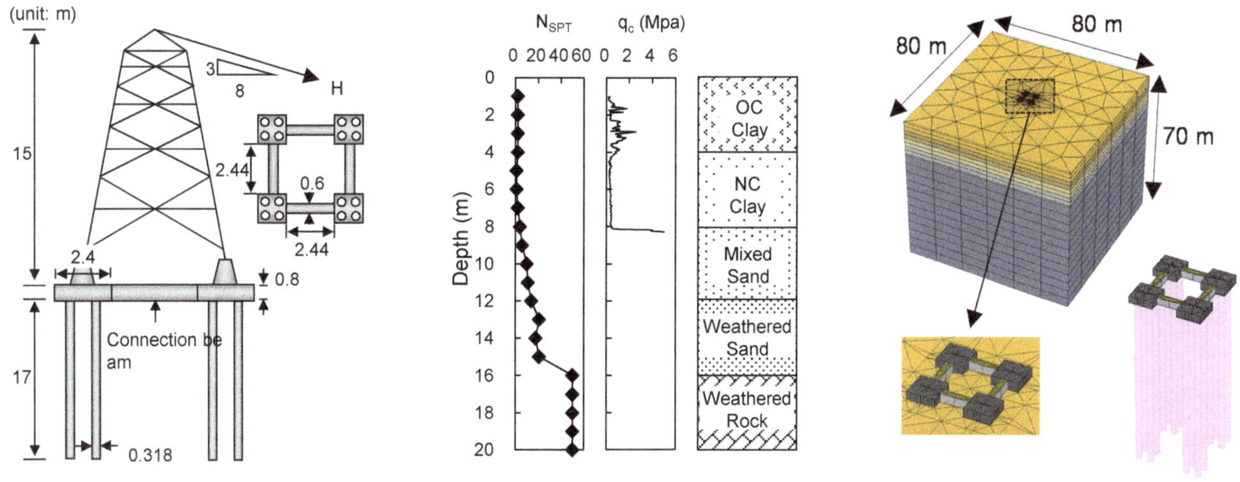

(a) (b) (c)

Figure 8. The prototype model test for connected foundation: (**a**) detailed configuration of prototype model; (**b**) site investigation results for the site; and (**c**) FE model configuration.

Figure 8b shows the depth profiles of soil layers, SPT, and CPT results at the test site. It is seen that the top 4 m of soil was a slight OC clay layer, below which an NC clay layer extended down to the depth of 8 m. Mixed and weathered sand layers were observed to depths of 13 to 16 m, beneath which was a weathered rock layer. The pile base was placed on this weathered rock layer. The layered soil condition was considered for modeling the prototype model load test in the finite element analysis. Figure 8c shows the configuration of finite element model for the prototype model load test and Table 2 shows the material properties used in the analysis.

Table 2. Material properties for the prototype model load test.

Material	Model	Depth (m)	γ_t^1 (kN/m^3)	E^2 (kN/m^2)	s_u^3 (kN/m^2)	ϕ'^4 (°)	υ^5
OC Clay	Mohr–Coulomb	0–4	18.7	2900	5–64	-	0.5
NC Clay	Mohr–Coulomb	4–8	18.7	2900	20	-	0.5
Mixed Sand	Mohr–Coulomb	8–12	19.5	2,200–5,200	-	27	0.3
Weathered Sand	Mohr–Coulomb	12–16	20.5	5,200–12,000	-	30	0.35
Weathered Rock	LE	16–70	25	300,000	-	-	0.25
Mat	LE	-	25	40,000,000	-	-	0.2
Pile	LE	-	75	205,000,000	-	-	0.3

γ_t^1 = total unit weight, E^2 = elastic modulus, s_u^3 = undrained shear strength, ϕ'^4 = friction angle, υ^5 = Poisson's ratio.

Figure 9 compares the results from the prototype model load test (PMT) and finite element analysis (FEA). As shown in Figure 9a,b, the compressive and uplift load responses of the connected foundations from the finite element analyses were in close agreement with the measured results from the field load test. Figure 9c shows the estimated and measured vertical displacement profiles at lateral load of H = 1200 kN. The profiles of vertical and differential settlements also show reasonable agreement between FEM and PMT results.

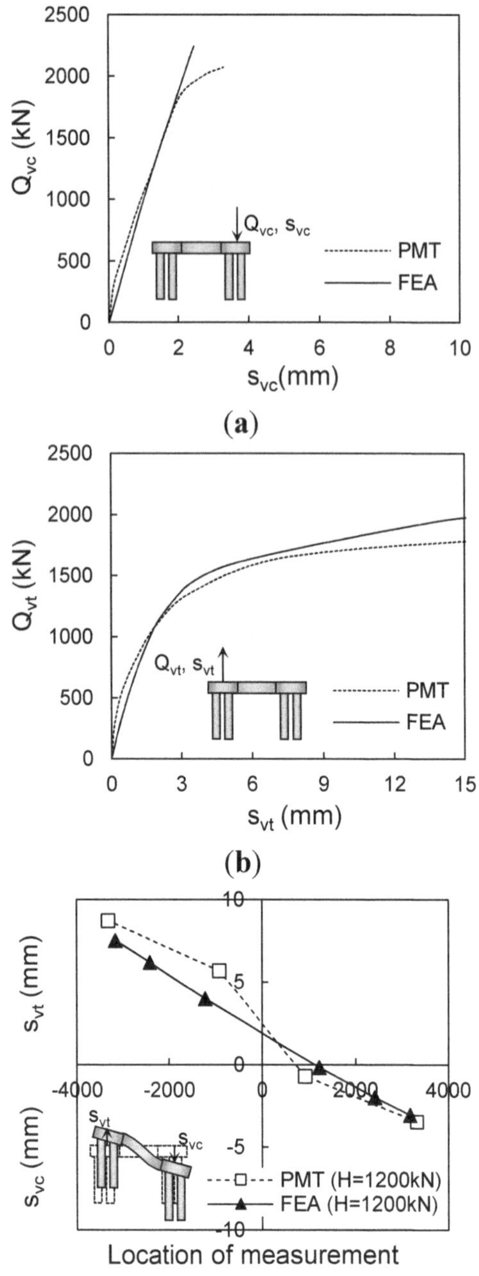

(c)

Figure 9. Compared load-displacement curves of prototype model test (PMT) and finite element analysis (FEM): (**a**) compressive load responses; (**b**) uplift load responses; and (**c**) vertical settlement profiles.

4. Parametric Study Analysis

4.1. Effects of Connection Beam Stiffness

Figure 10 shows the load-displacement curves obtained from the finite element analyses for unconnected and connected foundations. The results for different ratios of connection-beam stiffness (EI_c) to mat stiffness (EI_m) were all included in Figure 10. In Figure 10a,b, the uplift and compressive load-displacement curves were plotted for clay with I_R = 100. As shown in Figure 10, both the uplift and compressive load-carrying capacities increase as the connection beam stiffness increases. It is also noted that the amount of load-carrying capacity increase from the connected foundation was higher for the uplift foundation case than for the compressive foundation components. This is due to a higher confining effect on the uplift side where larger displacements and lower load carrying capacity are observed. When the uplift and compressive sides are connected, however, uplift and compressive displacements become similar due to the effect of the connection beam. This means that the use of connection beams would be more effective for the uplift foundations that usually control the design of transmission tower foundations.

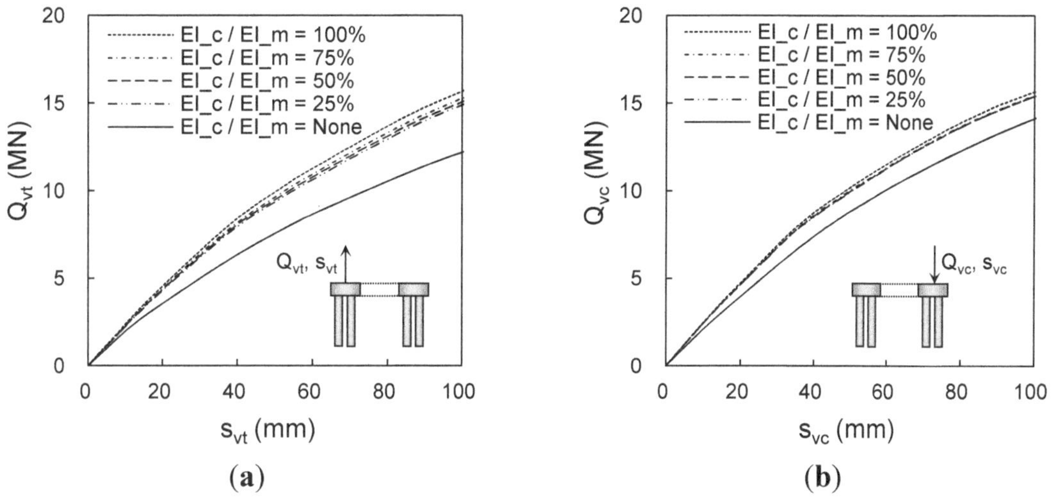

Figure 10. Load-displacement curves with different connection beam of conditions: (**a**) uplift foundation and (**b**) compressive foundation components.

Figure 11 shows the reductions in vertical displacement with relative connection-beam stiffness (EI_c/EI_m). All displacements in Figure 11 were measured at a load equal to the ultimate load capacity of the unconnected foundation. Figure 11a,b shows those for the uplift displacement (s_{vt}) and compressive settlement (s_{vc}), respectively. The values of s_{vt} and s_{vc} both decreased with increasing connection beam stiffness. The reduction effect increased with increasing EI_c/EI_m but the reductions were relatively small after EI_c/EI_m equal to 25%. The reductions in uplift displacement (s_{vt}) in Figure 11a were larger than those of compressive settlement (s_{vc}) in Figure 11b. Reductions in differential settlements for unconnected and connected foundations are shown in Figure 11c. Note that the variation of differential settlement occurrence with EI_c/EI_m follows the same tendency to those of tensile and compressive displacements. Approximately 21–25% reductions were observed, showing similar tendency and efficiency to those shown in Figure 11a,b.

Figure 12 shows the lateral load capacity H_u of tower structure and uplift and compressive load capacities of connected foundation for different EI_c/EI_m. As shown in Figure 12a, increases in H_u show a similar tendency to those in Figure 11. H_u increased by 25%, 26%, 28%, and 32% for EI_c/EI_m values of 25%, 50%, 75%, and 100%, respectively.

Figure 12b shows the transferred uplift and compressive loads of $Q_{vt,p}$ and $Q_{vc,p}$ on the piles at the rear uplift and front compressive sides, respectively. Note that $Q_{vc,p}$ and $Q_{vt,p}$ do not include the mat capacity and thus $Q_{vc,p}$ is different Q_{vc}. Both $Q_{vc,p}$ and $Q_{vt,p}$ decreased with increasing connection beam stiffness due to the load sharing effect of connection beams. For the unconnected foundation (EI_c/EI_m = 0), $Q_{vt,p}$ was larger than $Q_{vc,p}$ due to the contact resistance of the mat on the ground at compressive side.

For the connected foundation, however, the stiffer the connection beam with increasing EI_c/EI_m, the smaller the difference between $Q_{vc,p}$ and $Q_{vt,p}$. This is because connection beams confine the uplift and compressive foundations, together producing combined behavior.

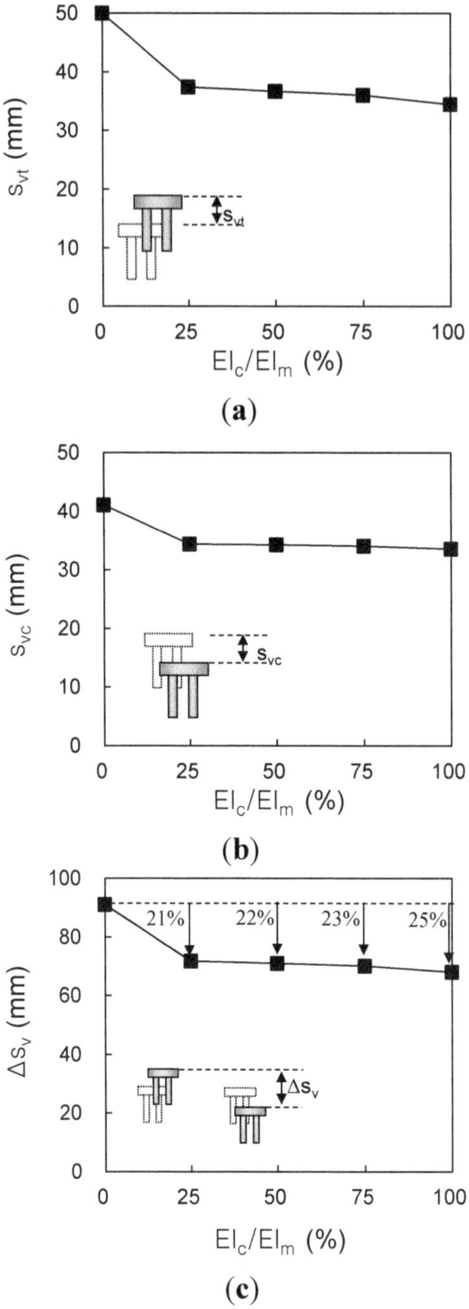

Figure 11. Reductions in vertical displacements with connection beam: (**a**) uplift displacement (s_{vt}); (**b**) compressive settlement (s_{vc}); and (**c**) differential settlement (Δs_v).

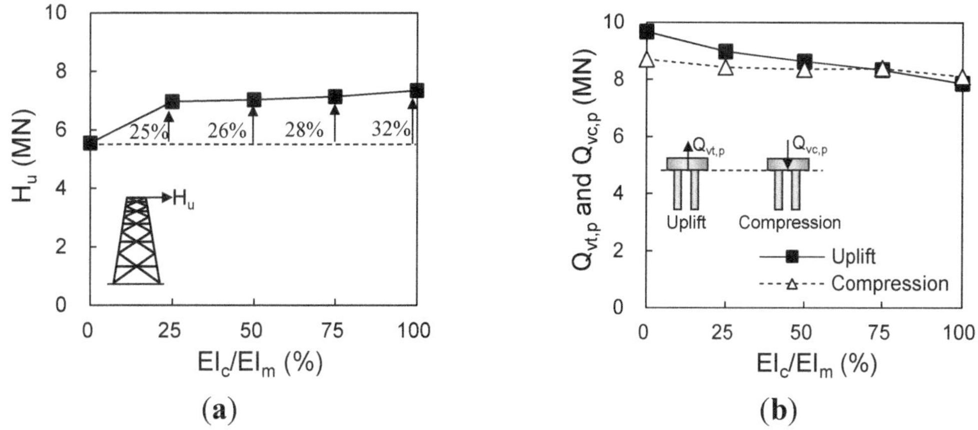

Figure 12. Load capacities of tower structure and connected foundation: (**a**) ultimate lateral load capacity of transmission tower (H_u) and (**b**) transferred uplift ($Q_{vt,p}$) and compressive ($Q_{vc,p}$) loads.

4.2. Effects of Soil Condition

To check the effect of soil condition on the performance of connected foundation, changes in differential settlement and H_u from connected foundations were obtained from the results of finite element analysis and plotted in Figure 13a,b with I_R, respectively. As I_R represents the combined characteristics of s_u and E_s, for which values increase with soil depth (z), cases with different values of s_u and E_s were considered and included in Figure 13. All results in Figure 13 were obtained for the 25% relative connection beam stiffness. As shown in Figure 13a, larger reductions in differential settlement are observed as I_R increases for both increasing E_s and decreasing su. For I_R = 100, the higher E_s case showed higher reduction. The results in Figure 13 indicate that reductions in differential settlement become larger as soil becomes stiffer with higher I_R condition. This is because stiffer clays with higher I_R produce less compressive settlement while changes in uplift displacements were smaller.

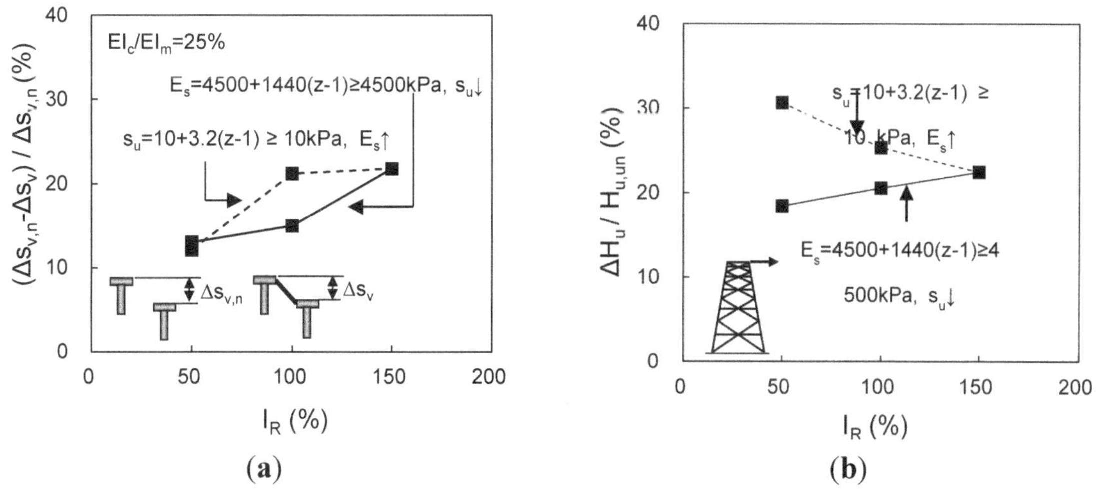

Figure 13. Effects of soil conditions on connected foundation: (**a**) reduction ratio of differential settlement and (**b**) increases in lateral load capacity of transmission tower.

Figure 13b shows the ratios of increase in lateral load capacity (ΔH_u) for the transmission tower with connected foundation to H_u for unconnected case. It is seen that $\Delta H_u/H_u$ decreases with increasing E_s and increases with decreasing s_u. These results indicate that $\Delta H_u/H_u$ increases more in weaker soil conditions with lower s_u than for higher E_s cases. From the results in Figure 13a,b, it can be summarized that the use of a connected foundation is more effective in soft clays with lower s_u. For the effect of soil stiffness, however, different tendencies were observed for differential settlement and lateral load capacity.

4.3. Effect of Bearing Rock Layer

Figure 14 shows the compressive and uplift load-displacement curves of connected foundations with and without a bearing rock layer. The compressive load-displacement curves for the with-rock layer condition are much higher than those for the without-rock layer condition. This can be attributed to increases in the end-bearing capacity of the piles that extend to the depth of the bearing rock layer. However, the uplift load-displacement curve demonstrates no major differences between with or without-rock layer conditions. This is because the existence of a rock layer does not contribute much to the uplift pile capacity that is mobilized on the pile shaft.

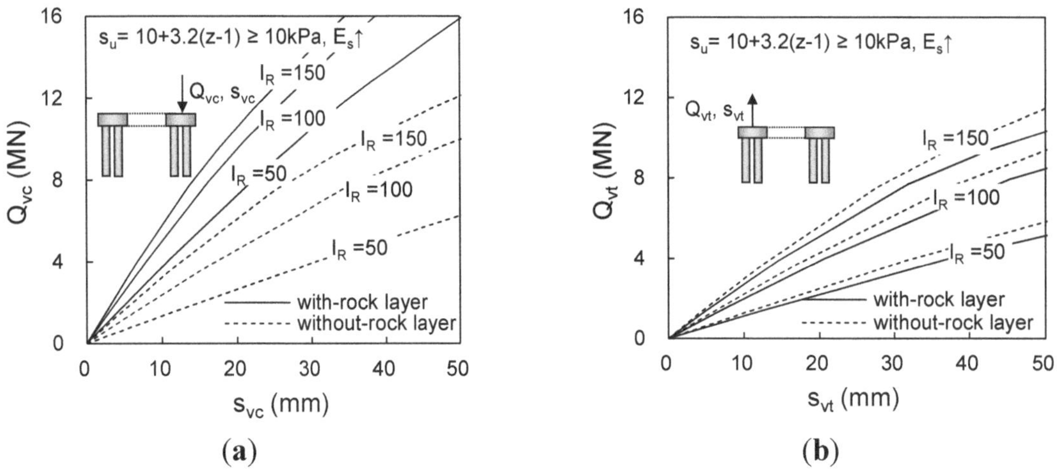

Figure 14. Load-displacement curves of connected foundations with and without bearing rock layer: (**a**) compressive and (**b**) uplift foundation.

Figure 15 shows the changes in differential settlement and lateral load capacity with and without a bearing rock layer. As shown in Figure 15a, smaller amounts of differential settlement were observed in all soil conditions for both unconnected and connected foundations when a bearing rock layer was present. This is due to restricted settlement of the pile base that is resting on the rock layer. It is also seen that reductions in differential settlement with the presence of connection beams were greater for the without-rock layer condition than for the with-rock layer condition.

Figure 15b shows that the lateral load capacity (H_u) of the transmission tower was higher in the absence of a rock layer than in the presence of a rock layer. When the foundation was not reinforced by connection beams, the values of H_u were similar for both with- and without-rock layer because each foundation behaves independently and the uplift capacity controls H_u anyway. When the foundation was reinforced by connection beams in a rock layer, the uplift displacement experienced a greater increase than the one that occurs in the absence of a rock layer because the overturning axis was closer to the compressive foundation, which was caused by decreasing compressive displacement. Therefore, the rate of H_u increase for connected foundations becomes lower when a bearing rock layer exists.

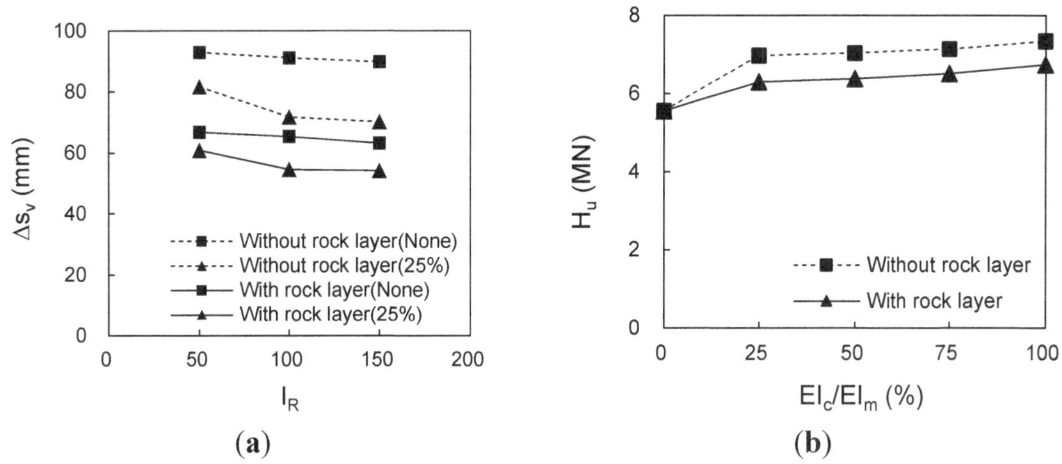

Figure 15. Differential settlement and lateral load capacity for different rock layer conditions: (**a**) differential settlement and (**b**) lateral load capacity of tower structure.

4.4. Bending Moment Distribution for Connection Beam

The connected foundation of the transmission tower is subjected to uplift and compressive loads on different sides, and bending moment occurs in the connection beam. The bending moment in the connection beam can cause structural damage, while this can be prevented or reduced by changing the shape of the beam. Figure 16a shows the distribution of bending moment along the connection beam in different soil conditions. As shown in Figure 16a, the bending moment was largest at ends and decreased with distance from the ends. It is seen that the distribution of bending moment is not symmetrical and shows the location of zero bending moment at the distance of 1/4 length from the uplift-side of the connection beam. Minor differences in the bending moment were observed for the results with different soil conditions.

In order to decrease the amount of bending moments of the connection beam, haunch-shaped connection beams were considered and analyzed in the finite element analysis. As shown in Figure 16b, the haunch shapes were defined by the ratio of D_m to D_c. The value of D_c was set equal to 0.61 m corresponding to 1/4 of connection beam length. The values of D_m were then changed to consider different haunch shapes. Three haunch-shaped connection beams were used in the finite element analysis. The considered D_m/D_c values were equal to 2/5, 3/5, and 4/5. Figure 16b shows the distribution of bending moments for the considered haunch shapes. The magnitude of bending moment of the connection beams decreased with increasing D_m/D_c showing 6.1, 4.7, and 4.1 kN·m for uplift side and 5.1, 4.1, and 3.4 kN·m for the compressive side with D_m/D_c equal to 2/5, 3/5, and 4/5, respectively. This confirms that introducing haunch-shaped connection beams is effective for increasing connection beam stability.

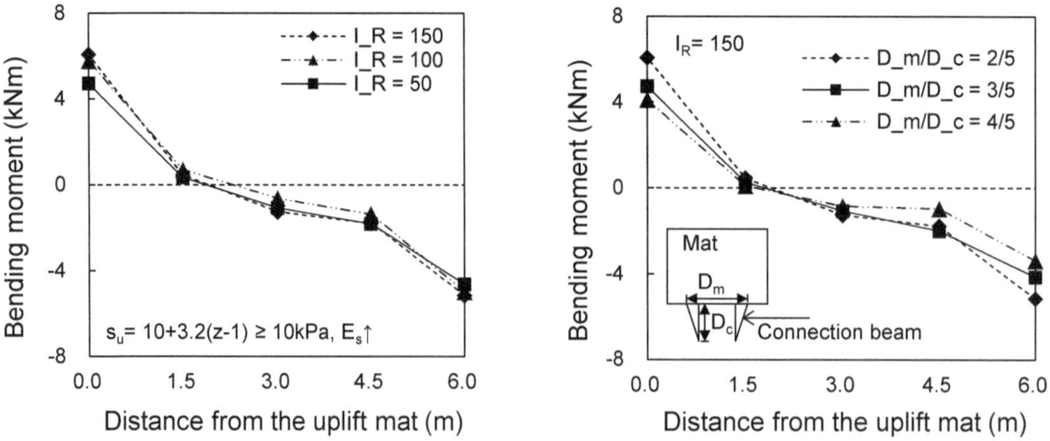

(a) **(b)**

Figure 16. Distribution of bending moment along connection beams: (**a**) bending moment distribution for different soil conditions and (**b**) bending moment distributions for different haunch shapes.

5. Summary and Conclusions

In the present study, the effects of connected foundation on the performance of electrical transmission tower structures embedded in soft ground were investigated with consideration of various connection-beam and soil conditions. For this purpose, a series of finite element analyses were performed and used to analyze the improved performance of transmission tower foundation. To assess the validity of the finite element analysis, the field load test results were compared with those from the finite element analyses.

The application of connected foundation produced increases in resistance for uplift side higher than for compressive side. This indicates that the use of connection beams would be more effective for uplift foundations that usually control the design of transmission tower foundations. Uplift displacement and settlement both decreased with increasing connection beam, stiffness showing the most effective reduction ratio at EI_c/EI_m equal to 25%. The reductions in uplift displacement were larger than in compressive settlement. The lateral load capacity of a tower structure increased similarly to other resistance components.

For the effect of soil condition, it was seen that increases in the lateral load capacity H_u of a tower structure decreases with increasing E_s and increases with decreasing s_u. This implies that the use of a connected foundation is more effective in soft clays with lower s_u. The compressive load carrying capacity for the with-rock layer condition was much higher than for the without-rock layer condition due to an increase in the end-bearing capacity of piles. Smaller amounts of differential settlement were observed in all soil conditions for both unconnected and connected foundations when a bearing rock layer was present. When the foundation was not reinforced by connection beams, the values of H_u were similar for both with- and without-rock layers. It was confirmed that introducing haunch-shaped connection beams is effective for increasing connection beam stability.

The results obtained from this study indicate that the connected foundation is an effective option to improve the resilience of electrical transmission tower infrastructures with the effect of increasing load carrying capacity and reducing differential settlements. It is particularly effective within soft grounds where extra safety margins are necessary as there is increasing possibility of damage with weaker soil conditions. In addition, extreme weather events due to climate change can cause harmful changes in subsoil condition with chances of additional settlements and unexpected reduction in the bearing capacity of foundations. The design strategy for optimizing connected foundations of a transmission tower system given in this study can be used to prepare more robust and sustainable transmission infrastructures with improved resilience against various climate change scenarios and unfavorable soil conditions.

Acknowledgments

This work was supported by a Power Generation & Electricity Delivery of the Korea Institute of Energy Technology Evaluation and Planning (KETEP) grant funded by the Korean Ministry of Knowledge Economy (No. 20101020200060). This work was also supported by the Basic Science Research Program through the National Research Foundation of Korea (NRF) grant funded by the Korean government (MSIP) (Nos. 2011-0030040 and 2013R1A1A2058863).

Author Contributions

All three authors significantly contributed to the scientific study and writing. Doohyun Kyung and Junhwan Lee contribute to the overall idea, planning, financing, analyzing, and writing of the manuscript; Youngho Choi performed finite element analyses of connected foundations.

Conflicts of Interest

The authors declare no conflict of interest.

References

1. Ebinger, J.; Walter, V. *Climate Impacts on Energy Systems: Key Issues for Energy Sector Adaptation*; World Bank: Washington, DC, USA, 2011.
2. Yates, D.; Luna, B.Q.; Rasmussen, R.; Bratcher, D.; Garre, L.; Chen, F.; Tewari, M.; Hansen, P.F. Assessing climate change hazards to electric power infrastructure: A sandy case study. *IEEE Power Energy Mag.* **2014**, *12*, 66–75.
3. Wilbanks, T.; Fernandez, S.; Backus, G.; Garcia, P.; Jonietz, K.; Kirshen, P.; Savonis, M.; Solecki, B.; Toole, L.; Allen, M.; *et al. Climate Change and Infrastructure, Urban Systems, and Vulnerabilities*; Technical report for the U.S. Department of Energy in Support of the National Climate Assessment, U.S. Department of Energy: Oak Ridge, TN, USA, 2012.
4. European Commission. Commission Staff Working Document: Adapting Infrastructure to Climate Change. In *An EU Strategy on Adaptation to Climate Change*; European Commission: Brussels, Belgium, 2013.
5. HM Government. *Climate Resilient Infrastructure: Preparing for a Changing Climate*; UK for The Stationery Office Limited on behalf of the Controller of Her Majesty's Stationery Office: London, UK, May 2011.

6. Neumann, J.E.; Price, J.C. *Adapting to Climate Change: The Public Policy Response−Public Infrastructure*; An Initiative of the Climate Policy Program at RFF, Resources for the Future: Washington, DC, USA, June 2009.
7. Konrad, J.; Samson, M. Hydraulic conductivity of kaolinite-silt mixtures subjected to close-system freeze and thaw consolidation. *Can. Geotech. J.* **2000**, *37*, 857–869.
8. Othman, M.A.; Besnson, C.H. *Effect of Freeze-Thaw on the Hydraulic Conductivity of Three Compacted Clays from Wisconsin*; Transportation Research Board: Washington, DC, USA, 1992; pp. 118–125.
9. Yarbasi, N.; Kalkan, E.; Akbulut, S. Modification of the geotechnical properties, as influenced by freeze-thaw, of granular soils with waste additives. *Cold Reg. Sci. Technol.* **2007**, *48*, 44–55.
10. Ausilio, E.; Conte, E. Influence of groundwater on the bearing capacity of shallow foundations. *Can. Geotech. J.* **2005**, *42*, 663–672.
11. Shahriar, M.A.; Sivakugan, N.; Das, B.M.; Urquhart, B.; Tapiolas, M. Water table correction factors for settlement of shallow foundations in granular soils. *Int. J. Geomech. ASCE* **2015**, *15*, 06014015.
12. Yasuhara, K.; Murakami, S.; Mimura, N.; Komine, H.; Recio, J. Influence of global warming on coastal infrastructural instability. *Sustain. Sci.* **2007**, *2*, 13–25.
13. IEEE. IEEE Guide for Transmission Structure Foundation Design and Testing, IEEE Standard 691–2001. In Proceedings of the IEEE Power Engineering Society and the American Society of Civil Engineers, New York, NY, USA, 2001.
14. Jang, S.H.; Kim, H.K.; Ham, B.W.; Chung, K.S. A study on the transmission tower foundation design and construction method-A focus of cylindrical foundation. *J. Korean Inst. Electr. Eng. (KIEE)* **2007**, *56*, 1031–1034.
15. KECA. *Handbook for Transmission Structure*; Korea Electrical Contractors Association: Seoul, Korea, 2003.
16. Morinaga, Y.; Kamiji, M.; Imoto, S.; Ogawa, S.; Iwamori, K. Transmission Tower Foundation in Japan. In Proceedings of the Transmission and Distribution Conference and Exhibition, Yokohama, Japan, 6–10 October 2002; Volume 3, pp. 2162–6165.
17. Kim, J.B.; Cho, S.B. The design and the full load test results of 765 kV tower foundation. In Proceeding of the Korean Institute of Electrical Engineers (KIEE) Fall National Conference, Seoul, Korea, November 1995; pp. 447–449.
18. Jeoung, S.S.; Ham, H.K.; Lee, D.S. Load transfer analysis of drilled shaft reinforced by soil nails. *J. Korean Geotech. Soc.* **2004**, *20*, 37–47.
19. Nam, D.S.; Kim, S.I.; Lee, J.H.; Yoon, K.S. Optimization of Reinforcement Effect of Large-diameter drilled deep foundation. *J. Korean Geotech. Soc.* **2003**, *19*, 207–216.
20. Yuan, G.L.; Li, S.M.; Xu, G.A.; Si, W.; Zhang, Y.F.; Shu, Q.J. The anti-deformation performance of composite foundation of transmission tower in mining subsidence area. *Prodedia Earth Planet. Sci.* **2009**, *1*, 571–576.
21. TEPCO. *Design Guideline for UHV Foundation*; Tokyo Electric Power Company: Tokyo, Japan, 1988.
22. Kyung, D.H.; Kim, D.H.; Lee, J.H. Improved mechanical performance of connected foundations for transmission tower structures in soft soils. *Eng. Struct.* **2014**, in submit.

23. KEPCO. *Design Standard for Transmission Tower Foundation*; DS-1110; Korea Electronic Power Corporation: Seoul, Korea, December 2011.
24. Yang, J.S.; Yang, Y.H.; Yan, L.; Zhang, H.L.; Hu, X.; Tang, P. Construction scheme choice of large-span tunnels under-passing high voltage transmission tower and its application. *Chin. J. Rock Mech. Eng.* **2012**, *31*, 1184–1191.
25. Wang, S.; Yang, J.; Yang, Y.; Zhang, F. Construction of large-span twin tunnels below a high-rise transmission tower: A case study. *Geotech. Geol. Eng.* **2014**, *32*, 453–467.
26. PLAXIS 3D. *Foundation PLAXIS 3D Foundation User Manual, Version 2.0*; Brinkgreve, R.B., Swolfs, W.M., Eds.; PLAXIS Inc.: Delft, The Netherlands, 2008.
27. Liang, F.Y.; Chen, L.Z.; Shi, X.G. Numerical analysis of composite piled raft with cushion subjected to vertical load. *Comput. Geotech.* **2003**, *30*, 443–453.
28. Hansbo, S. *Foundation Engineering*; Elsevier Science B.V.: Amsterdam, The Netherlands, 1994; pp. 89–91.
29. Jamiolkowski, M.; Ladd, C.C.; Germaine, J.T.; Lancellotta, R. New developments in field and laboratory testing of soils, theme lecture. In Proceedings of the 11th International Conference on Soil Mechanics and Found Engineering, San Francisco, CA, USA, 1985; Volume 1, pp. 57–153.
30. Mersi, G. Reevaluation of $s_{u(mob)}=0.22\sigma'_p$ using laboratory shear tests. *Can. Geotech. J.* **1989**, *26*, 162–164.
31. Skempton, A.W. Discussion on the planning and design of new Hong Kong airport. *Proc. Inst. Civ. Eng.* **1957**, *7*, 305–307.
32. Wroth, C.P. Interpretation of in-situ soil tests. *Geotechnique* **1984**, *34*, 449–489.
33. Coduto, D.P. *Foundation Design—Principles and Practices*; Prentice Hall: Englewood Cliffs, NJ, USA, 1994.
34. Terzaghi, K.; Peck, R.B. *Soil Mechanics in Engineering Practice*; John Wiley & Sons: New York, NY, USA, 1967.

Optimization and Analysis of a Hybrid Energy Storage System in a Small-Scale Standalone Microgrid for Remote Area Power Supply (RAPS)

Fengbing Li [1,2], Kaigui Xie [1,*] and Jiangping Yang [2]

[1] State Key Laboratory of Power Transmission Equipment & System Security, Chongqing University, Chongqing 400044, China; E-Mail: cqulifengbing@163.com

[2] State Grid Sichuan Electric Power Company Meishan Power Supply Company, Meishan 620010, China; E-Mail: yajia1987@126.com

* Author to whom correspondence should be addressed; E-Mail: kaiguixie@vip.163.com

Academic Editor: Neville Watson

Abstract: The analysis and application of hybrid energy storage systems (HESSs) in small-scale standalone microgrids for remote area power supply (RAPS) has received extensive attention. This application mode has its own characteristics which must be considered but have not been considered in the existing research. To reflect the common satisfaction of load demands and maximize the utilization of renewable energy in a standalone microgrid, a new index named effective rate of energy storage system (ESS) is proposed. To reflect the true work state of supercapacitor ESS (SC-ESS), the second-level data of field measurements is used in calculation and analysis. To further enhance the operational performance of the HESS, a coordinated control strategy based on state cooperation is adopted. To get a more reasonable and more credible HESS optimization model, the comparison of existing models and proposed model with different considerations on cost and life is provided. In addition, a comparative analysis of technical and economic characteristics improvements is presented for different ESS application schemes in practical projects.

Keywords: small-scale standalone microgrid; HESS; ESS effective rate; coordinated control strategy; cost and life; technical and economic optimization

---

1. Introduction

The utilization of renewable energy sources, such as wind-turbine power generation (WT), photovoltaic power generation (PV), *etc*., can effectively alleviate the increasingly serious problems of energy crisis and environmental pollution. However, there are still some remote areas not suitable for independently getting their power supply from large power grids or small conventional power generators due to economic, environmental and other factors. However, these areas are often rich in wind energy and solar energy which can be used as distributed generation and to constitute a small standalone RAPS system [1,2]. These small systems, called standalone microgrids, are an effective way of solving the problem of villages in remote areas without electricity. WT and PV can exist simultaneously or only partially in a standalone microgrid. Wind energy and solar energy themselves are both random and uncontrollable, so their output power can't satisfy the time-varying load demands. ESSs can not only provide voltage and frequency support in the form of a main power supply to maintain system stability, but also have many advantages such as environmental-friendliness, bidirectional flow of energy, and continuously adjustable charge-discharge power features. Therefore, ESSs have become an important part of such standalone microgrids [3–5].

Energy storage technology can be divided into energy type and power type. The former has a large storage capacity and short life cycle, while the latter has a large power density, fast response and long life cycle. In recent years, energy-type energy storage devices have been commonly used in standalone microgrids including renewable energy generation [6–10], but the face the challenge of large fluctuation in charge-discharge power, and even the challenge of frequent switching between charge-discharge status. This will accelerate the life loss of energy-type energy storage systems which are sensitive to the charge-discharge process, and severely degrade their performance. If the two types of energy storage are used as a whole, namely HESS, complementarity of technical characteristics can be achieved. The typical HESS, which consists of lithium-ion battery ESS (LB-ESS) and SC-ESS, is commonly used in RASP systems [11–17].

The parameters such as power rating and storage capacity of each sub-ESS, which can dirctly affect the overall technical and economic characteristics of HESS, should be optimized. In [18], the supercapacitor was increased to meet the peak load and reduce the battery capacity configuration in a PV independent power supply system. In a stand-alone power system with single source HESS described in [19], the power rating and storage capacity configurations for two ESSs have been analyzed, and the comparison between a HESS and a single energy storage system (SESS) is given. A method for configuring HESS capacities for a WT/PV autonomous multi-microgrids cluster has been presented [20] based on the maximum off-grid time of allowable continuous running, and the expected stable system operation time under extreme conditions. Capacity optimization of battery arrays and supercapacitor arrays are proposed to minimize one-time investment and operation costs for a WT/PV standalone system in [21]. The optimization results of HESS based on the total cost is given in [22] for a PV off-grid power system. The sum of capital cost and balancing cost is the objective function of the total cost optimization in a WT independent system described in [23], and the results show that HESS is better than SESS.

Through analysis, research on HESS optimization of independent microgrids should be focused on HESS control strategy, data interval, cost consideration factors, energy storage life quantization and system constraints. In addition to [19] using a first-order filter to allocate load power, HESSs in the above studies are all controlled by a simple rule, *i.e.*, batteries to meet the load energy demand and supercapacitors to meet the instantaneous pulse loads. The technological complementarities of HESS can't be fully utilized because there is no coordination between the two types of ESSs. In a small-scale RAPS stand-alone microgrid, the storage capacity of SC-ESS can't be too large based on the considerations of application requirements and economic cost. In general, the rated power discharge duration of a SC-ESS is no more than several hundred seconds. However, the data interval of WT, PV and load in these studies can be basically one hour, or even a month [21], and the true working states of SC-ESS are not reflected. The storage arrays are connected to the power supply system through a Power Convert System (PCS), so a HESS consists of several sub-ESSs storage array components and the corresponding PCSs. The optimization considering cost of storage arrays alone will affect the credibility of the study results. At the same time, the battery life quantification is considered only in [19], and battery life is simply computed as an equivalent number of charge-discharge cycles. There is no consideration of life quantization for supercapacitors and PCSs. Further, Loss of Power Supply Probability (LPSP) or Loss of Load Probability (LOLP) [20,22,23] are taken as system constraints in the above HESS optimizations of independent microgrids. The indexes only reflect the satisfaction of load power, and can't accurately reflect the system requirements when renewable generation is included.

Aiming at the insufficiencies of existing research, a new index is proposed in this paper to accurately reflect the satisfactory rate of net load in a standalone microgrid. In order to reflect the true work state of SC-ESS, field measurements data with small enough time intervals is used for calculation and analysis. Compared with considering only power allocation in HESS control, the necessity of using the coordinated control strategy based on state cooperation is analyzed. Optimization models are established to reflect the impact of HESS optimization without a comprehensive consideration of cost and life, and to analyze the merits of common ESS application schemes in practical projects.

2. A Small-Scale Standalone Microgrid and ESSs

2.1. System Structure of a Standalone Microgrid for RAPS

A RAPS system described above is a small-scale off-grid system consisting of loads, ESSs and distributed renewable energy generation. The system is a standalone microgrid taking ESSs as the main power supply for power balance and system stability. A system structure of the microgrid for AC applications considered in this paper is shown in Figure 1.

The load demands are the requirement of time-varying power from AC BUS to the load shown as P_{LD} in Figure 1. P_{DG} shown in Figure 1 is the sum of output powers of distributed generation sent to the AC BUS. The distributed generation consists of renewable energy generation sources, such as WT, PV and so on, which should be utilized as much as possible. The distributed generation sources can't maintain the system stability and power balance because their output powers are random, intermittent and uncontrollable. The balance node is ESSs which can absorb and output power satisfying the system requirement within the allowable range. HESS or SESS consisting of energy-type ESS can be used here. P_{out_ESS} shown in Figure 1 is the total charge-discharge power of ESSs, *i.e.*, the bi-directional power flow between ESSs and the AC BUS. At the same time, the sum of load demands and distributed generations can be regarded as the net load of this system P_{net_LD} shown in Figure 1. If the direction to AC BUS is defined as the positive direction, the values of P_{LD} and P_{DG} are negative and positive, respectively, and the values of P_{out_ESS} and P_{net_LD} both can be positive or negative.

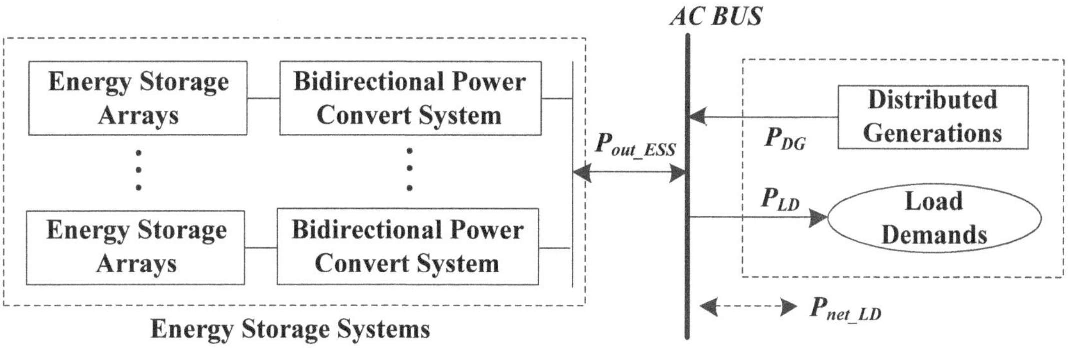

Figure 1. Configuration diagram of a standalone microgrid.

2.2. The Proposed Index Named ESS Effective Rate

In the existing research on HESS optimization in standalone microgrids, the indexes LPSP or LOLP are used as system constraints [20,22,23]. As some of the most common reliability indices, LPSP and LOLP are both indexes that show the system performance under a particular load. LPSP or LOLP actually are the ratio of load energy shortage and total load energy demand as shown as Equation (1), which ranges from 0 to 100%. In the discrete calculation with the same time interval, this value is equal to the ratio of load power shortage and total load power demand. When LPSP or LOLP is equal to 100%, the load is never satisfied and the system should be redesigned. When LPSP or LOLP is equal to 0, the system is able to supply the power to the load all the time:

$$L_{pro} = 100\% \times \sum_{n=1}^{N_T} \left\{ P_{loss\_LD(n)} \cdot \Delta T_{com} \right\} \Big/ \sum_{n=1}^{N_T} \left\{ P_{LD(n)} \cdot \Delta T_{com} \right\} \tag{1}$$

According to the positive direction as shown in Section 2.1, the value of load is negative because it absorbs energy from the AC bus, and the value of DG output is positive because it outputs energy to the AC bus. Therefore, the value of load energy shortage and total load energy demand are both negative. When $P_{LD} + P_{DG} \leq 0$, the DG output is equal to or less than the load demand. In this case, the microgrid is under the state of power requirement, and this can be cumulatively calculated in LPSP or LOLP. When $P_{LD} + P_{DG} > 0$, the DG output is more than the load demand, and the power supply of microgrid is in an overabundance with respect to the load demand. Although this state often occurs in actual operation, it can't be reflected in the calculation of LPSP or LOLP. In other words, the indexes LPSP or LOLP can only count the negative part of the net load which is the sum of load demand and DG output, as shown in Equation (2):

$$P_{loss\_LD(n)} = \min\left\{ P_{LD(n)} + P_{DG(n)}, 0 \right\} = -\left| P_{net\_LD(n)} \right| \tag{2}$$

However, in the standalone microgrid which is different from the traditional power grid, not only should the load demand be met, but also renewable energy generation sources should be utilized as much as possible. When all the renewable energy generation output power has been fully utilized, a new index is needed to show the system performance under the particular load and renewable energy generation situation. Therefore, the effective rate of ESS is proposed to reflect the degree to which the net load demand is satisfied. The effective rate R_{ESS} is computed using Equation (3). In order to get the power balance of a microgrid system, the sum of DG output, load power demand and ESS output should be equal to 0, so the reference value of ESS power output is calculated using Equation (4):

$$R_{ESS} = (1 - \frac{\sum_{n=1}^{N_T} \{ |P_{ref\_ESS(n)} - P_{out\_ESS(n)}| \cdot \Delta T_{com} \}}{\sum_{n=1}^{N_T} \{ |P_{ref\_ESS(n)}| \cdot \Delta T_{com} \}}) \times 100\% \tag{3}$$

$$P_{ref\_ESS(n)} = -(P_{LD(n)} + P_{DG(n)}) = -P_{net\_LD(n)} \tag{4}$$

The subscript "*ESS*" represents the current ESS, which can be SESS or HESS. The numerical range of R_{ESS} is [0, 100%]. In a standalone microgrid taking ESS as balance node, the smaller the difference between the actual output power of the ESS and the net load is, the bigger the value of R_{ESS} is, and the higher the requirements for ESS. Only when the two above are always exactly equal, R_{ESS} is 100%. Similar with the reliability requirement, the constraint of ESS effective rate R_{set} can be set for optimization analysis of ESS in a standalone microgrid. A penalty cost C_{pen_ESS} is needed because the requirement of effective rate must be satisfied in the optimization. In Equation (5), F is a fixed value which is much bigger than other costs:

$$C_{pen\_ESS} = \begin{cases} 0, & R_{ESS} \geq R_{set} \\ F, & R_{ESS} < R_{set} \end{cases} \tag{5}$$

2.3. General Mathematical Model of ESSs

In order to perform an optimization study of HESS, suitable mathematical models must be established. Although LB and SC have different technical characteristics, they are both energy storage technologies which can be charged and discharged as required through corresponding PCS. A general mathematical model available for LB-ESS and SC-ESS is presented in [24–30]. As shown in Equations (6)–(8), this model includes the computation of power output limits and state of charge (SOC) for ESSs. Moreover, the effect of self-discharge rate and charge-discharge efficiency is considered making the model more credible and effective:

$$P_{clmt\_ESS(n)} = -\min\{[S_{\max\_ESS} - (1-\sigma_{ESS})S_{ESS(n-1)}]\frac{E_{r\_ESS}}{\eta_{c\_ESS}\Delta T_{com}}, \ |P_{c\max\_ESS}|\} \tag{6}$$

$$P_{dlmt\_ESS(n)} = \min\{[(1-\sigma_{ESS})S_{ESS(n-1)} - S_{\min\_ESS}]\frac{\eta_{d\_ESS}E_{r\_ESS}}{\Delta T_{com}}, \ P_{d\max\_ESS}\} \tag{7}$$

$$S_{ESS(n)} = \begin{cases} (1-\sigma_{ESS})S_{ESS(n-1)} - P_{out\_ESS}\eta_{c\_ESS}\Delta T_{com} / E_{r\_ESS}, & (P_{out\_ESS} \leq 0) \\ (1-\sigma_{ESS})S_{ESS(n-1)} - \dfrac{P_{out\_ESS}\Delta T_{com}}{\eta_{d\_ESS}E_{r\_ESS}}, & (P_{out\_ESS} > 0) \end{cases} \tag{8}$$

3. Control Strategy of HESS

3.1. The Basic Control Strategy Considering Power Allocation Only

The existing work of HESS control in the standalone microgrid merely focuses on power allocation, and the methods used in [18–23] are simple rule and first-order filter. In the simple rule method, SC-ESS is used to meet pulse loads. However, the charge-discharge power of LB-ESS may still be changing rapidly, which will adversely affect the LB life. Although there are some other methods applicable to power allocation in all the research on HESS control, the first-order filtering method is the most common and widely accepted one. Through this method, the high-frequency parts of power fluctuations are borne by SC-ESS, and LB-ESS satisfies the rest, which is relatively flat. This can make full use of the advantages of SC and be conducive to delayed LB life loss. The basic control strategy considering power allocation only is the first-order filtering method shown as Equations (9) and (10):

$$P_{out\_SC}(s) = P_{HESS}(s) \cdot [sT_f / (1+sT_f)] \tag{9}$$

$$P_{out\_LB}(s) = P_{HESS}(s) - P_{out\_SC}(s) = P_{HESS}(s) / (1+sT_f) \tag{10}$$

3.2. The Coordinated Control Strategy Based on State Cooperation

In order to optimize the overall adjustment ability and improve the overall performance, the coordination between two HESS ESSs should be considered. A control strategy proposed in [31] considers the state coordination between two ESSs and can improve the overall performance of HESS. The strategy mainly includes four parts: the primary allocation of power commands, status adjustment of SC-ESS, coordination of overcharge and over-discharge protection and coordination of maximum power limit protection. The control flow of this coordinated control strategy based on state cooperation is shown in Figure 2.

The computation of primary allocation is same as Equations (9) and (10), where T_f is the initial value T_{f0}. When optimizing the overall adjustment ability, T_f can be repeatedly adjusted within the permissible range [T_{fmin}, T_{fmax}] in steps of ΔT_f. Figure 3 illustrates the principle of overall adjustment ability optimization as the core part of the coordinated control strategy, where SOC of SC-ESS is dynamically adjusted according to charge-discharge status of LB-ESS [32]. It shows that charging margin is reserved in SC-ESS when LB-ESS is discharging, and discharging margin is reserved in SC-ESS when LB-ESS is charging. Then there will be an enhanced response capability for instantaneous power changes, an optimization of charge and discharge process for SC-ESS, and an improvement of overall performance for HESS. Hence, a new index named as SOC coordinated response margin of SC-ESS is denoted by ΔS_{co_SC}, which is numerically equal to (S_{max_SC} -S_{LBd_SC}) or (S_{LBc_SC} -S_{min_SC}).

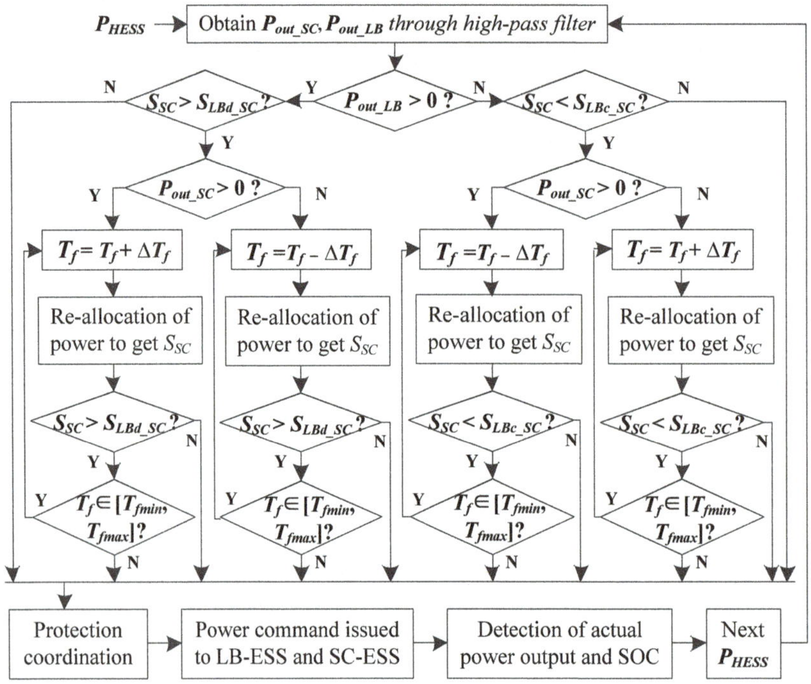

Figure 2. Flow chart of the coordinated control strategy for HESS.

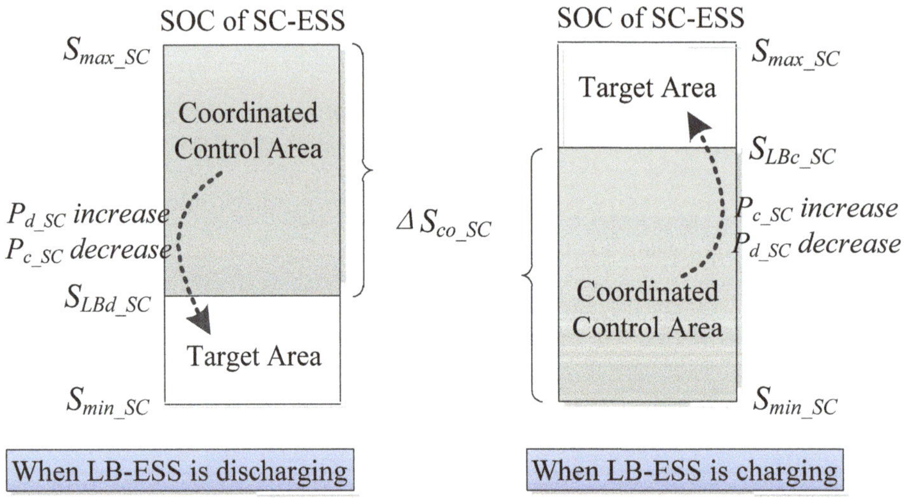

Figure 3. Coordination control diagram of overall adjustment.

4. Cost Calculation and Life Quantification of HESS

The costs of HESS primarily include the initial investment cost and the loss equivalent cost. The initial investment cost is the purchase and other expenses of HESS, which is a one-time investment cost. The loss equivalent cost is the depreciation expense of HESS within a certain time, which is a comprehensive cost, namely a form of life cycle cost. Owing to the different characteristics of different components, depreciation of HESS can't simply to use the service life, and the life of different components should be quantified separately.

4.1. Initial Investment Cost of HESS

The HESS in this paper consists of the LB arrays, the SC arrays and the two corresponding PCSs. The energy storage arrays typically consist of energy storage modules with series-parallel connection, which largely determines the ESS storage capacity. The total initial investment cost of LB storage arrays and SC storage arrays is computed by Equation (11). In practical applications, PCS usually is a standard equipment package which largely determines the power limits of ESS. The initial investment cost of PCSs is a fixed value under the determined level of rated power of ESSs. The initial investment cost of LB-ESS, SC-ESS and the whole HESS are computed using Equations (12), (13) and (14):

$$C_{initial\_HESSarr} = C_{unit\_LB} E_{r\_LB} + C_{unit\_SC} E_{r\_SC} \tag{11}$$

$$C_{initial\_LB} = C_{unit\_LB} E_{r\_LB} + C_{PCS\_LB} \tag{12}$$

$$C_{initial\_SC} = C_{unit\_SC} E_{r\_SC} + C_{PCS\_SC} \tag{13}$$

$$C_{initial\_HESS} = C_{initial\_LB} + C_{initial\_SC} \tag{14}$$

4.2. Life Quantification and Loss Equivalent Cost of Storage Arrays in LB-ESS

The useful life of the LB has an important impact on the overall performance of HESS, because the service lifetime of LBs is much shorter than that of other HESS components. Therefore, the quantification life and loss equivalent cost of LB are important indexes measuring the technical and economic characteristics of HESS. The LB lifetime is typically defined as cycle life or calendar life corresponding with the degenerative process from the nominal energy capacity to its 80% level [33–36]. A LB capacity degradation algorithm suitable for practical applications and irregular charge-discharge applications in particular is proposed in [33], as a function of the equivalent throughput cycles, the average SOC, the normalized SOC deviation and the operating temperature. The algorithm shown in Equation (15) validated by experiments and actual operation has repeatedly been cited or applied directly in [34–40]:

$$L_{array\_LB} = \sum_{m=1}^{M} \Delta D_{LB}(m) / D_{lmt\_LB}$$

$$\begin{cases} \Delta D_{LB}(m) = D_2 \cdot \exp[K_T (T_{LB} - T_{ref}) \cdot \dfrac{T_{a\_ref}}{T_{a\_LI}}] \\ D_2 = D_1 \cdot \exp[4K_{SOC}(S_{avg\_LB} - 0.5)] \cdot [1 - D_{LB}(m-1)] \\ D_1 = K_{co} N_{LB} \cdot \exp(\dfrac{S_{dev\_LB} - 1}{K_{ex}} \cdot \dfrac{T_{a\_ref}}{T_{a\_LB}}) + \dfrac{0.2\tau}{\tau_{life\_LB}} \end{cases} \tag{15}$$

where $D_{lmt\_LB} = 0.2$, $T_{ref} = 25$ °C, $T_{a\_ref} = T_{ref} + 273$, $T_{a\_LB} = T_{LB} + 273$, $K_T = 0.0693$, $K_{co} = 3.66\text{e-}5$, $K_{ex} = 0.717$, $K_{SOC} = 0.916$ [33]. The coefficients K_T, K_{co}, K_{ex}, K_{SOC} are typical values used as empirical constants of a particular LB, and can be amended in accordance with the lifetime data for different LBs. The degradation of LB capacity changes from 0 (when LB is new) to 1 (when the capacity of LB degenerates to 0). D_{lmt_LB} = 0.2 means that the actual capacity of LB degenerates to 80% of its nominal capacity. The cycle uses number or usage time length from 0 to D_{lmt_LB}, namely cycle life or calendar life of LB.

The loss equivalent cost considering life quantification of storage arrays in LB-ESS is computed as follows:

$$C_{loss\_LBarr} = L_{array\_LB} C_{array\_LB} \tag{16}$$

4.3. Overall Life Quantification and Total Loss Equivalent Cost of HESS

In order to more accurately reflect the performance of HESS, the life of each component should be quantified. The number of SC charge-discharge cycles is more than one million times, which is more than LB as shown in [15,41], soan approximate calculation model is used for lifetime quantization of SC arrays. The lifetime loss coefficient of storage arrays in SC-ESS is approximately the ratio of used cycles and its total cycles, which is computed by Equation (17). The loss equivalent cost considering life quantification of storage arrays in SC-ESS is computed with Equation (18):

$$L_{array\_SC} = N_{cycle\_SC} / N_{total\_SC} \tag{17}$$

$$C_{loss\_SCarr} = L_{array\_SC} C_{array\_SC} \tag{18}$$

Various PCSs used in LB-ESS and SC-ESS usually work uninterruptedly, and their power electronic devices have a limited service lifetime. Without taking into account any failure, the lifetime loss coefficient of PCS is approximately the ratio of running time and corresponding service lifetime. The coefficient is computed by Equation (19):

$$L_{loss\_PCS} = T / T_{life\_PCS} \tag{19}$$

The total loss equivalent cost of each ESS and HESS are computed as follows:

$$C_{loss\_LB} = C_{loss\_LBarr} + L_{loss\_PCS} C_{PCS\_LB} \tag{20}$$

$$C_{loss\_SC} = C_{loss\_SCarr} + L_{loss\_PCS} C_{PCS\_SC} \tag{21}$$

$$C_{loss\_HESS} = C_{loss\_LB} + C_{loss\_SC} \tag{22}$$

4.4. Impact Analysis Models of Different Cost and Life Considerations

The consideration of cost and life in optimization of HESS is not comprehensive in the existing research [18–23], and the optimized results may not be the best overall. In this paper, an overall optimization model shown by Equation (26) is established considering the life quantification and cost conversion of each HESS component. For comparative analysis, three different optimization objective functions based on different cost and life considerations, shown as Equations (23), (24) and (25), are set to indicate the lack of existing research. The total initial investment cost and total loss equivalent cost of HESS in four different optimized results are compared to analyze the changes of overall technical and economic characteristics of HESS. The HESSs all use the coordinated control strategy based on state cooperation, and satisfy the net load demand of a standalone microgrid for RAPS.

Optimization Objective 1, to minimize the total initial investment cost of storage arrays in HESS:

$$\min(C_{opt1\_HESS}) = \min(C_{initial\_HESSarr} + C_{pen\_ESS}) \tag{23}$$

Optimization Objective 2, to minimize the initial investment cost of the whole HESS:

$$\min(C_{opt2\_HESS}) = \min(C_{initial\_HESS} + C_{pen\_ESS}) \tag{24}$$

Optimization Objective 3, to minimize the loss equivalent cost of storage arrays of LB-ESS:

$$\min(C_{opt3\_HESS}) = \min(C_{loss\_LBarr} + C_{pen\_ESS}) \tag{25}$$

Optimization Objective 4, to minimize the total loss equivalent cost of the HESS:

$$\min(C_{opt4\_HESS}) = \min(C_{loss\_HESS} + C_{pen\_ESS}) \tag{26}$$

In all these optimization models, the parameters to be optimized are power ratings and storage capacities of LB-ESS and SC-ESS, and control parameters of HESS, *i.e.*, $Pr_{\_LB}$, $Er_{\_LB}$, $Pr_{\_SC}$, $Er_{\_SC}$, T_{f0} and ΔS_{co_SC}. The thing to note is that, the storage capacity of ESS is mainly affected by energy storage arrays, and the output power limits of ESS are mainly affected by PCS. The increase or decrease of module number in energy storage arrays can make an approximately continuous change in storage capacity of ESS. PCSs generally use standard equipment packages in the engineering applications. Then the rated output power of ESSs is usually one of the few optional specification values, such as 50, 100, 200, 250, 300, 400, 500 kW, *etc*.

The constraints of four optimization models are the output power limits and SOC operation range of LB-ESS and SC-ESS, and the allow range of SOC coordinated response margin, as shown in Equations (27)~(31):

$$P_{clmt\_LB} \le P_{out\_LB} \le P_{dlmt\_LB} \tag{27}$$

$$S_{min\_LB} \le S_{LB} \le S_{\max\_LB} \tag{28}$$

$$P_{clmt\_SC} \le P_{out\_SC} \le P_{dlmt\_SC} \tag{29}$$

$$S_{\min\_SC} \le S_{SC} \le S_{\max\_SC} \tag{30}$$

$$0 \le \Delta S_{co\_SC} \le (S_{\max\_SC} - S_{\min\_SC}) \tag{31}$$

The flowcharts of algorithm for impact analysis models of Different Considerations in Cost and Life are shown in Figure 4 as follows.

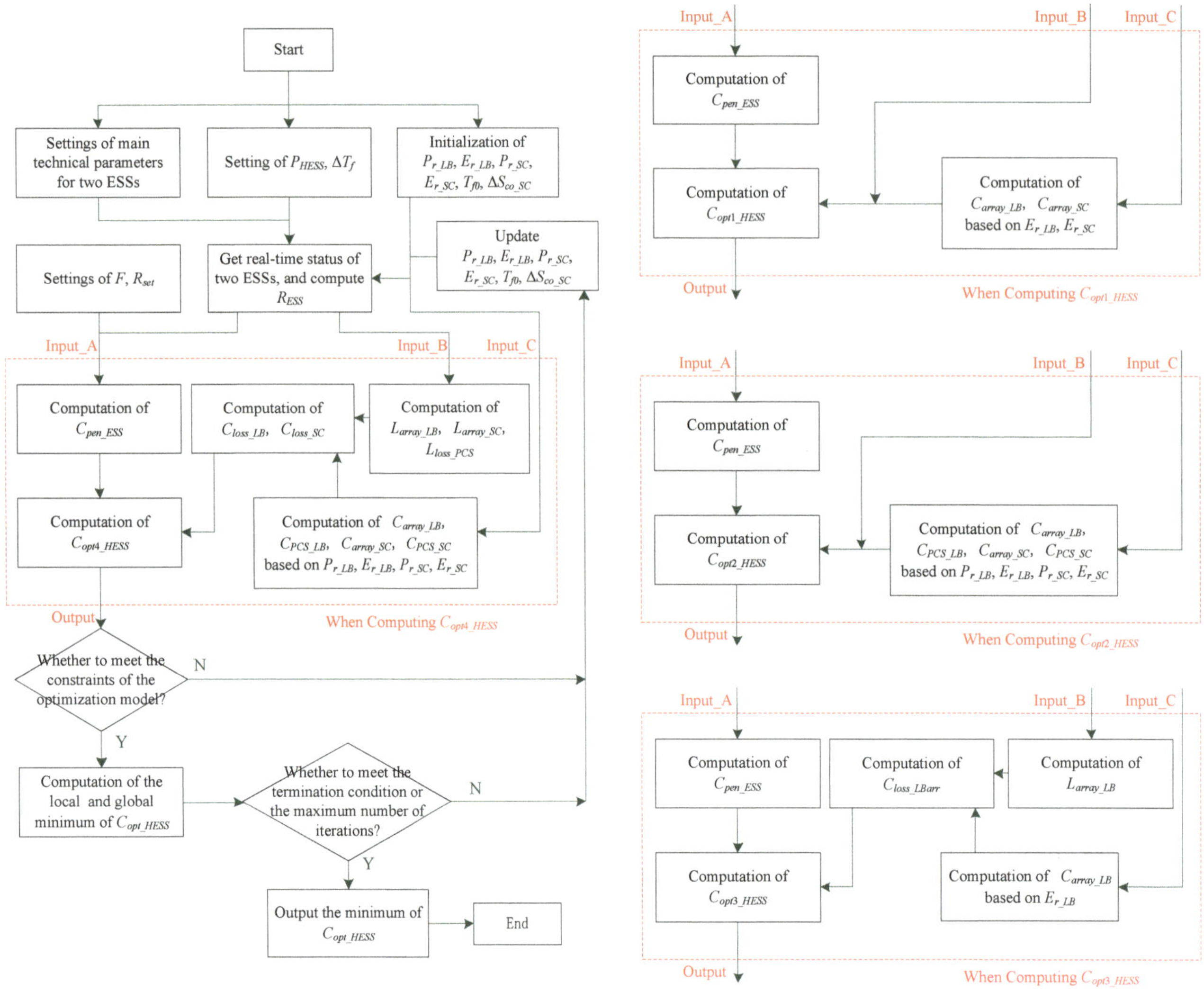

Figure 4. Flowcharts of algorithm for impact analysis models of Different Considerations in Cost and Life.

5. Comparative Analysis Models of ESSs under Different Application Schemes

In the practical engineering applications of standalone microgrids for RAPS, ESSs may be configured in three ways. They are SESS used alone, SESS newly increased based on the existing SESS, and HESS used directly. Optimization models aiming at minimizing the total loss equivalent cost of the current ESS are established to comparatively analyze the three application schemes.

5.1. SESS Used Only

When SESS is used, ESS usually adopts an energy-type ESS such as LB-ESS which can be controlled by a *Vf* pattern. The optimization model of SESS used alone is shown as Equation (32). The parameters to be optimized are P_{r_LB} and E_{r_LB}. The constraint conditions are shown by Equations (27) and (28):

$$\min(C_{opt\_SESS}) = \min(C_{loss\_LB} + C_{pen\_ESS}) \tag{32}$$

5.2. SC-ESS Increased Newly Based on the Existing SESS

The optimization model of Supercapacitor-Added ESS (SAESS) based on SESS of the existing project is shown by Equation (33). The power rating and storage capacity of LB-ESS are the same as when SESS alone is used, and will not be changed. The parameters to be optimized are P_{r_SC}, E_{r_SC}, T_{f0} and ΔS_{co_SC}. The constraint conditions are shown by Equations (27)~(31):

$$\min(C_{opt\_SAESS}) = \min(C_{loss\_LB} + C_{loss\_SC} + C_{pen\_ESS}) \tag{33}$$

5.3. HESS Used Directly

The optimization model of HESS used directly is shown by Equation (34). The parameters to be optimized are P_{r_LB}, E_{r_LB}, P_{r_SC}, E_{r_SC}, T_{f0} and ΔS_{co_SC}. The constraint conditions are shown in Equations (27)~(31):

$$\min(C_{opt\_HESS}) = \min(C_{loss\_LB} + C_{loss\_SC} + C_{pen\_ESS}) \tag{34}$$

In addition, as an auxiliary index for analysis, the total initial investment cost of ESSs under the three application schemes is computed using Equations (35), (36) and (37), respectively:

$$C_{total\_SESS} = C_{initial\_LB} \tag{35}$$

$$C_{total\_SAESS} = C_{total\_SESS} + C_{initial\_SC} \tag{36}$$

$$C_{total\_HESS} = C_{initial\_LB} + C_{initial\_SC} \tag{37}$$

6. Case Studies

6.1. System Demand and ESS Parameters

The net load curve of one day shown as Figure 5 is the second-level data of field measurements in a standalone microgrid project. The parameters are set as R_{set} = 99.9%, ΔT_f = 1, F = 10,000\$. The main parameters of LB-ESS and SC-ESS are provided in Table 1, where the matching PCSs typically have a lifetime of 10 years, and their initial investment costs are shown in Table 2. A single-stage PCS is used for LB-ESS, and a dual-stage PCS is used for SC-ESS.

Table 1. Main parameters setting of two ESSs.

Parameter type	LB-ESS	SC-ESS
Operating range of SOC	0.25~0.95	0.2~0.9
SOC threshold of over-charge protection	0.9	0.85
SOC threshold of over-discharge protection	0.3	0.25
Initial value of SOC	0.8	0.8
Charge and discharge efficiency	90%	95%
Self-discharge rate ($\%\cdot s^{-1}$)	0	0.00017
Unit capacity cost ($\$\cdot kWh^{-1}$)	655.7	157,377.0

Table 2. Initial investment cost of the matching PCS for LB-ESS and SC-ESS.

P_{r_PCS} (kW)	50	100	200	250	300	400	500
C_{PCS_LB} (10^4·\$)	1.00	1.97	3.77	4.61	5.41	6.89	8.20
C_{PCS_SC} (10^4·\$)	1.21	2.36	4.52	5.54	6.49	8.26	9.84

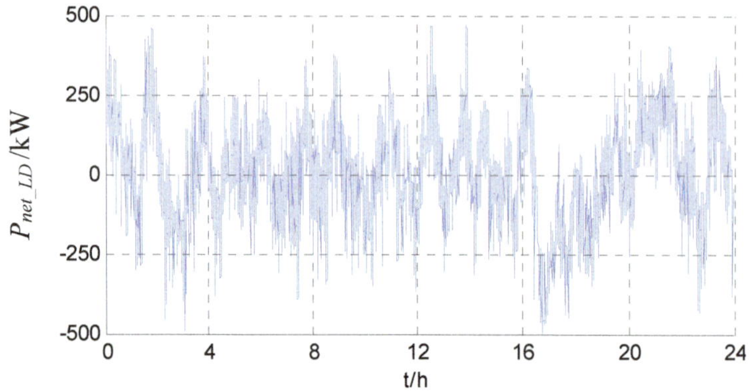

Figure 5. Net load curve of the standalone microgrid in one day.

6.2. Comparison of HESS Operation under Different Control Strategies

A case is presented to analyze the difference between the basic control strategy and the coordinated control strategy. Through the comparison of HESS operation under different control strategies, the strategy which will be used for further research can be determined. The operation of HESS using two strategies to satisfy the net load shown in Figure 4 should be compared through the case study. In order to maintain comparability, *i.e.*, the effectiveness of comparative analysis results for different control strategies, the constraint of ESS effective rate should always be fully met. Therefore, HESS should have a redundant configuration based on net load demand. Combined with analysis of Figure 4, the rated output power and rated storage capacity of LB-ESS and SC-ESS are set as 500 kW/1,000 kWh and 500 kW/10 kWh, respectively.

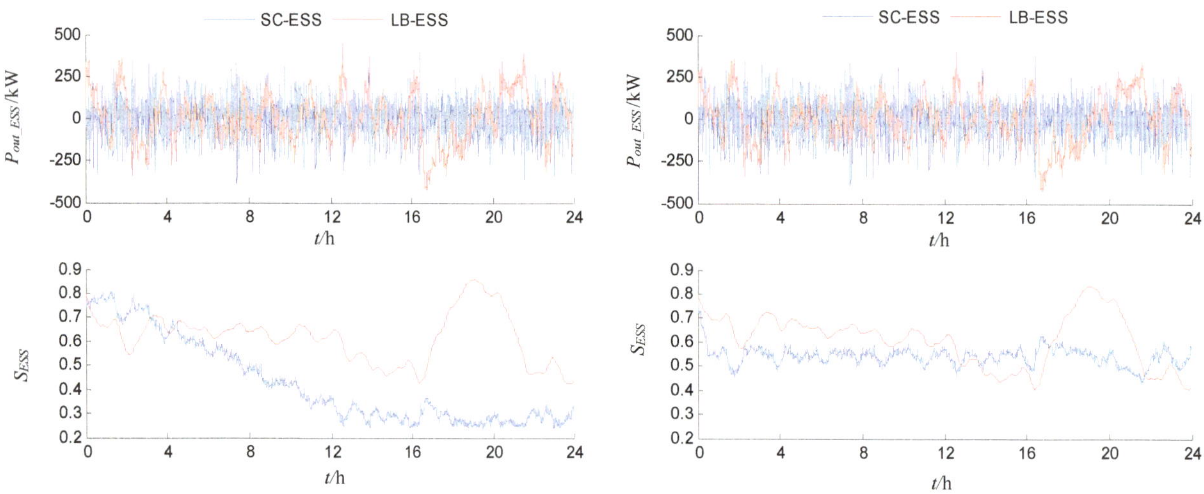

(**a**) When using basic control strategy (**b**) When using coordinated control strategy

Figure 6. Output power and SOC of two ESSs using two strategies when $T_f = 15$.

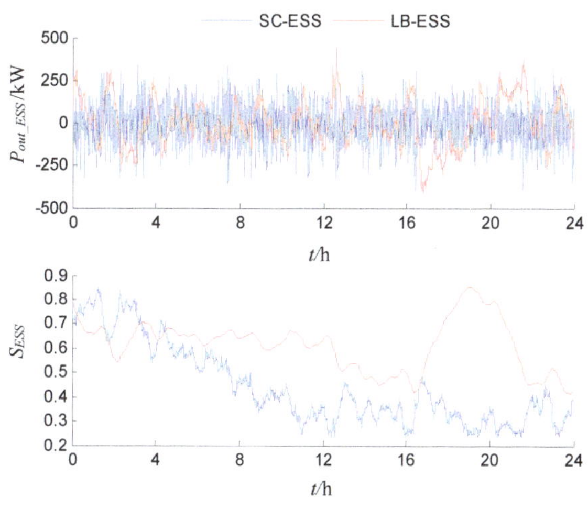

(**a**) When using basic control strategy

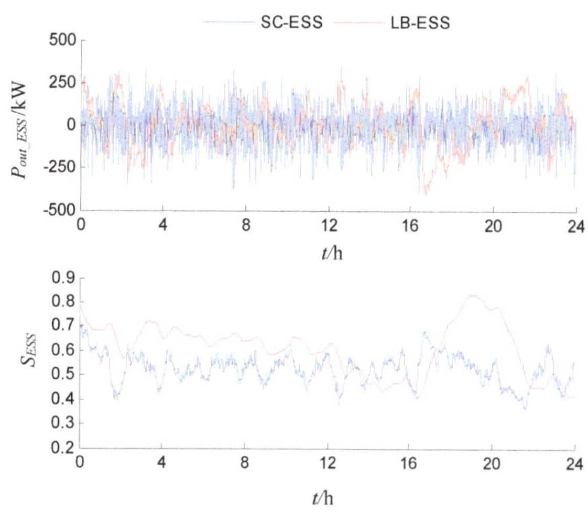

(**b**) When using coordinated control strategy

Figure 7. Output power and SOC of two ESSs using two strategies when $T_f = 30$.

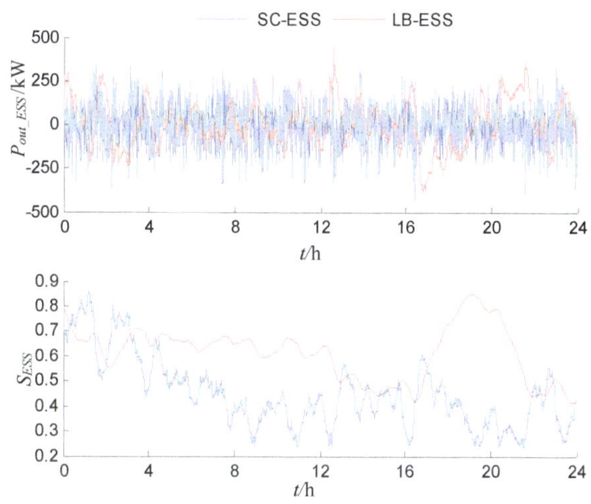

(**a**) When using basic control strategy

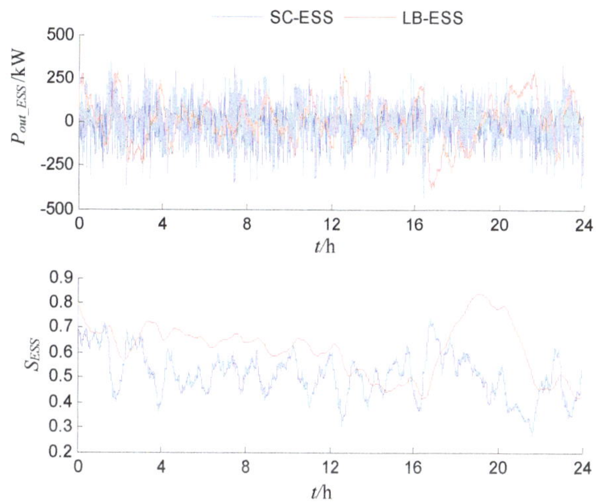

(**b**) When using coordinated control strategy

Figure 8. Output power and SOC of two ESSs using two strategies when $T_f = 45$.

The main control parameters of HESS using the two strategies are T_{f0} and ΔS_{co_SC}. The value of ΔS_{co_SC} is set as 0.35 the median of the margin indictor and remains unchanged, because it exists only in the coordinated control strategy. The changes of T_{f0} may result in changes of HESS operation under both control strategies. It is necessary to exclude the impact of comparative analysis results of different control strategies by the filter time constant. So the HESS operation curves using different control strategies are given as Figures 6–8 when T_{f0} is 15, 30 and 45, respectively.

The comparative analysis of Figures 6–8 shows that, although there are some differences in the details due to the variation in filter time constant, the variation trends of output power and SOC of LB-ESS and SC-ESS are the same. This conclusion is applicable to both when the basic control strategy is used and the coordinated control strategy is used. No matter which control strategy is used and how much the filter time constant is, the high-frequency parts of power fluctuations are borne by SC-ESS, and the LB-ESS satisfies the remaining non-high-frequency parts. The advantages of different types of ESSs are utilized reasonably.

At the same time, the SOC variation trends of SC-ESS are mainly related to control strategy. For all three cases where filter time constant changes, the SOC of SC-ESS when using the basic control strategy is close to its limit for a long time in the absence of an orderly control. This will result in a significant decrease in their ability to participate in the system power regulation. By contrast, state adjustment of SC-ESS using the coordinated control strategy keeps its SOC in a reasonable range all the time. Through the cooperation between two ESSs, a certain ability to regulate power is always maintained. Therefore, compared with the basic control strategy, the coordinated control strategy has a better performance for HESS operations.

6.3. Comparative Analysis Based on Different Cost and Life Considerations

After determining the HESS control strategy, the comparative analysis of the proposed optimization model and the existing optimization models can be undertaken. With regard to the four different cost and life considerations, the proposed optimization models are solved by a Particle Swarm Optimization (PSO) algorithm. The optimal results are given as Table 3, and the total initial investment cost and total loss equivalent cost of HESS based the corresponding optimal results are given in Table 4.

Table 3. Optimal results of four different considerations in cost and life.

Optimal results	Optimized parameters						Optimized objectives					
	P_{r_LB} (kW)	E_{r_LB} (kWh)	P_{r_SC} (kW)	E_{r_SC} (kWh)	T_{f0} (s)	ΔS_{co_SC}	$C_{initial\_HESSarr}$ ($)	$C_{initial\_HESS}$ ($)	C_{loss_LBarr} ($)	C_{loss_HESS} ($)	C_{pen_ESS} ($)	C_{opt_HESS} ($)
opt1	500	779.4	300	1.73	15	0.52	78.3×10^4	——	——	——	0	78.3×10^4
opt2	500	779.4	300	1.73	15	0.52	——	93.0×10^4	——	——	0	93.0×10^4
opt3	500	749.6	300	5.92	43	0.68	——	——	406.3	——	0	406.3
opt4	500	756.8	300	3.17	24	0.63	——	——	——	467.9	0	467.9

Table 4. Comparison of initial investment cost and loss equivalent cost based on optimal results.

Optimal results	Initial investment cost					Loss equivalent cost				
	C_{array_LB} (10^4·$)	C_{PCS_LB} (10^4·$)	C_{array_SC} (10^4·$)	C_{PCS_SC} (10^4·$)	$C_{initial\_HESS}$ (10^4·$)	C_{loss_LBarr} ($)	C_{loss_LB} ($)	C_{loss_SCarr} ($)	C_{loss_SC} ($)	C_{loss_HESS} ($)
opt1	51.1	8.2	27.2	6.5	93.0	443.2	465.7	25.2	27.0	492.7
opt2	51.1	8.2	27.2	6.5	93.0	443.2	465.7	25.2	27.0	492.7
opt3	49.2	8.2	93.2	6.5	157.0	406.3	428.7	44.8	46.6	475.3
opt4	49.6	8.2	49.9	6.5	114.2	415.9	438.3	27.8	29.6	467.9

The analysis combined with the optimization results in Table 3 and Table 4 shows that:

1. The initial investment cost of HESS is optimized in both Objective 1 and Objective 2, and the difference is whether to consider the cost of matched PCS. The optimized results are the same for both objectives, because PCSs in practical engineering applications are usually standard equipment packages. The rated output power of PCS has only a few optional specification values. This characteristic of non-continuous changes makes the initial investment cost of matched PCS relatively fixed. However, if this part of the cost has not counted in the project budget, a poor decision will be made. For example, an additional expense of 0.147 million

dollars will be produced in this case study. Therefore, to minimize the amount of capital in initial investment of HESS, optimization using Objective 2 is more reasonable.

2. The data comparing optimization when using Objective 2, Objective 3 and Objective 4 is analyzed to obtain the differences between minimizing the initial investment cost and minimizing the loss equivalent cost. The initial investment cost of whole HESS when using Objective 2 is 0.930 million dollars, which is 0.640 million dollars lower than using Objective 3 with a decline of 40.75%, and 0.212 million dollars lower than using Objective 4 with a decline of 18.55%. At the same time, the total loss equivalent cost of HESS when using Objective 2 is 492.7 dollars, which is 17.4 dollars more than using Objective 3 with an increase of 3.67%, and 24.8 dollars more than using Objective 4 with an increase of 5.29%. Compared with the cost when optimizing the loss equivalent cost, although the initial investment cost of whole HESS when optimizing the initial investment cost is lower, the total loss equivalent cost of HESS per unit time is significantly higher. As a result, from the point of life cycle cost, the overall performance of HESS is better when optimizing the loss equivalent cost.
3. The life loss and cost conversion of storage arrays in LB-ESS are considered in Objective 3, and the life loss and cost conversion of whole HESS are considered in Objective 4. The life of storage arrays in LB-ESS is much less than the life of other components in HESS, so it is the main factor in HESS performance. As shown in Table 4, the minimum total loss equivalent cost of HESS is 467.9 dollars, *i.e.*, the result when using Objective 4. When using Objective 3, the total loss equivalent cost of HESS is 475.3 dollars, which is only 7.3 dollars more than the optimal value with a difference of 1.57%. That is to say that, an approximately optimal value is obtained when using Objective 3. Meanwhile, the loss equivalent cost of storage arrays in LB-ESS is lowest when using Objective 3, which is 9.6 dollars lower than when using Objective 4 with a decline of 2.32%. The initial investment cost of whole HESS when using Objective 3 is 0.428 million dollars more than when using Objective 4 with an increase of 37.48%. Hence, the economic cost when using Objective 3 is difficult to accept, although the loss equivalent cost of storage arrays in LB-ESS can be lowest and an approximate minimum of total loss equivalent cost of HESS can be obtained.

In summary, taking the total loss equivalent cost of HESS as optimization objective is best. The technical and economic characteristics of HESS will be optimal only when the life quantification and cost conversion of all storage arrays and matched PCSs are taken into account. When considering storage arrays only, there will be a poor decision of initial investment in the project budget and an approximate optimal result of HESS overall performance. Although the optimization of HESS overall technical and economic characteristics is very important, the optimization of HESS initial investment is also important especially when considering capital as a decision-making factor in the project.

6.4. Comparative Analysis Based on Different ESS Application Schemes

Through the analysis for the same scheme of ESS application in Section 6.3, the overall optimization can be achieved only when the technical and economic impact of all devices in HESS are considered. In the practical engineering, there are some different ESS application schemes. The comparative analysis based on the different application schemes of ESSs are discussed in this section.

For the three schemes of SESS used alone, SC-ESS newly increased and HESS used directly, the proposed optimization models are solved by the PSO algorithm, and the optimal results are given as follows.

Table 5. Optimization results of ESSs in three schemes.

Optimal results	Optimized parameters						Loss equivalent cost			Initial investment cost		
	P_{r_LB} (kW)	E_{r_LB} (kWh)	P_{r_SC} (kW)	E_{r_SC} (kWh)	T_{f0} (s)	ΔS_{co_SC}	C_{loss_LB} ($)	C_{loss_SC} ($)	C_{loss_HESS} ($)	$C_{initial\_LB}$ (10^4·$)	$C_{initial\_SC}$ (10^4·$)	$C_{initial\_HESS}$ (10^4·$)
SESS used only	500	781.5	——	——	——	——	577.7	——	577.7	59.4	——	59.4
SC-ESS increased newly	——	——	300	3.58	28	0.54	456.7	30.6	487.4	——	62.8	122.3
HESS used directly	500	756.8	300	3.17	24	0.63	438.3	29.6	467.9	57.8	56.4	114.2

The analysis based on the sheet data above shows that:

1. As shown in the optimal results of the three schemes, the loss equivalent cost of LB-ESS is 577.7, 456.7 and 438.3 dollars, respectively. Compared with the loss equivalent cost of LB-ESS when SESS is used alone, this cost is decreased by 20.9% when SC-ESS is newly increased, and decreased by 24.1% when HESS is used directly. The data analysis above shows that, compared with SESS used alone, the utilization of SC-ESS can effectively reduce the life loss equivalent cost of LB-ESS.
2. Compared with SESS used alone, an additional initial investment cost of 0.628 million dollars is needed after SC-ESS is newly increased. However, the total loss equivalent cost of HESS in the simulation time is reduced from 577.7 dollars to 487.4 dollars, which is decreased by 15.6% based on the cost of SESS. This indicates that the initial investment cost is increased because of a newly increased SC-ESS, but the total loss of ESSs demanded by a standalone microgrid is effectively reduced in unit time, so the technical and economic characteristics are much better than SESS.
3. After overall optimization of HESS used directly, the total loss equivalent cost has been decreased to 467.9 dollars, which represents a significant decreased amplitude of 19.0% compared with SESS used alone. In addition, compared with optimization of SC-ESS increased newly, the overall optimization of HESS used results in a decrease of the total initial investment cost from 1.223 to 1.142 million dollars, *i.e.*, a savings 80.7 thousand dollars in the initial investment cost. Therefore, the technical and economic characteristics have a further enhance after overall optimization of HESS used directly.

Meanwhile, combined with the optimal results in Table 3, output power curves and SOC curves of various ESSs in three schemes are shown as follows.

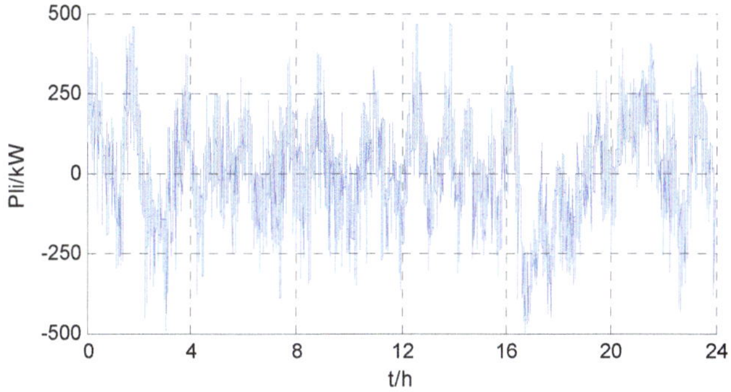

(**a**) SESS used only

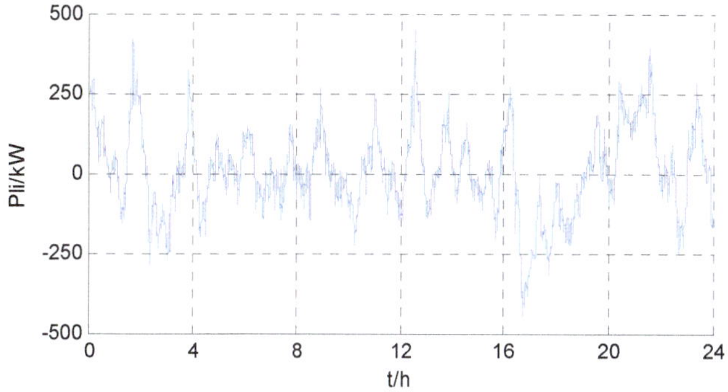

(**b**) SC-ESS increased newly

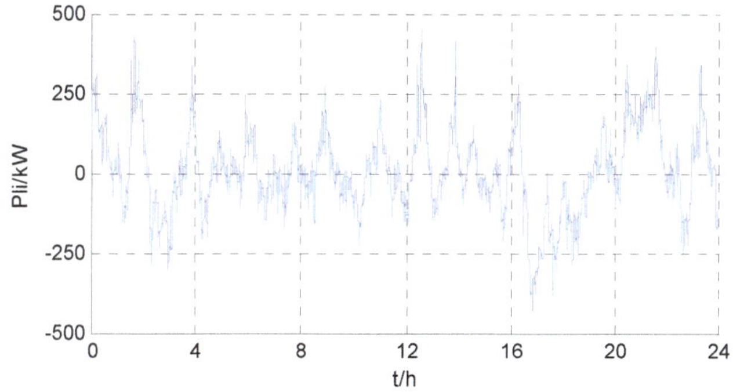

(**c**) HESS used directly

Figure 9. The output power curves of LB-ESS in the three schemes.

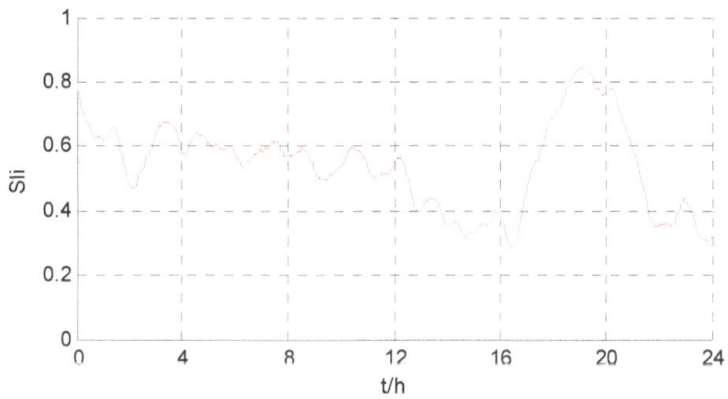

(**a**) SESS used only

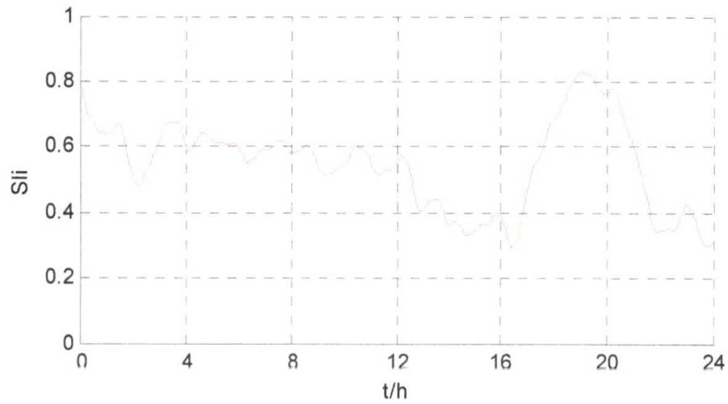

(**b**) SC-ESS increased newly

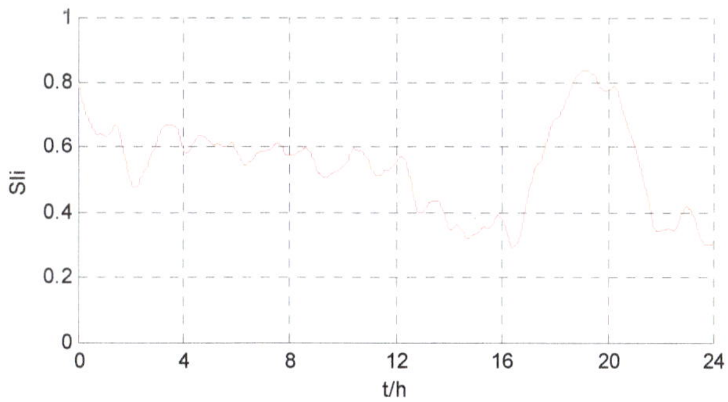

(**c**) HESS used directly

Figure 10. The SOC curves of LB-ESS in the three schemes.

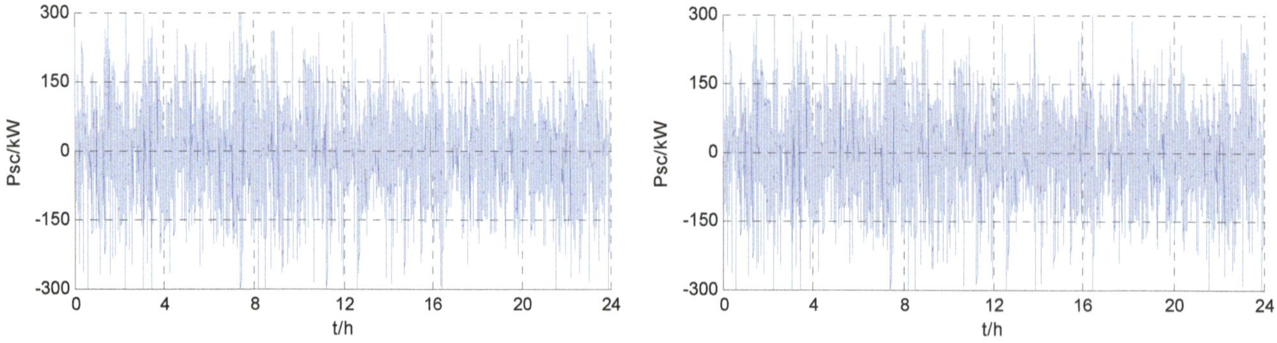

(**a**) SC-ESS increased newly (**b**) HESS used directly

Figure 11. The output power curves of SC-ESS in the latter two schemes.

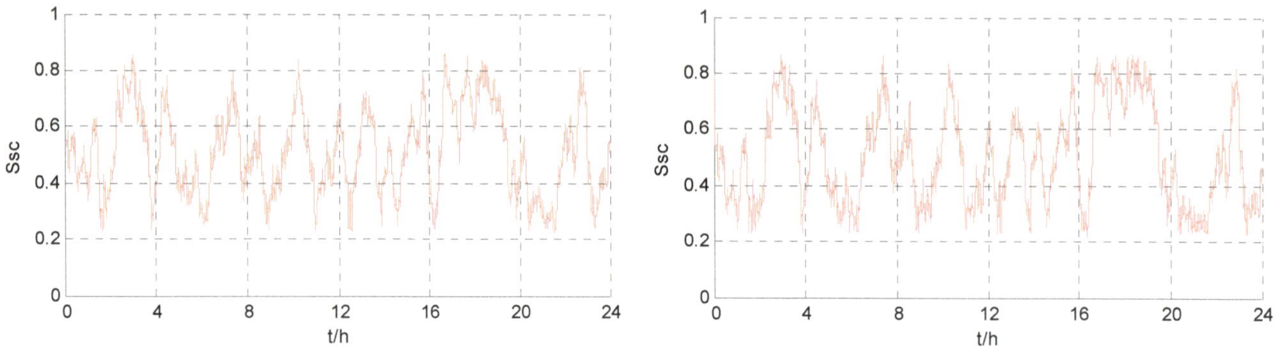

(**a**) SC-ESS increased newly (**b**) HESS used directly

Figure 12. The SOC curves of SC-ESS in the latter two schemes.

As shown in Figures 9–12, the output power and SOC of each ESS in the three schemes are all close to the permitted upper and lower limits of normal operation. By contrast, the output power and SOC of ESSs given in Figures 6–8 above can't do it, especially the SOC curves of SC-ESS. In other words, the required performances of this application mode have been achieved by minimum power rating and minimum storage capacity of the ESSs after optimization of the three schemes. The advantages of technical complementary characteristics between energy-type energy storage and power-type energy storage are given full play in the latter two schemes, and optimal configuration and operation of the whole ESSs is achieved. Therefore, the scheme parameter combinations of this day as shown in Table 3 can be accepted as optimal in a more intuitive expression.

7. Conclusions

A new index named ESS efficiency rate is proposed in this paper considering the characteristics of a small-scale standalone microgrid for RAPS. The index reflects the common satisfaction of load demands and maximized utilization of renewable energy. Compared with LOLP or LPSP, ESS efficiency rate is more suitable as the constraint condition of ESS optimization analysis model in the microgrid. At the same time, the second-level data of field measurement is used in optimization instead of hour-level data making the analysis results more credible. Because the real time curves of output power and SOC of SC-ESS can be obtained, although the duration of SC-ESS continuous discharge at rate power is very short.

With regard to the operating control of HESS, the coordinated control strategy based on state cooperation is based on the basic control strategy considering power allocation only. The increased measures in coordinated control strategy are status adjustment of SC-ESS, coordination of overcharge and over-discharge protection, and coordination of maximum power limit protection. A numerical example shows that, for different filter time constants, the power adjustment abilities of SC-ESS when using coordinated control strategy are all significantly better than when using the basic control strategy.

Three different optimization objective functions are established in the optimization of HESS. The impact of incomplete consideration of cost and life on overall performance of HESS is comparatively analyzed. A numerical example shows that the cost of matched PCSs must be taken in account in the optimization. Although the initial investment cost of HESS can be reduced when taking one-time investment cost as optimization objective, the usage cost per unit time is unacceptable and the optimized results when taking the loss equivalent cost of LB storage arrays as optimization objective are approximately optimal. Only when considering the life quantification and cost equivalent of the whole HESS, the optimized results can get the best technical and economic characteristics of HESS.

The different application schemes of ESS in practical projects are contrastively analyzed. The analysis results show that, although the initial investment cost of SC-ESS increased newly based on SESS increases, the total loss equivalent cost of ESSs in unit time decreases significantly. If HESS is used directly, the total loss equivalent cost of ESSs can be further reduced. The ESS characteristics of practical projects are considered, such as self-discharge of ESSs, PCSs to be standard equipment packages, and so on. The analysis results have guiding significance and provide reference values for ESS applications in existing and new projects.

Acknowledgments

This work has been supported by National High Technology Research and Development Program of China (863 Program) (2014AA052005), and State Grid Corporation of China.

Author Contributions

The manuscript is prepared by Fengbing Li and revised by Kaigui Xie and Jiangping Yang. All authors have read and approved the final manuscript.

Nomenclature

P_{DG}	Output power of distributed generations
P_{LD}	Power demand of loads
P_{net_LD}	Power demand of net loads
P_{out_ESS}	Output power of ESS
R_{ESS}	Effective rate of ESS
ΔT_{com}	Time step of computation
N_T	Number of discrete data points in time T with in interval ΔT_{com}
$P_{ref\_ESS(n)}$	Output power reference of ESS at the nth time interval
$P_{out\_ESS(n)}$	Actual output power of ESS at the nth time interval
R_{set_ESS}	Setting constraint of ESS effective rate

C_{pen_ESS}	Penalty cost without satisfying R_{set_ESS}
F	Fixed value much bigger than other cost
$P_{clmt\_ESS(n)}$	Charging power limit of ESS within the nth time interval
$P_{dlmt\_ESS(n)}$	Discharging power limit of ESS within the nth time interval
P_{cmax_ESS}	Maximum charging power of ESS
P_{dmax_ESS}	Maximum discharging power of ESS
E_{r_ESS}	Rated storage capacity of ESS
S_{max_ESS}	Maximum SOC of ESS
S_{min_ESS}	Minimum SOC of ESS
$S_{ESS(n)}$	SOC of ESS at the nth time interval moment
$S_{ESS(n-1)}$	SOC of ESS at the n-1th time interval moment
σ_{ESS}	Self-discharge rate of ESS
η_{c_ESS}	Charging efficiency of ESS
η_{d_ESS}	Discharging efficiency of ESS
P_{HESS}	Power command of HESS
P_{out_LB}	Output power of LB-ESS
P_{out_SC}	Output power of SC-ESS
T_f	Filter time constant
T_{f0}	Initial value of T_f in power allocation
ΔT_f	Adjustment step of T_f
T_{fmin}	Minimum of T_f Adjustment
T_{fmax}	Maximum of T_f Adjustment
S_{SC}	SOC of SC-ESS
S_{min_SC}	Minimum SOC of SC-ESS
S_{max_SC}	Maximum SOC of SC-ESS
S_{LBd_SC}	SOC control objective of SC-ESS when LB-ESS discharges
S_{LBc_SC}	SOC control objective of SC-ESS when LB-ESS charges
ΔS_{co_SC}	SOC coordinated response margin of SC-ESS
E_{r_LB}	Rated storage capacity of LB-ESS
C_{unit_LB}	Unit storage capacity cost of LB-ESS
E_{r_SC}	Rated storage capacity of SC-ESS
C_{unit_SC}	Unit storage capacity cost of SC-ESS
$C_{initial\_HESSarr}$	Total initial investment cost of storage arrays in HESS
C_{PCS_LB}	PCS cost of LB-ESS
$C_{initial\_LB}$	Initial investment cost of LB-ESS
C_{PCS_SC}	PCS cost of SC-ESS
$C_{initial\_SC}$	Initial investment cost of SC-ESS
$C_{initial\_HESS}$	Initial investment cost of whole HESS
L_{array_LB}	Lifetime loss coefficient of LB-ESS storage arrays
M	Number of time interval in the life quantification of LB-ESS storage arrays
$\Delta D_{LB(m)}$	Degenerate increment of LB-ESS storage capacity within the mth time interval
D_{lmt_LB}	Degenerate limit of LB-ESS storage capacity
D_1, D_2	Intermediate variable of ΔD_{LB}

τ	Duration of the mth time interval
S_{avg_LB}	SOC average value of LB-ESS
S_{dev_LB}	SOC normalized deviation of LB-ESS
N_{LB}	Equivalent throughput cycle of LB-ESS
T_{ref}	Reference temperature in degrees centigrade
T_{LB}	Operation temperature of LB-ESS storage arrays in degrees centigrade
T_{a_ref}	Absolute temperature of T_{ref}
T_{a_LB}	Absolute temperature of T_{LB}
τ_{life_LB}	Calendar lifetime estimate of LB-ESS storage arrays end of 80% initial capacity
K_T, K_{co}, K_{ex}, K_{SOC}	Empirical constant of specific LB-ESS storage arrays
C_{loss_LBarr}	Loss equivalent cost of LB-ESS storage arrays
L_{array_SC}	Lifetime loss coefficient of SC-ESS storage arrays
N_{cycle_SC}	Charge and discharge cycles of SC-ESS storage arrays
N_{total_SC}	Total charge and discharge cycles of SC-ESS storage arrays
C_{loss_SCarr}	Loss equivalent cost of SC-ESS storage arrays
L_{loss_PCS}	Lifetime loss coefficient of PCS
T	Running time of PCS
T_{life_PCS}	Service lifetime of PCS
C_{loss_LB}	Loss equivalent cost of LB-ESS
C_{loss_SC}	Loss equivalent cost of SC-ESS
C_{loss_HESS}	Loss equivalent cost of HESS
C_{opt1_HESS}	Considered cost of HESS under optimization objective 1
C_{opt2_HESS}	Considered cost of HESS under optimization objective 2
C_{opt3_HESS}	Considered cost of HESS under optimization objective 3
C_{opt4_HESS}	Considered cost of HESS under optimization objective 4
P_{r_LB}	Rated output power of LB-ESS
P_{r_SC}	Rated output power of SC-ESS
P_{cmax_LB}	Maximum charging power of LB-ESS
P_{dmax_LB}	Maximum discharging power of LB-ESS
S_{LB}	SOC of LB-ESS
S_{min_LB}	Minimum SOC of LB-ESS
S_{max_LB}	Maximum SOC of LB-ESS
P_{cmax_SC}	Maximum charging power of SC-ESS
P_{dmax_SC}	Maximum discharging power of SC-ESS
C_{opt_SESS}	Loss equivalent cost of SESS used only
C_{opt_SAESS}	Loss equivalent cost of SC-ESS increased newly
C_{opt_HESS}	Loss equivalent cost of HESS used directly
C_{total_SESS}	Total initial investment cost of SESS used only
C_{total_SAESS}	Total initial investment cost of SC-ESS increased newly
C_{total_HESS}	Total initial investment cost of HESS used directly
P_{r_PCS}	Rated power of PCS

Conflicts of Interest

The authors declare no conflict of interest.

References

1. Mendis, N.; Muttaqi, K.M.; Perera, S. Management of low- and high-frequency power components in demand-generation fluctuations of a DFIG-based wind-dominated RAPS system using hybrid energy storage. *IEEE Trans. Ind. Appl.* **2014**, *50*, 2258–2268.
2. Patrick, T.; Moseley, P.T. Energy storage in remote area power supply (RAPS) systems. *J. Power Sources* **2006**, *155*, 83–87.
3. Dragicevic, T.; Guerrero, J.M.; Vasquez, J.C. A distributed control strategy for coordination of an autonomous LVDC microgrid based on power-line signalling. *IEEE Trans. Ind. Electron.* **2014**, *61*, 3313–3326.
4. Wu, D.; Tang, F.; Dragicevic, T.; Vasquez, J.C.; Guerrero, J.M. Autonomous active power control for islanded AC microgrids with photovoltaic generation and energy storage system. *IEEE Trans. Energy Convers.* **2014**, *29*, 882–892.
5. Dragicevic, T.; Pandzic, H.; Skrlec, D.; Kuzle, I.; Guerrero, J.M.; Kirschen, D.S. Capacity optimization of renewable energy sources and battery storage in an autonomous telecommunication facility. *IEEE Trans. Sustain. Energy* **2014**, *5*, 1367–1378.
6. Kim, J.Y.; Kim, H.M.; Kim, S.K.; Jeon, J.H.; Choi, H.K. Designing an energy storage system fuzzy PID controller for microgrid islanded operation. *Energies* **2011**, *4*, 1443–1460.
7. Li, J.; Wei, W.; Xiang, J. A simple sizing algorithm for stand-alone PV/Wind/Battery hybrid microgrids. *Energies* **2012**, *5*, 5307–5323.
8. Marzband, M.; Sumper, A.; Ruiz-Alvarez, A.; Dominguez-Garcia, J.L.; Tomoiaga, B. Experimental evaluation of a real time energy management system for stand-alone microgrids in day-ahead markets. *Appl. Energy* **2013**, *106*, 365–376.
9. Marzband, M.; Sumper, A.; Dominguez-Garcia, J.L.; Gumara-Ferret, R. Experimental Validation of a Real Time Energy Management System for Microgrids in Islanded Mode Using a Local Day-Ahead Electricity Market and MINLP. *Energy Convers. Manag.* **2013**, *76*, 314–322.
10. Marzband, M.; Ghadimi, M.; Sumper, A.; Dominguez-Garcia, J.L. Experimental validation of a real-time energy management system using multi-period gravitational search algorithm for microgrids in islanded mode. *Appl. Energy* **2014**, *128*, 164–174.
11. Gao, L.; Dougal, R.A.; Liu, S. Power enhancement of an actively controlled battery/ultracapacitor hybrid. *IEEE Trans. Power Electron.* **2005**, *20*, 236–243.
12. Dougal, R.A.; Liu, S.; White, R.E. Power and life extension of battery ultracapacitor hybrids. *IEEE Trans. Components Packag. Technol.* **2002**, *25*, 120–131.
13. Li, W.; Joos, G.; Belanger, J. Real-time simulation of a wind turbine generator coupled with a battery supercapacitor energy storage system. *IEEE Trans. Ind. Electron.* **2010**, *57*, 1137–1145.
14. Thounthong, P.; Rael, S.; Davat, B. Control strategy of fuel cell and supercapacitors association for a distributed generation system. *IEEE Trans. Ind. Electron.* **2007**, *54*, 3225–3233.

15. Gee, A.M.; Roninson, F.V.P.; Dunn, R.W. Analysis of battery lifetime extension in a small-scale wind-energy system using supercapacitors. *IEEE Trans. Energy Convers.* **2013**, *28*, 24–33.
16. Mendis, N.; Muttaqi, K.M.; Perera, S. Management of battery-supercapacitor hybrid energy storage and synchronous condenser for isolated operation of PMSM based variable-speed wind turbine generating systems. *IEEE Trans. Smart Grid* **2014**, *5*, 944–953.
17. Xue, F.; Gooi, H.B.; Chen, S. Hybrid energy storage with multimode fuzzy power allocator for PV systems. *IEEE Trans. Sustain. Energy* **2014**, *5*, 389–397.
18. Singo, T.A.; Martinez, A.; Saadate, S. Design and implementation of a photovoltaic system using hybrid energy storage. In Proceedings of the IEEE 11th International Conference on Optimization of Electrical and Electronic Equipment, Brasov, Romania, 22–24 May 2008.
19. Li, C.; Yang, X.; Zhang, M.; Wang, H.; Ye, J.; Chen, J. Optimal configuration scheme for hybrid energy storage system of super-capacitors and batteries based on cost analysis. *Autom. Electr. Power Syst.* **2013**, *37*, 1–5.
20. Tian, P.; Xiao, X.; Ding, R.; Huang, X. A capacity configuring method of composite energy storage system in autonomous multi-microgrid. *Autom. Electr. Power Syst.* **2013**, *37*, 168–173.
21. Zhou, T.; Sun, W. Optimization of battery-supercapacitor hybrid energy storage station in wind/solar generation system. *IEEE Trans. Sustain. Energy* **2014**, *5*, 408–415.
22. Glavin, M.E.; Chan, P.K.W.; Hurley, W.G. Optimization of autonomous hybrid energy storage system for photovoltaic application. In Proceedings of the Energy Conversion Congress and Exposition, San Jose, CA, USA, 20–24 September 2009.
23. Etinski, M.; Schulke, A. Optimal hybrid energy storage for wind energy integration. In Proceedings of the IEEE International Conference on Industrial Technology (ICIT), Cape Town, South Africa, 25–28 February 2013.
24. Yu, W.; Liu, D.; Huang, Y. Operation optimization based on the power supply and storage capacity of an active distribution network. *Energies* **2013**, *6*, 6423–6438.
25. Wee, K.W.; Choi, S.S.; Vilathgamuwa, D.M. Design of a least-cost battery-supercapacitor energy storage system for realizing dispatchable wind power. *IEEE Tans. Sustain. Energy* **2013**, *4*, 786–796.
26. Sen, C.; Usama, Y.; Carciumaru, T.; Lu, X.; Kar, N.C. Design of a novel wavelet based transient detection unit for in-vehicle fault determination and hybrid energy storage utilization. *IEEE Trans. Smart Grid* **2012**, *3*, 422–433.
27. Jiang, Q.; Hong, H. Wavelet-based capacity configuration and coordinated control of hybrid energy storage system for smoothing out wind power fluctuations. *IEEE Trans. Power Syst.* **2013**, *28*, 1363–1372.
28. Kazempour, S.J.; Moghaddam, M.P.; Haghifam, M.R.; Yousefi, G.R. Electric energy storage systems in a market-based economy: Comparison of emerging and traditional technologies. *Renew. Energy* **2009**, *34*, 2630–2639.
29. Nguyen, M.Y.; Nguyen, D.H.; Yoon, Y.T. A new battery energy storage charging/discharging scheme for wind power producers in real-time markets. *Energies* **2012**, *5*, 5439–5452.
30. Yang, W.; Wang, X.; Li, X.; Liu, Z. An active power sharing method among distributed energy sources in an islanded series micro-grid. *Energies* **2014**, *7*, 7878–7892.

31. Li, F.; Xie, K.; Zhang, X.; Wang, K.; Zhou, D.; Zhao, B. Design of control strategy for hybrid energy storage system based on charge-discharge state of lithium battery. *Autom. Electr. Power Syst.* **2013**, *37*, 70–75.
32. Li, F.; Xie, K.; Zhang, X.; Zhao, B.; Chen, J. Optimization of coordinated control parameters for hybrid energy storage system based on life quantization. *Autom. Electr. Power Syst.* **2014**, *38*, 1–5.
33. Millner, A. Modeling lithium ion battery degradation in electric vehicles. In Proceedings of the IEEE Conference on Innovative Technologies for an Efficient and Reliable Electricity Supply (CITRES), Waltham, MA, USA, 27–29 September 2010.
34. Xie, Q.; Lin, X.; Wang, Y.; Pedram, M.; Shin, D.; Chang, N. State of health aware charge management in hybrid electrical energy storage systems. In Proceedings of the Design, Automation & Test in Europe Conference & Exhibition (DATE), Dresden, Germany, 12–16 March 2012.
35. Xie, Q.; Yue, S.; Pedram, M.; Shin, D.; Chang, N. Adaptive thermal management for portable system batteries by forced convection cooling. In Proceedings of the Design, Automation & Test in Europe Conference & Exhibition (DATE), 18–22 March 2013.
36. Lam, L.; Bauer, P. Practical capacity fading model for Li-ion battery cells in electric vehicles. *IEEE Trans. Power Electron.* **2013**, *28*, 5910–5918.
37. Vergara, C.R. Parametric interface for battery energy storage systems providing ancillary services. In Proceedings of the 3rd IEEE PES International Conference and Exhibition on Innovative Smart Grid Technologies (ISGT Europe), 14–17 October 2012.
38. Traube, J.; Lu, F.; Maksimovic, D.; Mossoba, J.; Kromer, M.; Faill, P.; Katz, S.; Borowy, B.; Nichols, S.; Casey, L. Mitigation of solar irradiance intermittency in photovoltaic power systems with integrated electric-vehicle charging functionality. *IEEE Trans. Power Electron.* **2013**, *28*, 3058–3067.
39. Grahn, P.; Munkhammar, J.; Widen, J.; Alvehag, K.; Soder, L. PHEV home-charging model based on residential activity patterns. *IEEE Trans. Power Syst.* **2013**, *28*, 2507–2515.
40. Xie, Q.; Wang, Y.; Kim, Y.; Pedram, M.; Chang, N. Charge allocation in hybrid electrical energy storage systems. *IEEE Trans. Comput. Aided Des. Integr. Circuits Syst.* **2013**, *32*, 1003–1016.
41. Onar, O.C.; Khaligh, A. A novel integrated magnetic structure based DC/DC converter for hybrid battery/ultracapacitor energy storage systems. *IEEE Trans. Smart Grid* **2012**, *3*, 296–307.

Designing Structure-Dependent MPC-Based AGC Schemes Considering Network Topology

Young-Sik Jang, JoonHyung Park and Yong Tae Yoon *

Seoul National University, Gwanak-ro 599, Gwanak-gu, Seoul 151-744, Korea;
E-Mails: helou3@snu.ac.kr (Y.-S.J.); cdove@snu.ac.kr (J.H.P.)

* Author to whom correspondence should be addressed; E-Mail: ytyoon@ee.snu.ac.kr

Academic Editor: Josep M. Guerrero

Abstract: This paper presents the important features of structure-dependent model predictive control (MPC)-based approaches for automatic generation control (AGC) considering network topology. Since power systems have various generators under different topologies, it is necessary to reflect the characteristics of generators in power networks and the control system structures in order to improve the dynamic performance of AGC. Specifically, considering control system structures is very important because not only can the topological problems be reduced, but also a computing system for AGC in a bulk-power system can be realized. Based on these considerations, we propose new schemes in the proposed controller for minimizing inadvertent line flows and computational burden, which strengthen the advantages of MPC-based approach for AGC. Analysis and simulation results in the IEEE 39-bus model system show different dynamic behaviors among structure-dependent control schemes and possible improvements in computational burden via the proposed control scheme while system operators in each balancing area consider physical load reference ramp constraints among generators.

Keywords: automatic generation control; line flow control; model predictive control; network topology

---

1. Introduction

Bulk power systems are usually composed of interconnected balancing areas (BAs) to minimize the operation cost from arbitrage trading and to increase the system reliability by importing electric power. In this case, each BA has its own automatic generation control (AGC) scheme in an energy management system (EMS), which is responsible for maintaining a nominal frequency and stabilizing inadvertent tie-line flows to scheduled flows. AGC signals are traditionally developed to minimize area control errors (ACEs), which represent the discrepancy between generation supply and total megawatt (MW) obligation in each BA [1]. Given that the benefits of interconnection among BAs depend highly on the dynamic performance in maintaining frequency and tie-line flows, designing an AGC process to minimize ACEs has attracted continued interest while power systems are growing in size.

Recently, the model predictive control (MPC)-based designing process in AGC was studied extensively because of its two major advantages [2]. First, this approach provides satisfactory control performance under dynamic constraints, such as load reference ramp constraints. Second, it allows systematic design of multi-input-multi-output (MIMO) systems. The conventional proportional integral derivative-based approach in AGC presents intractable subjects for admissible load reference control in real-time operation. In contrast, the MPC-based approach does not suffer from these problems because the actual control objectives and operating constraints can be represented explicitly in a single multi-horizon optimization.

Several MPC-based studies and formulations are available in the literature, and these approaches can be classified into three main categories, namely, centralized, distributed, and decentralized control structures. Mohamed [3] investigated robust load frequency control (LFC) against parameter uncertainties and load changes but only dealt with LFC control application in a single-area power system. Kong [4] proposed state contractive constraint-based MPC algorithms to guarantee the stability of the control scheme. Yousef [5] utilized the MPC technique to investigate the design of the LFC system and thus improve power system dynamic performance over a wide range of operating conditions. However, centralized MPC-based AGC [3–5] among independent bulk BAs is stated to be an impractical approach [6,7] because each BA should have its own AGC scheme, and optimizing AGC signals in a large-scale power system model consumes considerable computational time to obtain an admissible solution. Camponogara [6] introduced the coordination of optimization computations using iterative exchange of information among distributed BAs. Venkat [7] proposed a new object function that measures the system-wide effect of local control actions to build a reliable MPC-based controller. Nong [8] provided a distributed MPC with fuzzy modeling to deal with the valve limit on the governor. Ma [9] provided an MPC-based AGC that considers generation rate constraint and load reference ramp constraint. However, these distributed studies [6–9] assumed that the dynamic model in each area is modeled as a single-input-single-output (SISO) model, even when each BA has various local frequencies and multiple generators as the MIMO system [1]. Control signals in each BA should also be periodically updated because EMS is a discretized controller [10]. A few of the decentralized approaches are introduced or briefly mentioned in literature [7,11]. However, these control schemes are necessary to build a highly reliable and real-time communication system because each entity should communicate with other entities in real time to minimize the system-wide effects of local control actions and maximize its profits.

In this study, we develop a constrained MPC-based AGC in centralized and distributed control structures. In centralized MPC-based AGC, we develop a novel scheme to maintain designated line flows to strengthen its unique advantages. As loop flows and arrangements for parallel path compensation become increasingly important, this developed control scheme supports system operators by reducing the increase in power losses or the overloading of network elements. In distributed MPC-based AGC, the pseudo-frequency signal computed from ACEs is proposed. On the basis of this proposed signal, we develop individual and unsynchronized controllers to consider the system-wide effect of local control actions in each BA. In both control structures, the proposed MPC-based controllers provide the discretized AGC signals derived by minimizing a constrained quadratic programming (QP) based on the generator dynamic models in each BA. Moreover, we also consider computation burden because the real-time MPC-based controller in a bulk system can suffer from the computation capacity limits caused by dimensional complexity. According to the number of participating generators in AGC, computation time in the MPC-based approach is compared, and we develop a bulk-area partitioning scheme in accordance with the results to reduce computation time and ensure dynamic performance.

The remainder of this paper is divided into five sections: Section 2 introduces the coupling model of electric generator and power network. Section 3 develops MPC schemes in consideration of the generator dynamic model in power network. Section 4 describes a case study of the proposed control model and discusses the MPC-based approach for implementation in AGC. Finally, Section 5 gives the conclusions.

2. Generator and Power Network Dynamics

Studies on modeling the generator-load dynamic relationship between incremental mismatch power and frequency deviations exist. This section introduces the formulation of a coupling model of generator and power network dynamics to develop the discretized frequency-state transition function for the MPC-based approach.

2.1. Generator Dynamic Model

We assume that an electric generator is represented by a continuous-time model, such as that given in Equation (1) [10,12]. This model represents the characteristics of a drooping governor and a discretized frequency control. The discrete-time-state feedback control w_G^{ref}, which is the load reference setting through AGC, is periodically updated for every discretized sampling time T_u:

$$\begin{bmatrix} \dot{\omega}_G \\ \dot{P}_m \\ \dot{\alpha} \end{bmatrix} = \begin{bmatrix} -\frac{D}{M} & \frac{1}{M} & 0 \\ 0 & -\frac{1}{T_a} & \frac{K_t}{T_a} \\ -\frac{1}{T_g} & 0 & -\frac{r}{T_g} \end{bmatrix} \begin{bmatrix} \omega_G \\ P_m \\ \alpha \end{bmatrix} + \begin{bmatrix} -\frac{1}{M} \\ 0 \\ 0 \end{bmatrix} P_G + \begin{bmatrix} 0 \\ 0 \\ \frac{1}{T_g} \end{bmatrix} w_G^{ref}[k_u] \tag{1}$$
$$= A_{pri} z + B w_G^{ref}[k_u] + C P_G$$

In Equation (1), ω_G is the frequency, P_m is the mechanical power, and α is the valve position in the electric generator. P_G is the real (electrical) power output supplied by the generator. However, P_G is no longer an independent input when generators are connected because P_G depends on the constraints imposed by the interconnections with other generators (P_G), the power demands (P_L) at load bus, and the injected powers (F_G or F_L), which are power demands at generator bus or tie-line power flows.

For the machine model in Equation (1), a steady-state solution for w_G can be found solely in terms of P_G, the network constraint, w_G^{ref}, and the load reference control. The steady-state solution is given in References [13,14] as follows:

$$\omega_G[k] = (1-\sigma D)w_G^{ref}[k] - \sigma P_G[k] \quad (2)$$

where the droop constant is $\sigma = r/(rD + K_t)$. An equivalent definition for the droop constant is:

$$\sigma = -\left.\frac{\partial \omega_G[k]}{\partial P_G[k]}\right|_{w_G^{ref}[k]=0} \quad (3)$$

2.2. Power Network Model

By linearizing power network equations under the decoupling assumption (*i.e.*, $\partial P^N/\partial V = 0$), we obtain:

$$P^N = \begin{bmatrix} P_G + F_G \\ -P_L + F_L \end{bmatrix} = \begin{bmatrix} J_{GG} & J_{GL} \\ J_{LG} & J_{LL} \end{bmatrix} \begin{bmatrix} \delta_G \\ \delta_L \end{bmatrix} \quad (4)$$

where $J_{ij} = \partial P_i^N/\partial \delta_j$, $i, j \in \{G, L\}$; G and L represent a set of all generators and loads in the area, respectively. From Equation (4), we obtain:

$$\begin{aligned} P_G &= (J_{GG} - J_{GL}J_{LL}^{-1}J_{LG})\delta_G + J_{GL}J_{LL}^{-1}P_L + J_{GL}J_{LL}^{-1}F_L - F_G \\ &= K_P\delta_G + D_P P_L - F \end{aligned} \quad (5)$$

$$\dot{P}_G = K_P\dot{\delta}_G + D_P\dot{P}_L - \dot{F} \quad (6)$$

The network model in Equation (5) shows that the power output of the system generators is affected by disturbances (P_L and F) and a generator angle (δ_G). In Equation (6), $\dot{\delta}_G$ is the generator frequency ω_G.

2.3. Generator and Network Coupling Model

From Equations (1) and (6), the extended-state space (or full-scale dynamic model) can be expressed as [13]:

$$\begin{bmatrix} \dot{\overline{z}} \\ \dot{\overline{P}}_G \end{bmatrix} = \begin{bmatrix} A_{pri}^{bd} & C^{bd} \\ K_p e^{bd} & 0 \end{bmatrix} \begin{bmatrix} \overline{z} \\ \overline{P}_G \end{bmatrix} + \begin{bmatrix} B^{bd} \\ 0 \end{bmatrix} \overline{w}_G^{ref}[k_u] + \begin{bmatrix} 0 & 0 \\ -I & D_p \end{bmatrix} \begin{bmatrix} \dot{F} \\ \dot{P}_L \end{bmatrix} \quad (7)$$

where $\overline{z}$ and $\overline{P}_G$ are the concatenation representations of machine state spaces and the electric power output of generators respectively, $\overline{w}_G^{ref}$ is a concatenation representation of load reference setting, $e = [100]$, the matrices with superscript ^{bd} show the block diagonal matrices with each matrix (e.g., A_{pri}^{bd} = diagonal [A_{pri}^1,.. A_{pri}^i,.. A_{pri}^n], where the number i indicates each generator).

The electric power output of generators P_G is a natural choice for state variables. This set of state can be interpreted in terms of the interaction of the generators with the transmission system.

From the given full-scale dynamic model in Equation (7), the discretized causal model with n-generator frequency sampling time T_f can be expressed as [13]:

$$\omega_G[k+1] = (I + K_p \Omega T_f)^{-1}\{\omega_G[k] + (I - \Omega\Phi)u[k] + \Omega(f[k] - D_P d[k])\} \\ = A_w \omega_G[k] + B_w u[k] + d_w[k] \tag{8}$$

where Ω = diagonal $[\sigma_1, \sigma_2, \ldots \sigma_n]$; Φ = diagonal $[D_1, D_2, \ldots D_n]$; $u[k] = w_G^{ref}[k+1] - w_G^{ref}[k]$, $f[k] = F[k+1] - F[k]$; and $d[k] = P_L[k+1] - P_L[k]$. ω_G in Equation (8) is imperfectly matched with the initial (transient) behavior of w_G in Equation (7), but it captures the slow frequency dynamics reasonably well [13].

2.4. Line Power Flow Model

On the basis of Equation (7), the generator frequency dynamics are affected by power flows. Therefore, line power flows should be controlled among areas or subsystems to stabilize frequency. The power flow on a single line from *i* bus to *j* bus is given in [10] as follows:

$$P_{ij} = \frac{E_i E_j}{X_{ij}} \sin(\delta_i - \delta_j) \tag{9}$$

By linearizing an initial operating point represented by $\delta_i = \delta_{i0}$ and $\delta_j = \delta_{j0}$, we obtain $\Delta P_{ij} = J_{ij}\Delta\delta_{ij}$, where $\Delta\delta_{ij} = \delta_i - \delta_j$, and $J_{ij} = (E_iE_j/X_{ij})\cos(\delta_{i0} - \delta_{j0})$.

The line power flows P_f can be described as:

$$P_f = \begin{bmatrix} J_{fG} & J_{fL} \end{bmatrix} \begin{bmatrix} \delta_G \\ \delta_L \end{bmatrix} \tag{10}$$

The phase angles on load buses can be expressed in terms of the phase angles on generator buses and the demands on load buses:

$$\delta_L = -J_{LL}^{-1}(J_{LG}\delta_G + P_L) \tag{11}$$

Substituting Equation (11) into Equation (10) eliminates the load bus phase angles. Based on this result, Equation (12) can be obtained as follows:

$$P_f = (J_{fG} - J_{fL}J_{LL}^{-1}J_{LG})\delta_G - J_{fL}J_{LL}^{-1}P_L \\ - K_f\delta_G + D_f P_L \tag{12}$$

3. MPC-Based AGC

On the basis of Equation (8), centralized and distributed MPC-based approaches are proposed in consideration of load reference ramp constraints. Given that the AGC signal is a discretized signal in the field, this research develops a continuous dynamic generator model controlled by discretized controllers.

3.1. Centralized MPC-Based AGC

From Equation (8), the discrete-time frequency model with constraints is described as:

$$\omega_{k+1} = A\omega_k + Bu_k \tag{13}$$

$$\underline{u} \le u_k \le \overline{u} \tag{14}$$

where $\omega \in \mathbb{R}^n$ is the frequency vector, and $u \in \mathbb{R}^n$ is the control vector. In this case, n is the total number of generators in the power network. Equation (14) provides the information of each generator regarding a different and limited load reference ramp rate of each capacity per minute (*i.e.*, combustion turbines can develop ramps of approximately 13% to 28% of their capacity per minute [15]).

From Equation (13), the state predictions for an N-step horizon are expressed as follows:

$$\begin{bmatrix} \omega_{k+1} \\ \omega_{k+2} \\ \vdots \\ \omega_{k+N} \end{bmatrix} = \begin{bmatrix} A \\ A^2 \\ \vdots \\ A^N \end{bmatrix} \omega_k + \begin{bmatrix} B & 0 & \cdots 0 \\ AB & B & \cdots 0 \\ \vdots & \vdots & \ddots \vdots \\ A^{N-1}B & A^{N-2}B & \cdots B \end{bmatrix} \begin{bmatrix} u_k \\ u_{k+1} \\ \vdots \\ u_{k+N-1} \end{bmatrix}$$
$$= M_w \omega_k + C_u U_k \tag{15}$$

where M_w is $(N \times n) \times n$ matrix; C_u is $(N \times n) \times (N \times n)$; and U_k is the N-horizon control set.

The attempt to minimize a squared-weighted sum of state (regulated variables) x and controls u of the given Equation (15) is described by:

$$J_k = \sum_{i=0}^{N-1} \left(\left\| \omega_{k+i|k} \right\|_Q^2 + \left\| u_{k+i|k} \right\|_R^2 \right) + \left\| \omega_{k+N|k} \right\|_{Q_N}^2$$
$$= U_k^T H U_k + 2\omega_k^T F_w^T U_k + \omega_k^T G \omega_k$$
$$= U_k^T H U_k + 2F_{w,1}^T U_k + c_0 \tag{16}$$

where Q, Q_N, and R are weighting matrices, $Q' = \text{diagonal}\ [Q, Q,\ldots, Q_N]$, Q_N is a weighting matrix for terminal state, and $R' = \text{diagonal}\ [R, R, \ldots R]$. $H = C_u^T Q' C_u + R'$, $F_w = C_u^T Q' M_w$, $G = M_w^T Q' M_w + Q$, $F_{w,1} = \omega_k F_w$, and c_0 is a constant. H, F_w, and G can be computed offline if the system is time invariant.

According to the result of Equation (16), N-horizon optimal control gain K_{mpc} is $-H^{-1} F_{w,1}$ when the system is unconstrained, and the optimal control u_k^* is derived by the first horizon gain.

3.2. Line Flow Control in Centralized Structure

In centralized control, the most attractive and distinguishing advantage is to control line flows because loop flows and arrangements for parallel path compensation become increasingly important as the demand for transmission capacity increases at a faster rate than actual capacity [16]. Therefore, we propose the control scheme to stabilize line flows in centralized MPC-based AGC.

From Equation (12), the state transition function for line flows can be described as follows:

$$P_f[k_{line}+1] = P_f[k_{line}] + K_f(\delta_G[k_{line}+1] - \delta_G[k_{line}]) + D_f(P_L[k_{line}+1] - P_L[k_{line}])$$
$$= P_f[k_{line}] + K_f T_{line} w_G^{set}[k_{line}] + D_f d[k_{line}]$$
$$= A_f P_f[k_{line}] + B_f w_G^{set}[k_{line}] + D_f d[k_{line}] \tag{17}$$

where k_{line} is the discretized line flow control time step. Load reference set-point signal w_G^{set} should be considered in Equation (17).

In a typical control design, the set-point command (path) is chosen as a smooth function that has no/few high-frequency components. Therefore, the time step k_{line} of w_G^{set} should be larger than the frequency control update time k_u and the state prediction time step k.

Optimal load reference set-point signal w_G^{set*} in the infinite horizon can be obtained from minimizing a criterion, such as:

$$J_\infty = \sum_{k=0}^{\infty} P_f[k_{line}+1]'Q_f\, P_f[k_{line}+1] + w_G^{set}[k_{line}]' R_f w_G^{set}[k_{line}] \tag{18}$$

From Equation (18), w_G^{set*} can be described as:

$$w_G^{set*}[k_{line}] = -K_G^{set} P_f[k_{line}] \tag{19}$$

where:

$$K_G^{set} = (\mathrm{R} + \mathrm{B}_f^T PB_f)^{-1} \mathrm{B}_f^T PA_f \tag{20}$$

$$0 = A_f^T PA_f - P + Q_f - A_f^T P\mathrm{B}_f(\mathrm{R}_f + \mathrm{B}_f^T P\mathrm{B}_f)^{-1} \mathrm{B}_f^T PA_f \tag{21}$$

where P is the solution to the discrete algebraic Riccati equation [17,18].

To drive the system to some other set-point w_G^{set}, w_k can be replaced with:

$$w_k^s = \omega_k - w_G^{set} \tag{22}$$

The state transition function in Equation (13) can be expressed as:

$$\begin{aligned} w_{k+1}^s &= \mathrm{A}\left(\omega_k - w_G^{set}\right) + Bu_k - (I-A)w_G^{set} \\ &= \mathrm{A}w_k^s + Bu_k + \eta \end{aligned} \tag{23}$$

where $\eta = -(I - A)\, w_G^{set}$.

On the basis of Equation (23), the predictive states for system frequency with a non-zero set-point w_G^{set} can be re-expressed as follows:

$$w_{k+i}^s = A^i w_k^s + \sum_{j=1}^{i} A^{i-j} Bu_{k+j-1} + \sum_{j=1}^{i} A^{j-1}\eta \tag{24}$$

$$W = M_w w_k^s + C_u U_k + D_s \eta \tag{25}$$

where D_s is the $(N \times n) \times n$ matrix.

The attempt to minimize a squared-weighted sum of state (regulated variables) w and control u of the given Equation (26) is described by:

$$\begin{aligned} J_k &= \sum_{i=0}^{N-1}\left(\left\|w_{k+i|k}^s\right\|_Q^2 + \left\|u_{k+i|k}\right\|_R^2\right) + \left\|w_{k+N|k}^s\right\|_{Q_N}^2 \\ &= (M_w w_k^s + C_u U_k + D_s\eta)^T Q'(M_w w_k^s + C_u U_k + D_s\eta) + U_k^T R' U_k + w_k^{s,T} Q w_k^s \\ &= U_k^T H U_k + 2(F_{w,1}^T U_k + F_{s,1}^T) U_k + c_1 \end{aligned} \tag{26}$$

where $F_{s,1} = C_u^T Q' D_S\ \eta$; c_1 is a constant.

In consideration of input constraints, the optimization of Equation (26) can be expressed as a quadratic program optimization:

$$\begin{aligned} &\min\ \frac{1}{2} U_k^T H U_k + \left(F_{w,1}^T + F_{S,1}^T\right) U_k \\ &\text{subject to} \\ &A_{in} U_k \le b_{in} \end{aligned} \tag{27}$$

where A_{in} and b_{in} express the load reference ramp constraints among generators.

The control gain K_{mpc} is with $-H^{-1}$ ($F_{w,1}$ + $F_{S,1}$) when the constraints are not activated. This value can be varied because of the activation of constraints. Figure 1 shows the centralized MPC-based AGC system. In this figure, S is the function to select the control at the first horizon.

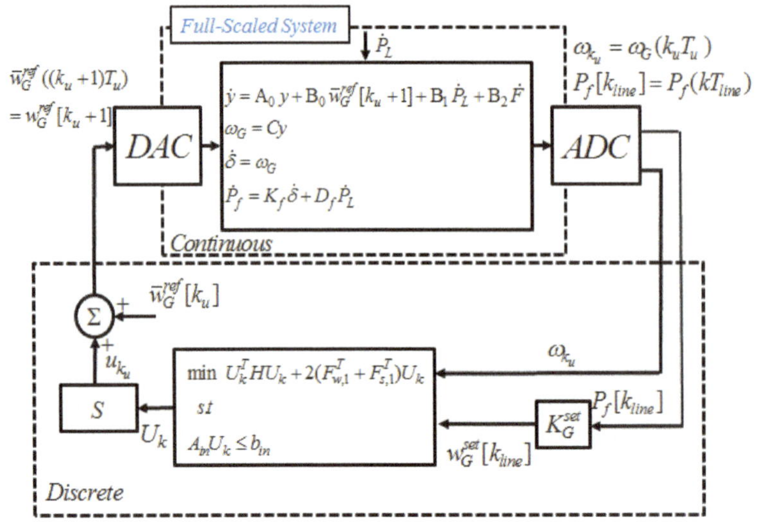

Figure 1. Centralized MPC-based AGC Considering Line-flows.

3.3. Distributed MPC-Based AGC

In the distributed MPC-based approach, each BA is assumed to be unaware of the state variables and control inputs in other BAs when its EMS computes AGC signals. Therefore, this paper provides individual AGC signals in each BA which mainly maintain their area frequency.

The major difference from centralized control is that the full-scale dynamic system in Figure 1 is affected from individual and unsynchronized AGC signals derived from each BA. Moreover, these signals in each BA are only responsible for maintaining a nominal frequency without any information for other area frequencies. This situation signifies that individual tie-line flow control is impossible in distributed MPC-based AGC because the angles δ in Equation (10) are affected by the power deviations in other areas, and these inadvertent tie-line flows eventually changes the frequencies in each BA.

To maintain frequency and tie-line flows at the same time, AGC signals in distributed BAs are traditionally computed by minimizing ACE, which can be described as:

$$\Delta ACE_i = \beta_i \Delta w_{i,k}^{r} + \Delta P_{i,k}^{net} \tag{28}$$

where i is the area number, β is the frequency bias factor, $w_{i,k}^{r}$ is a representative frequency at area i, and $P_{i,k}^{net}$ is the net inadvertent tie-line flow. When ACE is minimized, the tie-line control action in each BA is not a topological problem and is another form of active power balancing in each BA. Based on this scheme, EMS considers the system-wide effect of local control actions in each BA.

Following the same principle as ACE, we propose the pseudo-frequency at i-BA, which is described by:

$$\Delta w_{i,k}^{ps} = \Delta w_{i,k} + \mathrm{H}^{bd} \Delta P_{i,k}^{net} \tag{29}$$

where H^{bd} is the block diagonal matrices with $1/\beta$. The pseudo-frequency substitutes the measured frequency w_k in Equation (13) while EMS computes MPC-based AGC signals. Figure 2 shows the distributed MPC-based AGC system. In the figure, T_i is the function to compute $H^{bd}P_{i,f}^{net}$ from individual tie-line flows in i-BA.

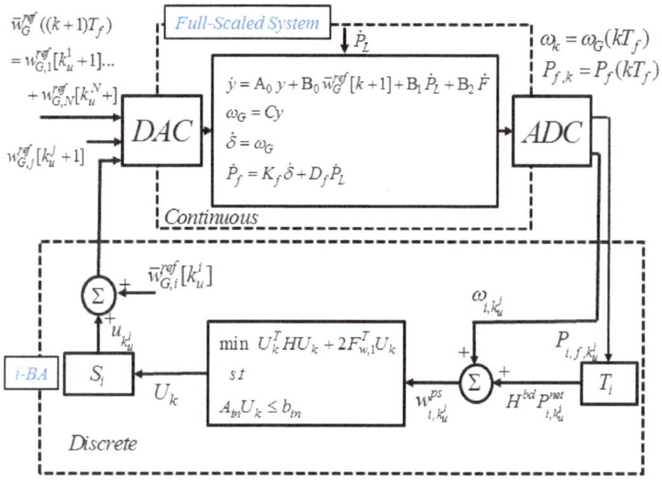

Figure 2. Distributed MPC-based AGC.

3.4. Bulk-Area Partitioning for MPC-Based AGC

Reference [6] showed that the average time for optimization in the MPC-based approach increases exponentially as the prediction horizon increases, but the desirable dynamic performance is only obtained when the prediction horizon is large enough. The average time for optimization also increases exponentially as the number of generators increases.

In the field, the numbers of participating generators exist in AGC, and the AGC signals should be rapidly computed for approximately 2 s [10]. Therefore, considering the computation time in the MPC-based approach is important in the design process.

The present research proposes bulk-area partitioning in bulk BAs. In the proposed scheme, each subsystem in a bulk BA has its own AGC scheme in EMS, such as a BA, and the net tie-line flows among subsystems is recomputed by the upper controller in its BA. The net of tie-line flows among subsystems is described as:

$$\Delta P_{i_s,k}^{net} = \underbrace{\sum_{j=1}^{n_A} \Delta P_{ij}[k]}_{\substack{Original \\ Net\ Tie-line\ Flows}} + \underbrace{\left(\Delta P_{m,i_s}^{net}[k_m] - \frac{1}{n_i} \sum_{k=1}^{n_i} \Delta P_{m,i_k}^{net}[k_m] \right)}_{Partitioning\ Control\ in\ BA\ i} \tag{30}$$

where n_A is the number of BAs, n_i is the number of subsystems in BA i, $P_{i_s j}$ is the power flow from subsystem i_s to BA (or subsystem) j, k_m is the discretized mechanical power sampling time step, and $\Delta P_{m,i_k}^{net}$ is the net of deviated mechanical power output in subsystem i_k. According to this control scheme, each subsystem equally reacts to the discrepancy between the generation supply and the total MW obligation in its BA. In this case, the function T_i in Figure 2 includes the control process for Equation (30).

4. Illustrative Example

Simulation is conducted on a modified New England 39-bus system, as shown in Figure 3; per-unit bus voltage, bus angle, and transmission line data given in [19,20] are used. Table 1 lists the parameters and load reference constraints for the generator dynamic models at each bus. Table 2 lists the load disturbance information.

This simulation demonstrates the results of both centralized and distributed MPC-based AGC. The MPC-based operation results are analyzed from the perspectives of computational burden and dynamic performance.

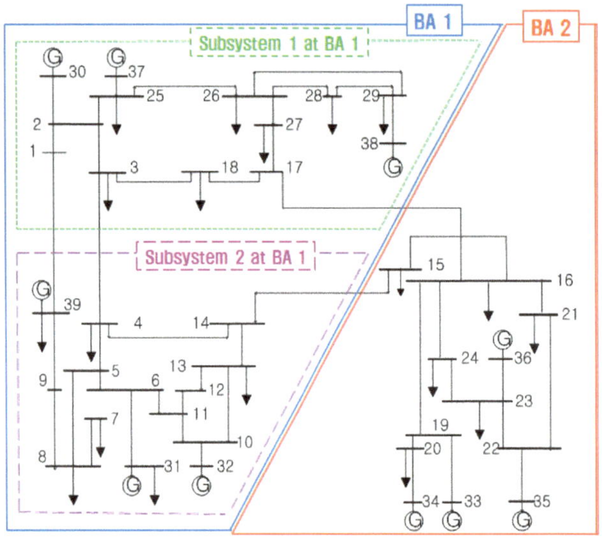

Figure 3. IEEE 39 Bus Systems with Two Bas.

Table 1. Parameters and load reference ramp constraints for generators.

Generator Bus	M	D	T_g	T_a	K_t	r	$\bar{u}$ (10^{-4})	$\underline{u}$ (10^{-4})
# 30 (Gen1)	2	5	0.25	0.2	250	19	3	−3
# 31 (Gen2)	3	5	0.25	0.2	250	19	3	−3
# 32 (Gen3)	2	5	0.25	0.2	250	19	2	−2
# 33 (Gen4)	2	5	0.25	0.2	250	19	2	−2
# 34 (Gen5)	3	5	0.25	0.2	250	19	3	−3
# 35 (Gen6)	2	5	0.25	0.2	250	19	3	−3
# 36 (Gen7)	2	5	0.25	0.2	250	19	3	−3
# 37 (Gen8)	2	5	0.25	0.2	250	19	3	−3
# 38 (Gen9)	3	5	0.25	0.2	250	19	2	−2
# 39 (Gen10)	2	5	0.25	0.2	250	19	2	−2

Table 2. Step load deviation events.

Event time [s]	Bus number	Magnitude [pu]
1	28	0.2
50	13	-0.1
100	22	0.15

4.1. Centralized MPC-Based AGC Systems

Table 3 lists the data and parameters for the centralized MPC-based control simulation. In Figure 4, each case shows that frequencies and line flows are maintained by the given control scheme. In this simulation, the proposed centralized controller maintains line flows P_{17-16} and P_{14-15}. Therefore, the 39-bus system naturally divides into two areas, BA1 and BA2 (Figure 3). We use a two-area notation rather than the original single-area notation to compare the centralized MPC-based results with the distributed MPC-based results.

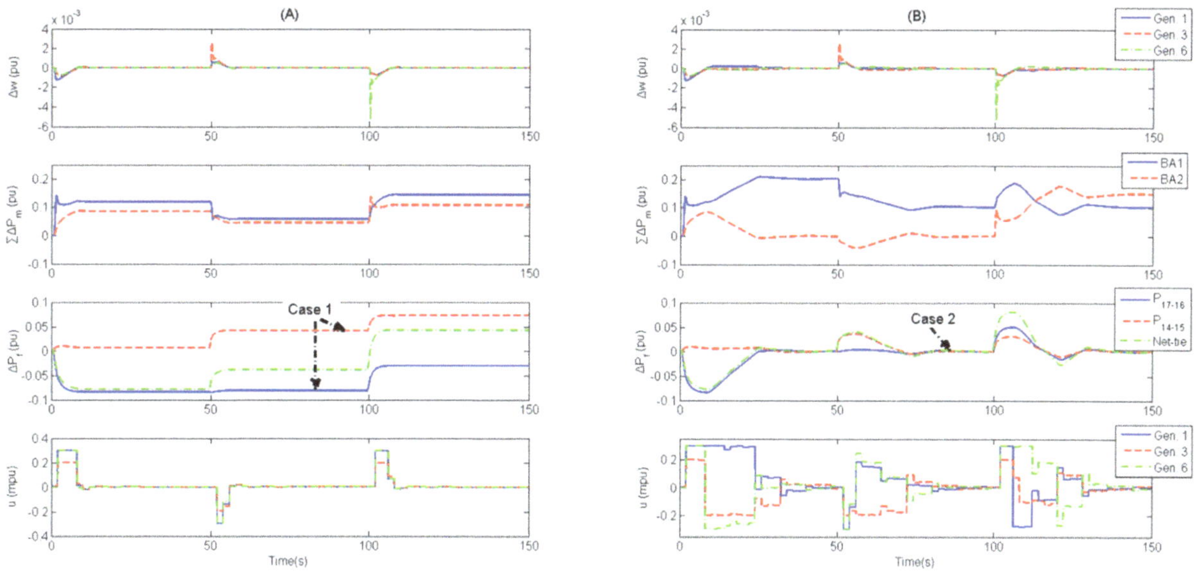

Figure 4. Centralized MPC-based AGC: (**A**) Without Line Flow Control (Original Case) (**B**) With Line Flow Control (Proposed).

Table 3. Data and parameters for centralized MPC-based AGC.

Description	Parameter	Value
Discretized Minimum Step Time [s]	T_f	1
AGC Update Time [s]	T_u	2
Line Control Update Time [s]	T_{line}	6
	Q	$10^2 I$
	Q_N	$10^2 I$
Weighting Matrix	R	I
	Q_f	I_{line}
	R_f	I
Prediction Horizon	N	20
Frequency Bias Factor	β	200

- I is a 10-by-10 identity matrix. - I_{line} is a 2-by-2 identity matrix.

Centralized control provides successful dynamic performance on stabilizing frequency. Figure 4A shows that each frequency at generator buses is rapidly maintained. Given that the centralized controller solves the optimization problem in consideration of the entire system dynamic behavior, its solution provides better dynamic performance than the distributed MPC-based solution. The specific results will be provided in the following section. In Figure 4B, each BA is responsible for its discrepancy between the generation supply and the total MW obligation through line flow control. This centralized controller adjusts the load references among generators to maintain the frequency every 2 s and to stabilize the inadvertent line flows every 6 s. The results show that BA1 and BA2 at 150 s produce powers of 0.1 and 0.15 pu respectively as the load deviations in each BA. This line control scheme secures state controllability while the number of designated lines is fewer than the generators involved in the line flow control. In the simulation, the controllability in Equation (17) is satisfied because two line flows are controlled by 10 generators. On the basis of this control property, the system operator can design AGC to maintain frequency in consideration of designated and important line flows.

This line control scheme helps the system operator reduce wheeling, which is a loop flow caused by network topology. In the field, wheeling increases the power losses or the overloading of network elements. Therefore, wheeling must be considered to minimize operation cost and secure system reliability. However, most of the previous MPC-based schemes, which did not consider network topologies, could not provide any method to minimize the loop flows such as the results at Case1 in Figure 4A. Based on the proposed control scheme, designated lines are effectively adjusted as shown at Case2 in Figure 4B, and it can help the system operators to reduce the power losses or the overloading of network elements.

4.2. Distributed MPC-based AGC Systems

Tables 4 and 5 list the data and parameters for distributed MPC-based control simulation. All the conditions are assumed to be the same as in Table 3, except for discretized control update times and frequency bias factors at each BA. In case of bulk-area partitioning, BA1, as shown in Figure 3, has two subsystems with their own frequency controllers, and the mechanical power outputs in each subsystem are controlled by an upper controller in BA1 every 4 s. Each BA has its own controller. These distributed controllers in BA1 and BA2 adjust the load references among generators to maintain frequency and to stabilize net-tie line flows for control update times of 3 and 2 s respectively. As shown in Figure 5, the load reference of Gen.6 is more smoothly updated than those of other generators in BA1.

Wheeling is an intractable and uncontrollable problem in a distributed control structure. In Figure 5, the net tie-line flow is maintained at zero, but the individual tie-line flows are different from the initial tie-line flows. This circumstance indicates that this control scheme suffers from the increase in power losses in the intervening systems or the overloading of network elements. Therefore, the load references should be recomputed among generators in BAs while wheeling makes significant effects.

Table 4. Data and parameters for MPC-based AGC in each BA.

Description	Parameter	BA1	BA2
Discretized Minimum Step Time [s]	T_f	1	1
AGC Update Time [s]	T_u	3	2
	Q	$10^2 I_1$	$10^2 I_2$
Weighting Matrix	Q_N	$10^2 I_1$	$10^2 I_2$
	R	I_1	I_2
Prediction Horizon	N	20	20
Frequency Bias Factor	β	120	80

- I_1 is a 6-by-6 identity matrix; - I_2 is a 4-by-4 identity matrix.

Table 5. Data and Parameters for Bulk-Area Partitioning in BA1.

Description	Parameter	SB1	SB2
Mechanical power sampling time [s]	T_m	4	
Weighting Matrix	Q	$10^2 I_{sb1}$	$10^2 I_{sb2}$
	Q_N	$10^2 I_{sb1}$	$10^2 I_{sb2}$
	R	I_{sb1}	I_{sb2}

- *SB* is subsystem; - I_{sb1} and I_{sb2} are 3-by-3 identity matrices.

In the bulk-area partitioning control scheme, each subsystem in BA1 equally generates the discrepancy between the generation supply and the total MW obligation. In Figure 5, although the mechanical power outputs among subsystems are different and do not become equal at Case 1, the results at Case 2 have the same mechanical power values in each subsystem. These results show that each subsystem shares its power generation capacities. Although this research does not consider the discriminative power sharing scheme in Equation (30), such a scheme should be cautiously considered in the future when renewable power sources and microgrids are briskly utilized in power system because these entities can produce large disturbances (P_L and F) to increase their profits.

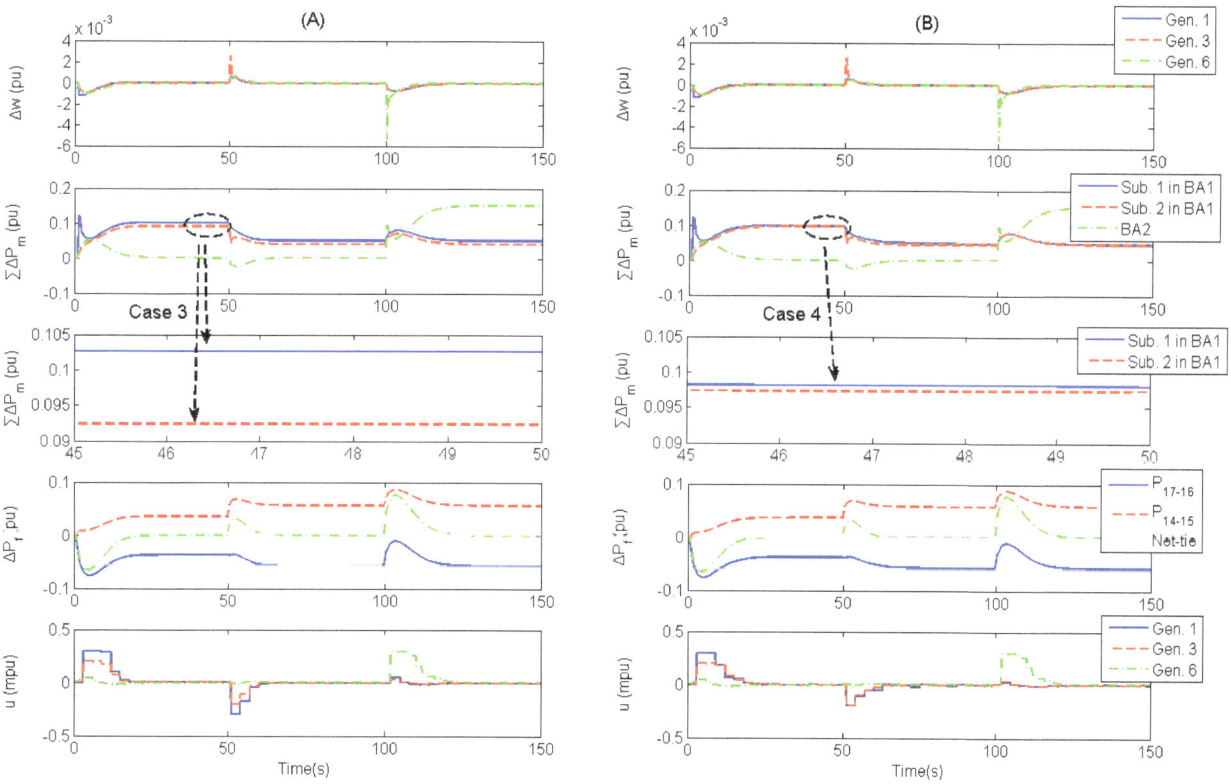

Figure 5. Distributed MPC-based AGC: (**A**) Without Partitioning (Original Case); (**B**) With Partitioning (Proposed Case).

Bulk-area partitioning dramatically reduces the computation time and ensures dynamic performance. In this simulation, each subsystem has three generators, half of the original six generators in BA1. Therefore, the object function in Equation (27) is much rapidly solved. In Figure 6, the results show the computation times in accordance with the horizon sizes and the number of states by using an Intel Core i5 CPU 750 @2.67 GHz and 2 GB of RAM. The prediction horizon size (N) affects the dynamic performance in stabilizing frequency and tie-line flows. Therefore, its size is manually chosen in accordance with the user-defined (or system) dynamic performance requirements. However, the number of generators is fixed and cannot be changed to a small value. In the MPC-based scheme, the numbers of generators in bulk areas will lead to computational burden, although simply modified SISO-based schemes, such as proportional–integral-based AGC, do not suffer greatly from this problem. The proposed bulk-area partitioning can mitigate this difficulty in solving a massive QP problem at a time by dividing the original state function. In addition, the results of the proposed scheme only present a slight difference in dynamic performance compared with the results in the (A) original case. This situation signifies that the bulk-area partitioning control scheme can provide substantial and realistic measures for suffering from "curse of dimensionality" as a desired result.

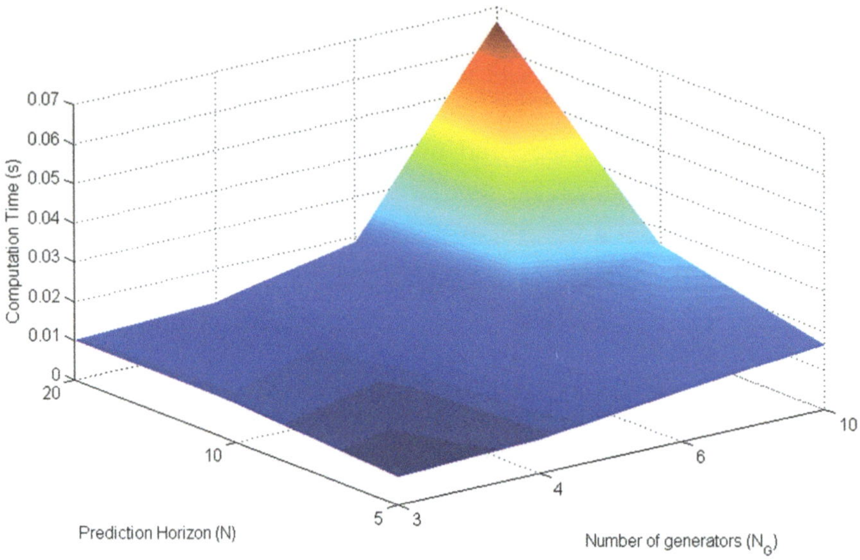

Figure 6. Average computation times according to operation conditions.

4.3. Discussion on Structure-Dependent AGC Scheme

Based on the tested simulation results, structure-dependent MPC-based AGC can be considered, as shown in Table 6. Each scheme has unique advantages and disadvantages. Given that centralized control considers the entire system dynamic behavior, this scheme in the original case provides improved settling time results. However, centralized controllers in bulk-power systems have severe problems (Table 6). Therefore, under the control design process, system operators should cautiously analyze whether or not to satisfy the system requirements in designated MPC-based AGC structure.

Centralized line flow control schemes poorly stabilize frequencies in comparison with the results in the original case. However, the difference can be ignored when the line control update time T_{line} is set to a large value. These results show a tradeoff between dynamic performances in stabilizing frequency and line flows. Therefore, the system operator should carefully choose T_{line} in consideration of the system operation requirements and economics.

Table 6. Structure-dependent MPC-based AGC.

Considerations for implementation	Centralized structure		Distributed structure	
	Original	Proposed	Original	Proposed
Wheeling	X	O	X	X
Computation Time	X	X	Δ	O
Practicality	X	X	O	O
Settling time (T_{st}) [s]	7.60	24.60	15.05	16.01

- X: severe, Δ: less severe, O: least severe; - T_{st}: The maximum time required for every generator frequency curve to stay within a range of certain percentage 0.01% (0.1 mpu) of the nominal frequency from the first disturbance 0.2 pu at Bus 28.

The proposed structure-dependent MPC-based AGC schemes can help system operators prepare for future power systems. Due to increasing profit-seeking entities such as wind-farm producers and microgrid operators, sharing the responsibility of deviated system frequency (or power imbalance) among entities can be critical and will be a very important issue in future. Based on proposed control structures, it is possible to consider how to share the responsibility of the deviated system frequency or design the operation rules. For instance, a system operator in a centralized structure should be responsible for stabilizing frequency by himself. However, he can reduce the computational burden as well as his responsibility via the proposed bulk-area partitioning control scheme since owners (operators) in subsystems are required to build dispatchable power resources to meet the contraction like Equation (30).

On the basis of the research findings, we propose to establish a centralized organization among distributed BAs to control individual tie line flows. In the proposed control scheme, line flow control signals are separately computed from the signals to stabilize the frequency, as shown in Figure 1. This situation signifies that the centralized line controller is a different control system and is interpreted as an additional organization. Therefore, founding a centralized organization among distributed BAs can help system operators to increase profits from the interconnection when system-wide wheeling causes significant effects. The limitation of computation capacity is not a significant problem in this control scheme because T_{line} is usually larger than T_u.

5. Conclusions

This paper presented an MPC-based AGC considering network topology. Based on the proposed MPC-based control process, the optimal AGC signal was obtained by solving a constrained QP, which revealed a significant advantage in terms of optimization procedure. In this simulation, the time for one step was approximately 70 ms when the MPC-based approach considered 20 prediction horizons and 10 generators.

This research mainly dealt with structure-dependent control schemes. To strengthen the advantages and minimize the disadvantages, different approaches were developed from previous MPC-based studies. In the centralized control structure, this research proposed designated line flow control to strengthen its unique advantages. The results showed that the proposed control effectively reduces inadvertent line flows in its area, signifying that wheeling is a manageable problem in this control scheme. In the distributed control structure, this research mainly considered bulk BAs from a computational perspective because the number of generators dramatically increases the computation time. Therefore, bulk-area partitioning was developed, and this proposed scheme provides only a slight difference in dynamic performance compared with original case result in the distributed structure. This research also proposed to establish a centralized organization among distributed BAs to manage individual tie-line flows when wheeling makes system-wide effects. This organization can help system operators to increase the profits from tie-line interconnections among BAs. In future works, we plan to obtain economic MPC-based AGC with performance constraints. This study will help system operators to reduce the operation cost of generators for AGC.

Nomenclature

The main symbols that appear in this paper are defined below. Other symbols are defined in the text as they appear.

Constants:

V	Bus voltage [pu]
δ	Bus voltage angle [radian]
P	Initial bus power generation or demand [pu/100 MW]
T_a	Time constant of the turbine [s]
T_g	Time constant of governor [s]
T_f	Discretized minimum time step [s]
T_u	Discretized automatic generation control update time [s]
T_m	Discretized mechanical power sampling time step [s]
T_{line}	Discretized line power flow control update time step [s]
T_{st}	Frequency settling time [s]
D	Damping constant
M	Moment of inertia of the generator [s]
K_t	Parameter in linearization for turbine characteristics
r	Parameter in linearization for governor droop characteristics
J	Jacobian matrix of power evaluated at angle
β_i	Frequency bias factor in *i*-balancing area
$\overline{u}$	Maximum generation control ramp rate [pu/s]
$\underline{u}$	Minimum generation control ramp rate [pu/s]

Variables:

w_G^{ref}	Load reference [pu]
w_G^{set}	Load frequency set point [pu]
ω_G	Generator frequency [pu]
P_m	Turbine mechanical power [pu]
α	Governor-controlled valve opening [pu]
F	Mapped injected-power [pu]
P_f	Tie-line power flow [pu]
y	Concatenated state variables in full-scale dynamic model

Indices:

k	Discretized minimum time step [$k = kT_f$]
k_u	Discretized frequency control update time step [$k_u = k_uT_u$]
K_{line}	Discretized tie-line flow control update time step [$k_{line} = k_{line}T_{line}$]
k_m	Discretized mechanical power sampling time step [$k_m = k_mT_m$]

Author Contributions

The main idea of this paper is suggested by Young-Sik Jang. He completed the mathematical modeling, simulations and wrote the first draft of the paper. JoonHyung Park contributed to the mathematical modeling and thoroughly reviewed the paper. Yong Tae Yoon coordinated the main theme of this paper and thoroughly reviewed the paper.

Conflicts of Interest

The authors declare no conflict of interest.

References

1. Zhang, G. *EPRI Power System Dynamics Tutorial*; Report 1016042; Electric Power Research Institute: Palo Aoto, CA, USA, 2009.
2. Rossiter, J.A. *Model-Based Predictive Control: A Practical Approach*; CRC: Boca Raton, FL, USA, 2003.
3. Mohamed, T.H.; Bevrani, H.; Hassan, A.A.; Hiyama, T. Model predictive based load frequency control design. In Proceedings of the 16th International Conference of Electrical Engineering. Busan, Korea, 11–14, July 2010; pp. 1–6.
4. Kong, L.F.; Xiao, L. A new model predictive control scheme-based load-frequency control. In Proceedings of the IEEE International Conference on Control and Automation, Guangzhou, China, 30 May–1 June 2007; pp. 2514–2518.
5. Yousef, A.M. Model predictive control approach based load frequency controller. *WSEAS Trans. Syst. Cont.* **2011**, *6*, 265–275.
6. Camponogara, E.; Jia, D.; Krogh, B.H.; Talukdar, S. Distributed model predictive control. *IEEE Control Syst. Mag.* **2002**, *22*, 44–52.
7. Venkat, A.; Hiskens, I.; Rawlings, J.; Wright, S. Distributed MPC strategies with application to power system automatic generation control. *IEEE Trans. Control Syst. Technol.* **2008**, *16*, 1192–1206.
8. Nong, H.; Liu, X. Nonlinear distributed MPC strategy with application to AGC of interconnected power system. In Proceedings of the Control and Decision Conference (CCDC), Qingdao, China, 25–27 May 2013.
9. Ma, M.; Chen, H.; Liu, X.; Allgöwer, F. Distributed model predictive load frequency control of multi-area interconnected power system. *Int. J. Electr. Power Energy Syst.* **2014**, *62*, 289–298.
10. Kundur, P. *Power System Stability and Control*; McGraw-Hill: New York, NY, USA, 1994; pp. 581–613.

11. Mohamed, T.H.; Bevrani, H.; Hassan, A.A.; Hiyama, T. Decentralized model predictive based load frequency control in an interconnected power system. *Energy Convers. Manag.* **2011**, *52*, 1208–1214.
12. Ilic, M.; Zaborszky, J. *Dynamics and Control of Large Electric Power Systems*, 1st ed.; John Willey and Sons Inc.: Hoboken, NJ, USA, 2000; pp. 352–370.
13. Liu, X. Structure Modeling and Hierarchical Control of Large Scale Electric Power System. Ph.D. Dissertation, Department of Electrical Engineering and Computer Science—Massachusetts Institute of Technology, Cambridge, MA, USA, 1994.
14. Eidson, B. Estimation and Hierarchical Control of Market-driven Electric Power Systems. Ph.D. Dissertation, Department of Electrical Engineering and Computer Science—Massachusetts Institute of Technology, Cambridge, MA, USA, 1995.
15. Makarov, Y.V.; Ma, J.; Lu, S.; Nguyen, T.B. *Assessing the Value of Regulation Resources Based on Their Time Response Characteristics*; CERTS Report PNNL-17632; Pacific Northwest National Laboratory: Richland, WA, USA, 2007.
16. Wood, A.J.; Wollenbergy, B.F. *Power Generation, Operation, and Control*; Wiley: New York, NY, USA, 1984.
17. Phillips, C.L.; Nagle, H.T. *Digital Control System*, 2nd ed.; Prentice Hall: Upper Saddle River, NJ, USA, 1990; pp. 356–372.
18. Bertsekas, D.P. *Dynamic Programming and Optimal Control*, 1st ed.; Athena Scientific: Belmont, MA, USA, 1995; Volume 1, pp. 130–157.
19. Pai, M.A. *Energy Function Analysis for Power System Stability*; Kluwer Academic Publishers: Boston, MA, USA, 1989.
20. Zimmerman, R.D.; Murillo-Sanchez, C.E.; Gan, D. *MATPOWER 5.0.0 User's Manual*. Available online: http://www.pserc.cornell.edu//matpower/ (accessed on 17 December 2014).

Dynamic Power Management for Portable Hybrid Power-Supply Systems Utilizing Approximate Dynamic Programming

Jooyoung Park [1,*], Gyo-Bum Chung [2], Jungdong Lim [1] and Dongsu Yang [3]

[1] Department of Control and Instrumentation Engineering, Korea University, Sejong 339-700, Korea; E-Mail: parkj@korea.ac.kr

[2] Department of Electronic & Electrical Engineering, Hongik University, Sejong 339-701, Korea; E-Mail: gbchung@hongik.ac.kr

[3] LG Electronics, Seoul 135-860, Korea; E-Mail: dongsu1234.yang@lge.com

* Author to whom correspondence should be addressed; E-Mail: parkj@korea.ac.kr

Academic Editor: William Holderbaum

Abstract: Recently, the optimization of power flows in portable hybrid power-supply systems (HPSSs) has become an important issue with the advent of a variety of mobile systems and hybrid energy technologies. In this paper, a control strategy is considered for dynamically managing power flows in portable HPSSs employing batteries and supercapacitors. Our dynamic power management strategy utilizes the concept of approximate dynamic programming (ADP). ADP methods are important tools in the fields of stochastic control and machine learning, and the utilization of these tools for practical engineering problems is now an active and promising research field. We propose an ADP-based procedure based on optimization under constraints including the iterated Bellman inequalities, which can be solved by convex optimization carried out offline, to find the optimal power management rules for portable HPSSs. The effectiveness of the proposed procedure is tested through dynamic simulations for smartphone workload scenarios, and simulation results show that the proposed strategy can successfully cope with uncertain workload demands.

Keywords: hybrid power-supply systems; dynamic power management; approximate dynamic programming

---

1. Introduction

Recently, mobile technologies have advanced to a remarkable extent, and power consumption from mobile devices has become a serious concern in portable electronics such as smartphones, tablet PCs, and electric vehicles. The most widely used power-supply systems for portable electronic devices are electrochemical-battery-based. However, battery technology has progressed very slowly and suffers from several drawbacks such as the high current load that rapidly depletes the battery's energy and its intrinsic low power density characteristic, resulting in inefficiency when handling high load fluctuations. One of the promising methods that can compensate for some of the battery's drawbacks is based on the use of supercapacitors [1]. The supercapacitor is an emerging technology that is currently experiencing rapid development and is particularly attractive for portable systems. This is because the supercapacitor has a higher power density and faster charge/discharge capability compared to a power supply based only on batteries. Owing to the strengths of supercapacitors that can compensate for the weaknesses of batteries, portable hybrid power-supply systems (HPSSs) employing both batteries and supercapacitors are now considered to be an important alternative to conventional battery-only power supply systems [2,3]. Because the lifetime of an HPSS can be significantly extended by judiciously controlling the level of each component's power consumption, optimization of the power flows in HPSSs is a key issue that should be solved for intelligent power management of HPSSs [4]. In this paper, the problem of dynamically managing power flows in portable HPSSs employing batteries and supercapacitors is considered, and we propose an approximate dynamic programming (ADP)-based solution. In recent years, stochastic optimal control theory has been applied to various engineering problems. As is well known, a general stochastic optimal control strategy can be derived, in principle, utilizing the concept of the state value function, which is defined as the minimum total expected cost achieved by an optimal control policy from the given state. This value-function-based method is called dynamic programming (DP) and provides an important theoretical foundation for various optimal control problems. However, most real-world optimal control problems cannot be solved by DP exactly, except in very simple cases, and many studies rely on ADP, which finds an approximate solution to the stochastic optimal control problem utilizing an approximate value function, in order to obtain a good suboptimal control policy [5–9]. The main objective of this paper is to consider an ADP-based solution to dynamic power management (DPM) for portable HPSSs employing batteries and supercapacitors. More specifically, we consider a mathematical formulation of DPM for an HPSS in the form of a stochastic optimal control problem and establish an ADP-based procedure for solving the resulting controller synthesis problem. Simulation results show that this procedure works well when applied to DPM for smartphones under various load profiles.

The remainder of this paper is organized as follows. In Section 2, a preliminary background is provided regarding the mathematical formulation of the DPM problem for portable HPSSs employing batteries and supercapacitors. In Section 3, we present the framework for stochastic optimal control and describe the ADP-based controller synthesis procedure for the DPM problem. In Section 4, the effectiveness of the ADP-based procedure is illustrated by using actual workload scenarios for mobile phones. Finally, in Section 5, concluding remarks are presented.

2. Problem Formulation

This section describes the portable HPSS, and presents a mathematical model in the form of a state-space description.

2.1. System Configuration and Related Background

Recently, the optimization of power flows in HPSSs has become an important issue for the purposes of energy savings, lifetime lengthening, and so forth. For the current state-of-the-art works on the subject, the reader is referred to papers such as [10–15]. Romaus *et al.* [10] addressed the problem of controlling the power flows of a hybrid energy storage system combining batteries and double layer capacitors, and proposed the use of self-optimization methods that can consider various operating conditions and take into account several partly conflicting objective functions. As an application domain, the reference paper [10] considered an autonomous rail-bound vehicle. Chen *et al.* [11] proposed a scheduling strategy for a hybrid energy storage system for a wind farm based on predictive wind speed and the particle swarm optimization (PSO) algorithm. The strengths of the proposed scheduling strategy are that the optimization results can not only satisfy the requirements of output power quality, but also have good prediction accuracy for the wind power output. Choi *et al.* [12] presented an optimal energy management scheme based on the so-called the multiplicative-increase-additive-decrease (MIAD) policy for hybrid energy storage systems with the objectives of minimizing the magnitude/fluctuation of the battery current, and minimizing the supercapacitor energy loss. Gee *et al.* [13] analyzed and implemented a battery lifetime extension strategy for a battery/supercapacitor hybrid energy storage system, in which current filtering using a low pass filter was proposed to divert the high-frequency component of the HPSS system charge/discharge current to the supercapacitor in real time. The results of [13] showed the potential improvement in battery lifetime of HPSS in a wind power system without applying any optimization methods. Blanes *et al.* [14] considered the implementation of a hybrid energy storage system using ultracapacitor, and investigated a simple FPGA control method with a bidirectional buck-boost converter for an electrical vehicle HPSS system. In the method in [14], the battery supplies the low frequency part and the supercapacitor supplies the high-frequency part of the motor current, which reduces the battery stress and softens the battery peak current. Min *et al.* [15] presented a linear programming (LP) based method to address the problem of minimizing the installation cost while meeting the grid connection requirement, together with a quadratic programming-based approach to transmit the optimal operational signals to battery energy storage system and electric double layer capacitor modules.

This paper is concerned with an ADP-based DPM of an HPSS employing batteries and supercapacitors, which are used for portable and low-power applications. A typical example is shown in Figure 1. In this example, the HPSS consists of a single battery, two supercapacitors, two DC-DC converters, and five bidirectional switches. As is well known, a battery is currently the most widely used device that supplies energy to portable systems and has a high-energy and low-power density. However, the intrinsic low-power density of a battery often results in inefficiency, especially when handling high load fluctuations. On the other hand, supercapacitors have a low-energy and high-power density. Thus, batteries and supercapacitors can compensate for each other's weaknesses quite well [16]. In the portable HPSS in Figure 1, the battery and supercapacitors can operate in the charging and discharging modes, respectively, or all together. DC-DC converter #1 between the battery/supercapacitors and the load of the HPSS regulates the load voltage at a constant value. DC-DC converter #2 between the battery and the supercapacitors controls the charging current, which is calculated by the DPM unit. Because energy loss may occur during the process of DC-DC conversion, the converter topologies need be carefully selected [17]. Switches connect the components of the portable HPSS following the operation strategy provided by DPM. The DPM monitors the system at each sampling time step and determines the optimal decision to lengthen the lifetime of the HPSS, thus improving the energy conversion efficiency.

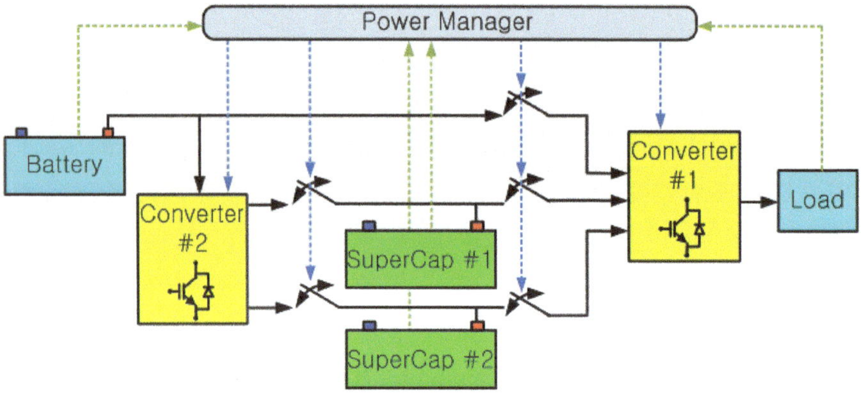

Figure 1. Structure of a typical portable hybrid power supply system.

2.2. State Equation and Performance Index

In this section, we describe the DPM of a portable HPSS employing batteries and capacitors within the framework of a stochastic optimal control problem. Throughout this paper, we consider an HPSS with the structure shown in Figure 1, which has a single battery and two supercapacitors. Generalization of the model and the corresponding solution procedure for a case with any multiple number of batteries and supercapacitors is straightforward. The steps for deriving the state equation and the performance index (PI) for the system are based on the results of [18] with some modifications. The details of these modifications will be specified later in this section.

Generally, dynamical systems in discrete-time stochastic optimal control problems are described by the state-space model in the form of:

$$x(t+1) = f(x(t), u(t), w(t)), \; t = 0,1,..., \tag{1}$$

where $x(t) \in X$ is the state vector, $u(t) \in U$ is the control input vector; and $w(t)$ is the external disturbance, all at time t. In the DPM problem presented in this paper, $w(t)$ is the load demand during a single time interval $(t, t + 1)$. The state vector that we consider for the DPM problem contains the amounts of charge of the supercapacitors and battery as well as the change in the battery's charge, *i.e.*,

$$x(t) = \begin{bmatrix} r_{cap,1}(t) \\ r_{cap,2}(t) \\ r_{bat}(t) \\ \Delta r_{bat}(t) \end{bmatrix}, \tag{2}$$

where $r_{cap,i}(t)$ is the charge state of the ith supercapacitor at time t, $r_{bat}(t)$ is the charge state of the battery at time t, and $\Delta r_{bat}(t) = r_{bat}(t-1) - r_{bat}(t)$ is the difference in the charge state of the battery. Note that in this paper, we assume that the state-of-charge (SOC) values of the supercapacitors and the battery are all available. In practical implementations, these SOC values can be estimated by e.g., extended Kalman filters [19]. The control input vector we consider for the DPM problem contains information regarding the proportion of the charge that each component (*i.e.*, the ith supercapacitor or battery) supplies to meet the workload and the level of the battery charge transferred to each supercapacitor. As in [18], an upper bound R_{th} is placed on the amount of charge that can be transferred from a battery to each supercapacitor within a time interval so that the battery charge may not be exhausted during the process of charging the supercapacitors. More specifically, the control input vector consists of the following five entries, all in the range of [0, 1]:

$$u(t) = \begin{bmatrix} a_{1x}(t) \\ a_{1y}(t) \\ a_{2x}(t) \\ a_{2y}(t) \\ a_{bat}(t) \end{bmatrix} \tag{3}$$

Here, $a_{ix}(t)$, $i = 1, 2$, is the normalized level of the ith supercapacitor being charged by the battery; thus, the amount of charge transferred from the battery to the ith supercapacitor during the time interval $(t, t+1)$ is $R_{th} \cdot a_{ix}(t)$. The other entries, $a_{1y}(t)$, $a_{2y}(t)$, and $a_{bat}(t)$, are the proportions of the charges that the two supercapacitors and battery supply to meet the workload $w(t)$ in the time interval $(t, t+1)$. Because these entries are all proportions, they satisfy:

$$a_{1y}(t) + a_{2y}(t) + a_{bat}(t) = 1 \text{ for } t = 0, 1, \ldots \tag{4}$$

With the above definitions of the state and control inputs, the state transition from time step t to $(t+1)$ can be expressed as follows [18]:

$$r_{cap,1}(t+1) = r_{cap,1}(t) - a_{1y}(t)w(t) + a_{1x}(t)R_{th} \tag{5}$$

$$r_{cap,2}(t+1) = r_{cap,2}(t) - a_{2y}(t)w(t) + a_{2x}(t)R_{th}, \tag{6}$$

$$r_{bat}(t+1) = r_{bat}(t) - a_{bat}(t)w(t) - a_{1x}(t)R_{th} - a_{2x}(t)R_{th}, \tag{7}$$

$$\begin{aligned} \Delta r_{bat}(t+1) &= r_{bat}(t) - r_{bat}(t+1) \\ &= a_{bat}(t)w(t) + a_{1x}(t)R_{th} + a_{2x}(t)R_{th}. \end{aligned} \tag{8}$$

Note that with the above definitions of the state vector and control input vector, these state transitions can be compactly written in the following state equation form:

$$\begin{aligned} x(t+1) &= Ax(t) + B(t)u(t) \\ &= Ax(t) + (B_1 + B_w w(t))u(t), \end{aligned} \tag{9}$$

where

$$A = \begin{bmatrix} 1 & 0 & 0 & 0 \\ 0 & 1 & 0 & 0 \\ 0 & 0 & 1 & 0 \\ 0 & 0 & 0 & 0 \end{bmatrix} \tag{10}$$

$$B_1 = R_{th} \begin{bmatrix} 1 & 0 & 0 & 0 & 0 \\ 0 & 0 & 1 & 0 & 0 \\ -1 & 0 & -1 & 0 & 0 \\ 1 & 0 & 1 & 0 & 0 \end{bmatrix} \tag{11}$$

and

$$B_w = \begin{bmatrix} 0 & -1 & 0 & 0 & 0 \\ 0 & 0 & 0 & -1 & 0 \\ 0 & 0 & 0 & 0 & -1 \\ 0 & 0 & 0 & 0 & 1 \end{bmatrix} \tag{12}$$

In addition, note that $u(t)$ should satisfy the following constraints from the definitions of the control inputs:

$$\begin{bmatrix} 0 & 1 & 0 & 1 & 1 \end{bmatrix} u(t) = 1, \tag{13}$$

$$0_{5\times1} \leq u(t) \leq 1_{5\times1} \tag{14}$$

In real portable HPSS applications, exhausting the charge of any component during operation is not desirable. Considering this practical constraint, one can additionally place constraints on the operating ranges of the battery and supercapacitors as follows:

$$R_{\min}^{cap} \leq r_{cap,1}(t) \leq R_{\max}^{cap} \tag{15}$$

$$R_{\min}^{cap} \leq r_{cap,2}(t) \leq R_{\max}^{cap} \tag{16}$$

$$R_{\min}^{bat} \leq r_{bat}(t) \tag{17}$$

As mentioned before, the state equation is based on the HPSS model of Mirhoseimi and Koushanfar in [18]. The main differences between these two models are as follows:

- In [18], it is assumed that the charge values of the batteries and supercapacitors take only discrete values. In this study, we omit this assumption; thus, $r_{cap,i}(t)$ and $r_{bat}(t)$ are all real-valued.
- In the model in [18], supercapacitors are constrained to not be simultaneously charged by the battery and discharged by the load. In this paper, we omit this constraint.
- In the model in [18], a decision for assigning the source to the workload is carried out such that only one electronic energy supply source can transfer the required charge to the load. Hence, the control inputs in [18] are all binary numbers, and only one member of $a_{iy}(t)$ and $a_{bat}(t)$ is one. In this study, we omit this constraint. As a result, $a_{iy}(t)$ and $a_{bat}(t)$ are nonnegative real numbers satisfying $\sum_i a_{iy}(t) + a_{bat}(t) = 1$.
- In the model in [18], it is assumed that at most one supercapacitor can be charged by the battery. This assumption is omitted in this paper.

In order to solve the DPM problem within the framework of stochastic optimal control, we have to choose a PI that can ensure improved energy savings after optimization. A widely used choice for the PI in a stochastic optimal control problem is the expected sum of the discounted stage costs, *i.e.*,

$$J = E\left[\sum_{t=0}^{T} \gamma^t \ell\left(x(t), u(t)\right)\right] \tag{18}$$

where $\gamma \in (0, 1)$ is the discount factor; $l(\cdot,\cdot)$ is the stage cost function; and T is the final time whose value is ∞ in infinite-horizon problems. By minimizing this PI over all admissible state feedback control policies ϕ_t: $X \rightarrow U$, the optimal control problem can be solved. For the performance index in the optimization of power flows for HPSSs, one may resort to some ad-hoc stage cost functions (e.g., [18]). However in this paper, it turns out that a traditional objective function which is more clear and widely used in the field of linear quadratic optimal control works well. More precisely, we derive an ADP-based solution to the dynamic power management of HPSSs with the purpose of minimizing the battery charge consumption rate and maintaining the charge level of supercapacitors while keeping the control efforts reasonably low. For the purpose, the stage cost function $l(\cdot,\cdot)$ of the DPM task is chosen as follows:

$$\begin{aligned} l\left(x(t), u(t)\right) &= w_1 \Delta r_{bat}(t) \\ &+ w_2 \sum_{i=1}^{2} \left\{ \left(r_{cap,i}(t) - 0.5 R_{\max}^{cap} \right)^2 \right\} \\ &+ w_3 \left\{ \sum_{i=1}^{2} a_{ix}^2(t) \right\} + w_4 \left\{ \left(\sum_{i=1}^{2} a_{iy}^2(t) \right) + \beta_{bat} \left(a_{bat}^2(t) \right) \right\}, \end{aligned} \tag{19}$$

where the weight values (w_1, w_2, w_3, w_4, and β_{bat}) are obtained empirically via a tuning process. In the DPM in [18], the Q-learning method [20] was used to derive an approximation for the optimal power management strategy, and the stage function was defined as

$$R\left(x(t), u(t)\right) = -e^{\Delta r_{bat}(t)} - e^{\left(\sum_i r_{cap,i}(t)/R_{\max}^{cap}\right)} - \lambda e^{E\left[\Delta r_{bat}(t)\right]}. \tag{20}$$

Here, we adopt the main idea of this stage reward function with the following significant modifications. First, representing the stage costs in a quadratic form is important for solving the DPM problem in the context of constrained linear quadratic stochastic optimal control. Hence, our stage cost function utilizes second order polynomial terms instead of exponential functions. Second, in order to minimize the battery charge consumption rate and maintain the charge level of supercapacitors with the control efforts kept reasonably low, we employ the cost terms involving $\Delta r_{bat}(t)$, $(r_{cap,i}(t) - 0.5R_{\max}^{cap})^2$, $i = 1, 2$, and the control-input-related quadratic terms, $a_{ix}^2(t)$, $a_{iy}^2(t)$, $i = 1, 2$, and $a_{bat}^2(t)$. Finally, note that our stage cost Equation (19) deals with the control effort directly with weighted summation of the quadratic input terms, whereas in the objective Equation (20) of the reference paper [18], consideration of the control efforts can be treated indirectly via the expectation operation.

3. Approximate Dynamic Programming Approach to Dynamic Power Management

3.1. Preliminaries

General stochastic optimal control problems can be analyzed, in principle, utilizing the concept of a state value function. This value-function-based method is called DP, and a variety of topics on stochastic optimal control and DP are well-addressed by [5–8]. A large class of stochastic optimal control problems deal with the dynamics of the form in Equation (1) and are concerned with finding a state-feedback control policy:

$$u(t)=\phi_t(x(t)),\ t=0,1,\ldots \tag{21}$$

that can optimize a given PI function. As mentioned before, a widely used choice for the PI is the expected sum of the discounted stage cost, *i.e.*, $J=E\left[\sum_{t=0}^{\infty}\gamma^t\ell(x(t),u(t))\right]$. By minimizing this PI over all admissible control policies ϕ_t: $X\rightarrow U$, the optimal value of J can be found. This minimal PI value is denoted J^*, and the optimal state-feedback function achieving the minimal value is denoted by ϕ_t^*. The state value function is defined as the minimum total expected cost achieved by an optimal control policy from the given initial state $x(0) = z$, *i.e.*,

$$V^*(z)=\inf_{\phi_t} E\left(\sum_{t=0}^{\infty}\gamma^t\ell(x(t),\phi_t(x(t)))\mid x(0)=z\right) \tag{22}$$

Note that when the initial condition $x(0)$ is a fixed vector x_0, the optimal PI value can be expressed as $J^* = V^*(x_0)$. According to stochastic optimal control theory [6,7], the state value function in Equation (23) is the fixed point of the Bellman equation:

$$V^*(z)=\inf_v\left(l(z,v)+\gamma E\left[V^*(f(z,v,w))\right]\right). \tag{23}$$

In its operator form, the Bellman equation can be written as:

$$V^*=TV^* \tag{24}$$

where T is the Bellman operator (see e.g., [5]), whose domain and codomain are both function spaces mapping X onto $R\cup\{\infty\}$, defined as:

$$(Tg)(z)=\inf_v(\ell(z,v)+\gamma E\left[g(f(z,v,w))\right] \tag{25}$$

As is well known, the Bellman equation or its operator equation version cannot be solved exactly, except in simple special cases [5,6], and an alternative strategy when finding the exact state value function is the use of ADP relying on an approximate state value function $\hat{V}:X\rightarrow R$. In the DPM problem discussed in this paper, we are concerned with an ADP solution utilizing convex quadratic functions. This class of quadratic functions is a good choice for the approximate state value function [5,8] because these functions are convenient and powerful for handling various optimization steps appearing in the ADP procedure.

3.2. ADP-Based Solution Procedure

From the above preliminary steps, the DPM of the portable HPSS can now be represented as the following dynamic stochastic control problem:

$$\begin{aligned} &\min_{u(\cdot)} && E\left[\sum_{t=0}^{\infty} \gamma^t l\big(x(t),u(t)\big)\right] \\ &\text{subject to} && x(t+1) = Ax(t) + \big(B_1 + B_w w(t)\big)u(t), \\ &&& \text{constraints } (13),(14),(15),(16),(17), \\ &&& \text{and } x(0) = x_0, \end{aligned} \tag{26}$$

where $l(\cdot,\cdot)$ is the quadratic stage cost function specified in Equation (19). To solve the above DPM problem for the portable HPSS, we propose an ADP-based solution procedure utilizing the approximate value function (AVF) policy approach of O'Donoghue, Wang, and Boyd [9]. In this AVF policy approach [9], the convex quadratic function:

$$\hat{V}_i(x) = x^T P_i x + 2p_i^{\ T} + q_i,\ i = 0,\ldots,M \tag{27}$$

is used for approximating the state value functions, and by letting the parameters (*i.e.*, P_i, p_i, and q_i) satisfy the iterated Bellman inequalities:

$$\hat{V}_i \le T\hat{V}_{i+1},\ i = 0,\ldots,M \tag{28}$$

with $\hat{V}_M = \hat{V}_{M+1}$, it is ensured that $\hat{V}_0$ is a lower bound of the optimal state value function V^* [5,9]. Further, by optimizing this lower bound $\hat{V}_0$ via convex optimization, the ADP approach finds a suboptimal approximate state value function $\hat{V}_0, \hat{V}_1, \cdots, \hat{V}_M$ and the corresponding ADP policies $\hat{\phi}_0, \hat{\phi}_1, \cdots, \hat{\phi}_M$. In the following, we derive an ADP-based solution to the dynamic management problem for the portable HPSS.

First, note that our stage cost function at time t can be expressed using the state $x(t)$ and input $u(t)$ as follows:

$$\begin{aligned} l\big(x(t),u(t)\big) &= w_1 \Delta r_{bat}(t) + w_2 \sum_{i=1}^{2} \big(r_{cap,i}(t) - 0.5 R_{\max}^{cap}\big)^2 \\ &+ w_3 \big(a_{1x}^2(t) + a_{2x}^2(t)\big) \\ &+ w_4 \big(a_{1y}^2(t) + a_{2y}^2(t) + \beta_{bat} a_{bat}^2(t)\big). \end{aligned} \tag{29}$$

Then, the general stage cost term can be written in the following matrix–vector form:

$$\begin{aligned} l(x,u) &= \begin{bmatrix} u \\ x \\ 1 \end{bmatrix}^T L \begin{bmatrix} u \\ x \\ 1 \end{bmatrix} \\ &= \begin{bmatrix} u \\ x \\ 1 \end{bmatrix}^T \begin{bmatrix} L_{11} & L_{12} & L_{13} \\ L_{12}^T & L_{22} & L_{23} \\ L_{13}^T & L_{23}^T & L_{33} \end{bmatrix} \begin{bmatrix} u \\ x \\ 1 \end{bmatrix}, \end{aligned} \tag{30}$$

where

$$L_{11} = \begin{bmatrix} w_3 & 0 & 0 & 0 & 0 \\ 0 & w_4 & 0 & 0 & 0 \\ 0 & 0 & w_3 & 0 & 0 \\ 0 & 0 & 0 & w_4 & 0 \\ 0 & 0 & 0 & 0 & \beta_{bat} w_4 \end{bmatrix}, \tag{31}$$

$$L_{12} = 0_{5\times 4}, \tag{32}$$

$$L_{13} = 0_{5\times 1}, \tag{33}$$

$$L_{22} = \begin{bmatrix} w_2 & 0 & 0 & 0 \\ 0 & w_2 & 0 & 0 \\ 0 & 0 & 0 & 0 \\ 0 & 0 & 0 & 0 \end{bmatrix}, \tag{34}$$

$$L_{23} = \begin{bmatrix} -0.5 w_2 R_{\max}^{cap} \\ -0.5 w_2 R_{\max}^{cap} \\ 0 \\ 0.5 w_1 \end{bmatrix}, \tag{35}$$

$$\text{and } L_{33} = 0.5 w_2 \left(R_{\max}^{cap} \right)^2 . \tag{36}$$

Next, we define the derived matrix variable G_i, $i = 1, \ldots, M$ to satisfy the following:

$$\begin{bmatrix} u \\ x \\ 1 \end{bmatrix}^T G_i \begin{bmatrix} u \\ x \\ 1 \end{bmatrix} = E\left[\hat{V}_i \left(Ax + \left(B_1 + B_w w \right) u \right) \right]. \tag{37}$$

Note that the right hand side of Equation (37) is the expectation of the convex quadratic function $\hat{V}_i$ evaluated at the next state. Since the expectation can be expressed as a quadratic function with respect to u and x, we use the derived matrix variable G_i [8] for notational convenience. Here, the expectation of the right-hand side of Equation (37) can be further expanded as follows:

$$\begin{aligned} & E\left[\begin{bmatrix} Ax + \left(B_1 + B_w w \right) u \\ 1 \end{bmatrix}^T \begin{bmatrix} P_i & p_i \\ p_i^T & q_i \end{bmatrix} \begin{bmatrix} Ax + \left(B_1 + B_w w \right) u \\ 1 \end{bmatrix} \right] \\ & = \begin{bmatrix} u \\ x \\ 1 \end{bmatrix} E\left[\begin{bmatrix} B_1 + B_w w & A & 0 \\ 0 & 0 & 1 \end{bmatrix}^T \begin{bmatrix} P_i & p_i \\ p_i^T & q_i \end{bmatrix} \begin{bmatrix} B_1 + B_w w & A & 0 \\ 0 & 0 & 1 \end{bmatrix} \right] \begin{bmatrix} u \\ x \\ 1 \end{bmatrix}. \end{aligned} \tag{38}$$

Furthermore, by further evaluating the right-hand side of Equation (38), one can obtain

$$G_i = \begin{bmatrix} G_{i,11} & G_{i,12} & G_{i,13} \\ \left(G_{i,12} \right)^T & G_{i,22} & G_{i,23} \\ \left(G_{i,13} \right)^T & \left(G_{i,23} \right)^T & G_{i,33} \end{bmatrix}, \tag{39}$$

where

$$G_{i,11} = B_1^T P_i B_1 + B_1^T P_i B_w \overline{w} + B_w^T P_i B_1 \overline{w} + B_w^T P_i B_w \overline{w^2}, \tag{40}$$

$$G_{i,12} = B_1^T P_i A + B_w^T P_i A \overline{w} \tag{41}$$

$$G_{i,13} = B_1^T p_i + B_w^T p_i \overline{w} \tag{42}$$

$$G_{i,22} = A^T P_i A \tag{43}$$

$$G_{i,23} = A^T p_i \tag{44}$$

$$G_{i,33} = q_i \tag{45}$$

In Equations (40)–(42), $\overline{w}$ and $\overline{w^2}$ are the first and second moments of the external workload demands, respectively.

With the convex quadratic approximation $\hat{V}_i(x) = x^T P_i x + 2p_i^T + q_i$, $i = 0,\ldots,M$, the Bellman inequalities $\hat{V}_{i-1} \le T\hat{V}_i$, $i = 1,\ldots,M$, for the DPM problem can be written as

$$\begin{bmatrix} u \\ x \\ 1 \end{bmatrix}^T \left(L + \gamma G_i - S_{i-1}\right) \begin{bmatrix} u \\ x \\ 1 \end{bmatrix} \ge 0,\ i = 1,\ldots,M, \tag{46}$$

where S_{i-1} is the matrix variable defined by

$$S_{i-1} = \begin{bmatrix} 0 & 0 & 0 \\ 0 & P_{i-1} & p_{i-1} \\ 0 & p_{i-1}^T & q_{i-1} \end{bmatrix}. \tag{47}$$

Note that the constraints in Equations (13)–(17) are all given in the form of an affine equality or affine inequalities with respect to the state and control inputs. Thus, these constraints can be compactly expressed via the affine form $Du(t) + Fx(t) + H$. More specifically, the equality constraint in Equation (13) can be written as

$$D^{(1)}u(t) + F^{(1)}x(t) + H^{(1)} = 0, \tag{48}$$

where $D^{(1)} = \begin{bmatrix} 0 & 1 & 0 & 1 & 1 \end{bmatrix}$, $F^{(1)} = 0_{1\times4}$, and $H^{(1)} = -1$.

Similarly, the remaining inequality constraints can be represented by

$$D^{(k)}u(t) + F^{(k)}x(t) + H^{(k)} \le 0,\ k = 2,3,4,5, \tag{49}$$

where

$$D^{(2)} = -I_{5\times5}, \tag{50}$$

$$F^{(2)} = 0_{5\times4}, \tag{51}$$

$$H^{(2)} = 0_{5\times1}, \tag{52}$$

$$D^{(3)} = I_{5\times5}, \tag{53}$$

$$F^{(3)} = 0_{5\times4}, \tag{54}$$

$$H^{(3)} = -1_{5\times1}, \tag{55}$$

$$D^{(4)} = 0_{3\times5}, \tag{56}$$

$$F^{(4)} = \begin{bmatrix} -1 & 0 & 0 & 0 \\ 0 & -1 & 0 & 0 \\ 0 & 0 & -1 & 0 \end{bmatrix}, \tag{57}$$

$$H^{(4)} = \begin{bmatrix} R_{\min}^{cap} \\ R_{\min}^{cap} \\ R_{\min}^{bat} \end{bmatrix}, \tag{58}$$

$$D^{(5)} = 0_{2\times5}, \tag{59}$$

$$F^{(5)} = \begin{bmatrix} 1 & 0 & 0 & 0 \\ 0 & 1 & 0 & 0 \end{bmatrix}, \tag{60}$$

$$\text{and } H^{(5)} = \begin{bmatrix} -R_{\max}^{cap} \\ -R_{\max}^{cap} \end{bmatrix}. \tag{61}$$

Note that these constraints can be further modified in accordance with the requirements considered in the design stage. One of the key steps in this solution procedure is to find $\hat{V}_t$ satisfying the Bellman inequalities in Equation (28) under the constraints in Equations (46) and (48). For this, we resort to the S-procedure [21]. On the basis of the S-procedure, we can ensure that the Bellman inequalities be satisfied under the constraints in Equations (46) and (48) by requiring that the following linear matrix inequalities (LMIs) hold true:

$$L + \gamma G_i - S_{i-1} + \sum_k \xi_i^{(k)} \Xi^{(k)} \geq 0, \tag{62}$$

$$diag\left(\xi_i^{(k)}\right) \geq 0, \ \forall k, \ i = 1, \ldots, M, \tag{63}$$

where $\xi_i^{(k)}$ represents the S-procedure multipliers [21], and

$$\Xi^{(k)} = \begin{bmatrix} 0 & 0 & D^{(k)}/2 \\ 0 & 0 & F^{(k)}/2 \\ \left(D^{(k)}/2\right)^T & \left(F^{(k)}/2\right)^T & H^{(k)} \end{bmatrix}. \tag{64}$$

By combining the above steps, one can obtain a DPM design procedure. The following show an illustrative design procedure:

Controller synthesis procedure:

Preliminary steps:

1. Choose the parameters of the problem: γ, λ, M, R_{th}, $R_{\min}^{cap}$, $R_{\max}^{cap}$, and $R_{\min}^{bat}$.
2. Estimate the 1st and 2nd moments of the external load demands, $\overline{w}$ and $\overline{w^2}$, from the training data.

Main steps:

1. Initialize the decision-making time $t = 0$, and choose $x(0)= x_0$.
2. Compute the stage cost matrix L in Equation (30) and the constant matrix $\Xi^{(k)}$ in Equation (64).
3. Define the LMI variables:

   (1) Define the basic LMI variables: P_i, p_i, and q_i in Equation (27).
   (2) Define the derived LMI variables: G_i in Equation (39) and S_{i-1} in Equation (47).
   (3) Define the S-procedure multipliers: $\xi_i^{(k)}$ in Equation (63).

4. Find the approximate state value functions, $\hat{V}_i$, by solving the following convex optimization problem:

$$\begin{aligned}
&\min \quad \hat{V}_0(x_0) = x_0^T P_0 x_0 + 2p_0^T x_0 + q_0 \\
&\text{subject to} \\
&\qquad L + \gamma G_i - S_{i-1} + \sum_{k=1}^{3} \xi_i^{(k)} \Xi^{(k)} \geq 0, \ i = 1,\ldots,M, \\
&\qquad S_{M-1} = S_M, \\
&\qquad P_i \geq 0, \ i = 0,\ldots,M, \\
&\qquad diag\left(\xi_i^{(k)}\right) \geq 0, \ i = 1,\ldots,M, \ k = 2,3.
\end{aligned} \tag{65}$$

5. Obtain the ADP controllers on the basis of

$$\Phi_t(x) = \underset{u}{\operatorname{argmin}}(\ell(x,u) + \gamma E V_{t+1}(f(x,u,w))) \ \text{ for } t = 0,\ldots,M \tag{66}$$

$$\Phi_t(x) = \underset{u}{\operatorname{argmin}}(\ell(x,u) + \gamma E V_{T+1}(f(x,u,w))) \ \text{ for } t > M \tag{67}$$

4. Simulation Results and Trajectories

In this section, we first describe a portable HPSS example to illustrate the proposed ADP-based DPM procedure. Afterwards, we conduct simulations and provide discussions for simulation results.

4.1. An Illustrative Example

We conducted simulations for a portable HPSS example [18] to illustrate the proposed ADP-based DPM procedure. For the example, we considered the workload scenarios of Figure 2, which are essentially based on the iPhone battery current measurements in [18,22]. The tasks considered in the scenarios include airplane mode (40 mA), default mode (100 mA), working internet with 3G network (310 mA), working game application "Tap-Tap" (220 mA), working 3D game application "GTI Racing" (400 mA), working application "YouTube" with WiFi (260 mA), working application "YouTube" (350 mA), and voice phone (170 mA) [18].

The portable HPSS in our example consists of two 200 [F] supercapacitors with $R_{\max}^{cap} = 400$ [C] and a single 4 [V] battery with $r_{bat}(0) = 1000$ [C]. The initial state conditions for the HPSS were assumed as follows:

$$\begin{aligned}
x(0) &= \left[r_{cap,1}(0), r_{cap,2}(0), r_{bat}(0), \Delta r_{bat}(0)\right] \\
&= [400, 400, 1000, 0]^T.
\end{aligned} \tag{68}$$

Following the data treatment of [18], we rescaled the source and demand charges such that at each instant t, the load charge demand $w_{load}(t)$ was transformed into $w(t) = w_{load}(t)/10$. We performed simulations based on the ten scenarios of Figure 2, and evaluated the resultant controllers. In the evaluation stage, we used a 10-fold cross-validation method for the iPhone workload data of Figure 2. For the simulation results, we considered all possible 10 different partitions of the training and test subsets. For the parameters of the problem, we chose the following:

$$\gamma = 0.99,\ R_{th} = 1,\ R_{\min}^{cap} = 0,\ R_{\max}^{cap} = 40, R_{\min}^{bat} = 0,\ M = 150,$$
$$w_1 = 5, w_2 = 0.2,\ w_3 = 0.5,\ w_4 = 5,\ \beta_{bat} = 5,$$

where the weight values, w_1, w_2, w_3, w_4, and β_{bat} were obtained empirically via a tuning process based on a subset of the training data. Simulation results show that the above set of weight values yielded reasonably good results. To develop a more disciplined design guideline for the weight values is an important topic that can be covered in follow-up works.

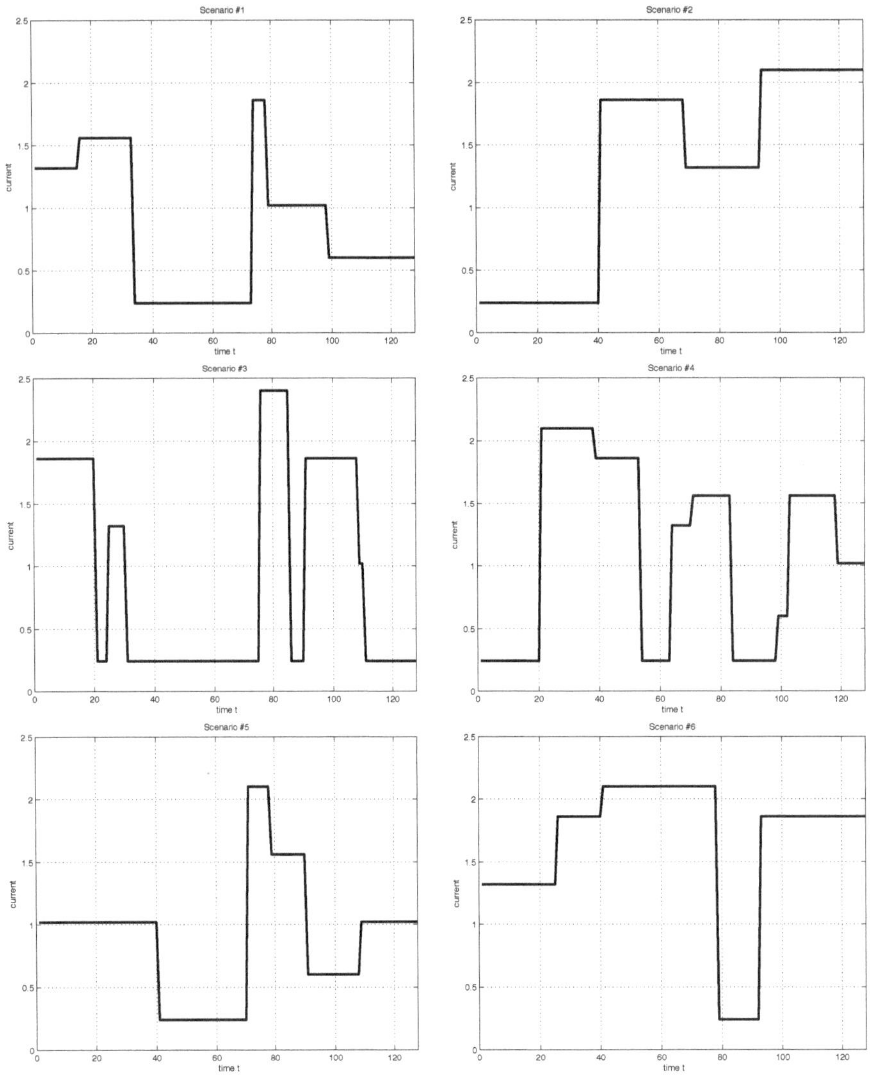

Figure 2. *Cont.*

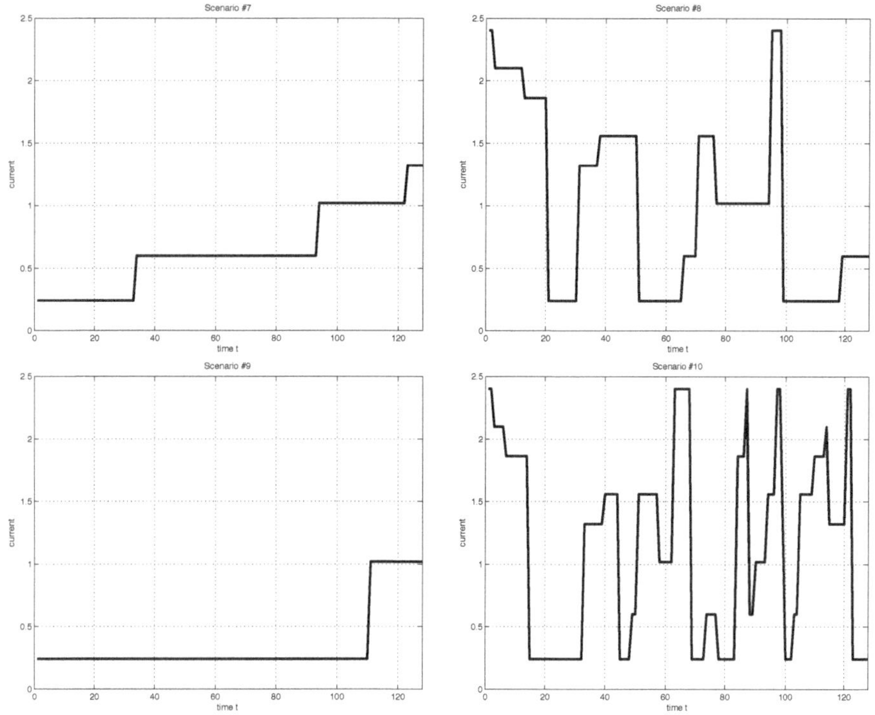

Figure 2. Considered workload scenarios, which are essentially based on the iPhone battery current measurements of [18].

4.2. Discussions and Performance Comparison

Figure 3 is the simulation results of the 10-fold cross-validation, and describes the charge depletion patterns of the battery and the supercapacitors in the HPSS. The exact meaning of the pictures in the figure is as follows: the *i*th picture shows the charge-depleting patterns obtained by the proposed ADP-based DPM procedure for the *i*th numerical experiment, in which the *i*th workload scenario of Figure 2 was used as the test set, and the other nine workload scenarios were used as the training set for estimating the first and second moments of $w(t)$. In the pictures, the solid line represents the charge value of the battery, $r_{bat}(t)$, and the dashed line is for the sum of supercapacitor charges, $r_{cap,1}(t) + r_{cap,2}(t)$. Note that in the sixth picture, the scenario consists of heavy loads (such as playing games, browsing the Internet, and watching YouTube videos), and the charges in the battery and the supercapacitors were completely depleted before the end of the considered time horizon. Figure 3 shows that the proposed ADP-based DPM procedure resulted in reasonable dynamic charge management for the HPSS. From the simulation results, one can see that when the demand charge changed abruptly, then the control actions for the supercapacitors reacted promptly. It is well known that in general, the battery alone cannot handle the situation of rapidly changing loads effectively due to its low power density. By comparison, Figure 3 shows that the HPSS utilizing the ADP-based DPM was capable of handling such situation reasonably well. We believe that this capability is ensured by the stage cost function of Equation (19), which is defined with the purpose of minimizing the battery charge consumption rate and maintaining the charge level of supercapacitors appropriately for speedy reaction to fluctuating loads. Note that our ADP-based DPM procedure can be applied to a significantly larger class of stage cost functions as long as they are quadratic-program (QP) representable [23] (*i.e.*, $l(\cdot,\cdot)$ is convex quadratic plus a convex piecewise linear function, possibly including linear equality and inequality constraints).

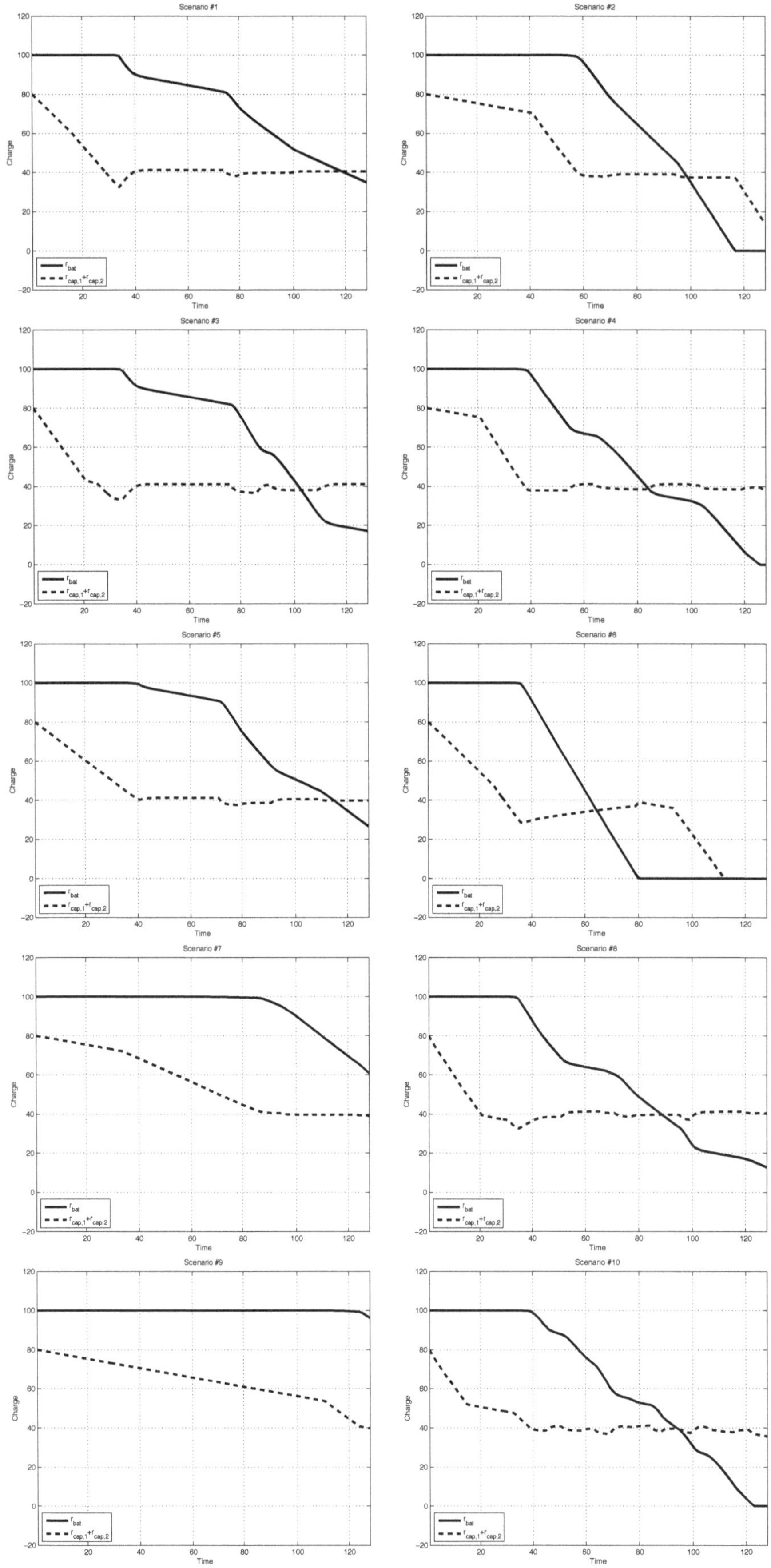

Figure 3. Charge-depleting patterns of the battery and supercapacitors resulting from the ADP-based approach of this paper.

For the sake of performance comparison, we considered the reinforcement learning (RL) based DPM technique of the reference paper [18]. RL is a branch of machine learning which is related to stochastic optimal control. Recently, RL methods [20] have been successfully applied to various decision-making problems, including the field of DPM for HPSSs [18]. One of the main differences of the RL methods for stochastic optimal control is that in their solution procedures, they use the training data on the state transitions together with function approximation instead of exact mathematical analysis. In the reference paper [18], Q-learning [24], which is one of the most widely used value-function-based RL methods, was utilized to derive an RL-based DPM strategy. In Q-learning, given the state transition data ($x(t)$, $u(t)$, $r(t)$, $x(t + 1)$), where $r(t)$ is the reward signal defined as the negative cost (*i.e.*, $r(t) = -l(x(t), u(t))$), the state-action value function Q [20,24] is updated via $Q(x(t), u(t)) \leftarrow Q(x(t), u(t)) + \alpha[r(x(t), u(t)) + \gamma \max_u Q(x(t+1),u) - Q(x(t), u(t))]$ In our numerical experiments for performance comparison, the discount rate of Q-learning was set to be the same (*i.e.*, $\gamma = 0.99$). Function approximation for state-action value function Q was performed via the tile coding [20]. Also for the learning rate and the ε value for exploration via the ε-greedy policy [20], we used α =0.1 and $\varepsilon = 0.1$ (For detailed meaning of α and ε in Q-learning, the reader is referred to [20]). Figure 4 shows the learning curve (the solid line) resulting from Q-learning, which describes the average total cost *vs.* the iteration over a set of 10 simulation runs, together with the total cost of the proposed ADP-based DPM approach (the dashed line). As shown in the figure, Q-learning took about 500 iterations of training scenarios to have a convergent behavior. Figure 5 shows the simulation results of the Q-learning-based DPM strategy, and describes the charge depletion patterns of the battery and the supercapacitors in the HPSS. Also, Figure 6 reports the average stage cost sums over a set of ten simulation runs for the two methods. As shown in Figures 5 and 6, the proposed ADP-based DPM strategy yielded better performance than the Q-learning-based DPM [18].

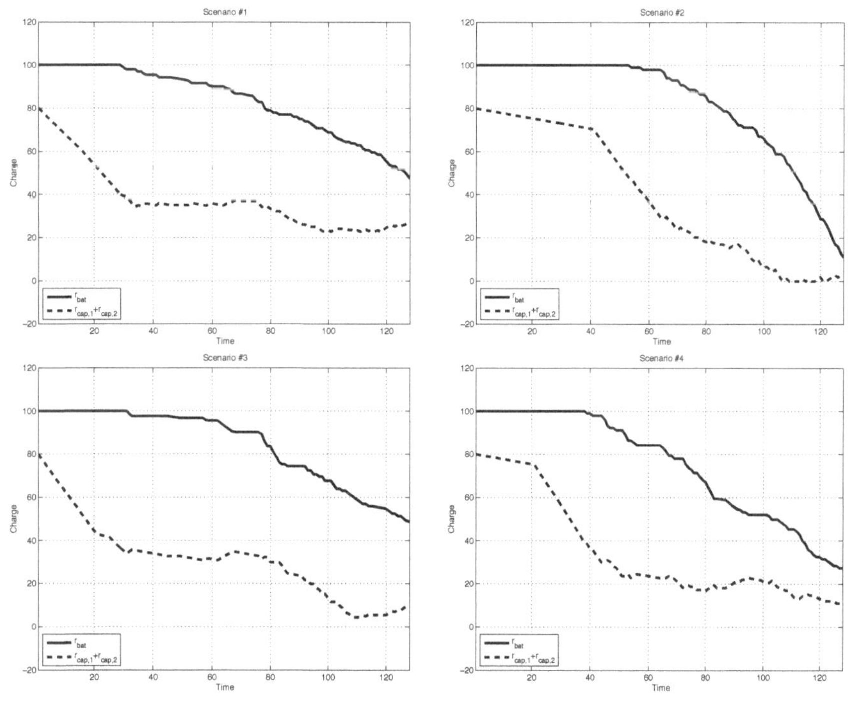

Figure 4. *Cont.*

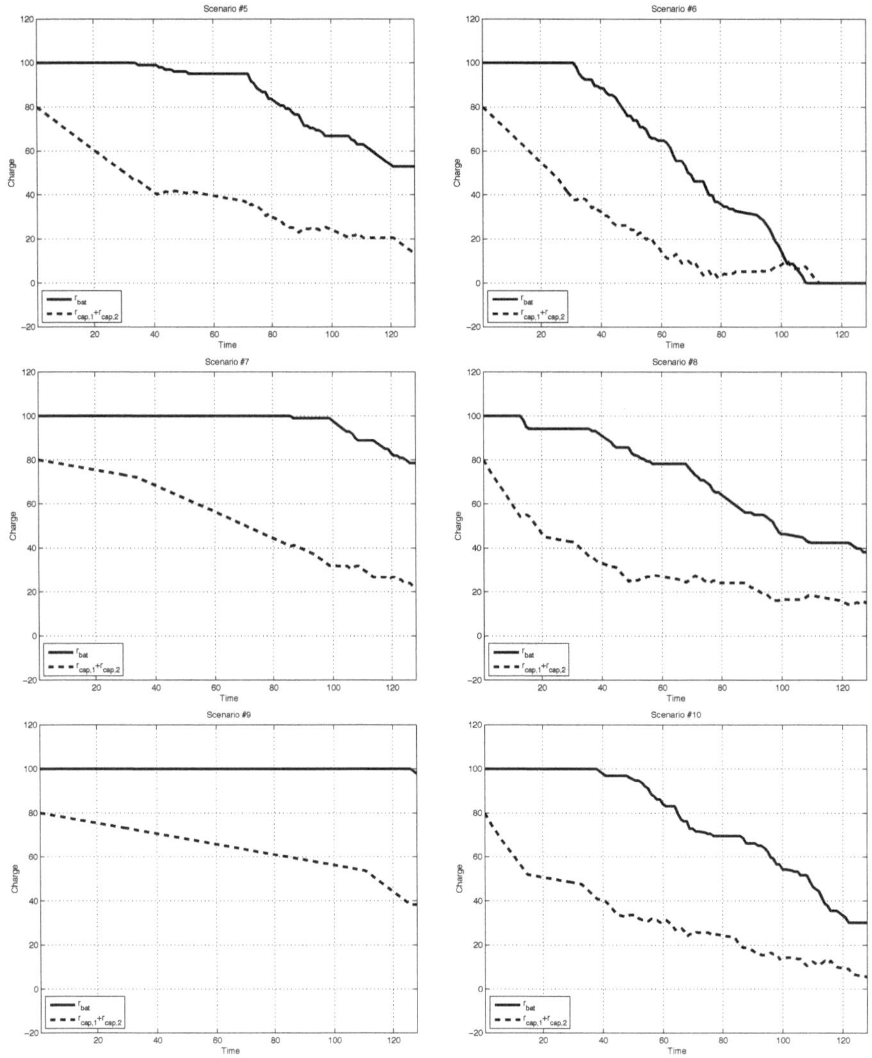

Figure 4. Charge-depleting patterns of the battery and supercapacitors resulting from the Q-learning-based DPM.

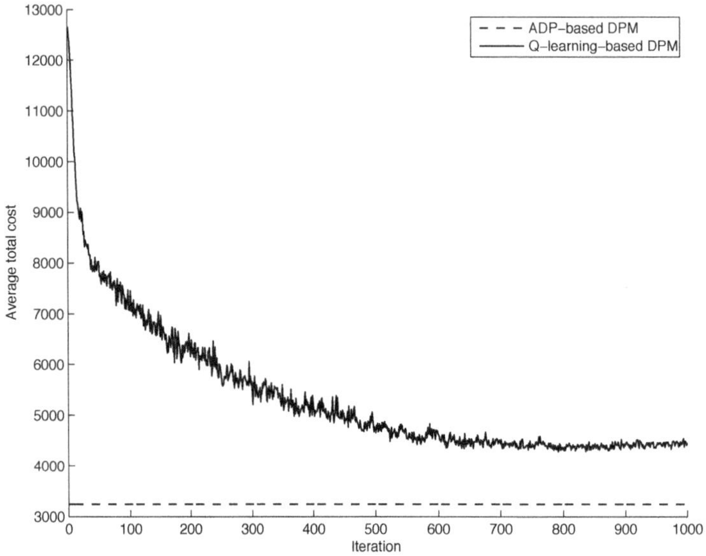

Figure 5. Learning curve (the average total cost *vs.* the iteration over a set of 10 simulation runs): Q-learning (dashed) and ADP (solid).

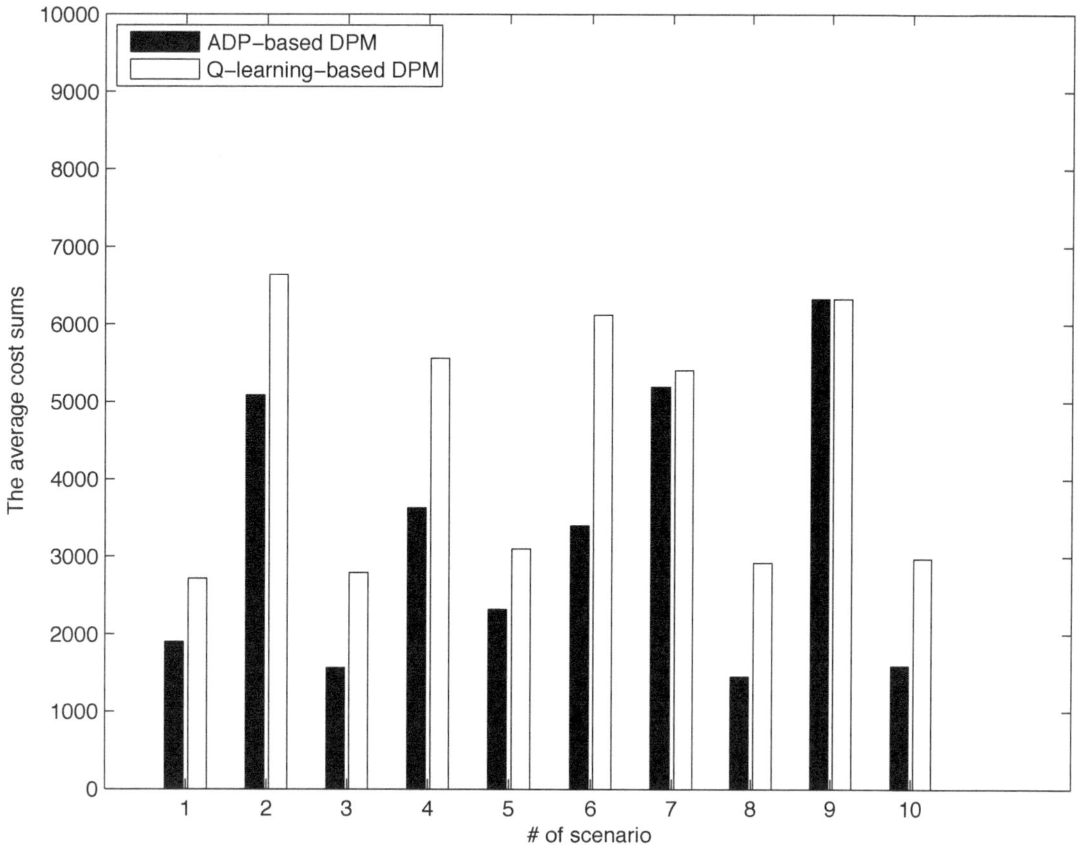

Figure 6. Performance comparison for ADP and Q: Average total cost for each scenario.

Finally, we think that the main significance of the ADP-based DPM approach of this paper and the Q-learning-based approach is in their capacity for customization. More specifically, they can customize the DPM for a (similar) user group, affecting the dynamic power management policy based on the first and second moments of the workloads and/or the past workload scenarios. We expect that this aspect will be covered in future works with sufficiently large training data sets.

5. Conclusions

Managing power flows in portable HPSSs has become an important issue with the advent of a variety of mobile systems and hybrid energy technologies, particularly when supercapacitors are used, which can compensate for the various weaknesses of conventional battery-based power supplies. The problem of optimally managing the power flows for HPSSs thus becomes more complicated. In this paper, the DPM problem for portable HPSSs with combined supercapacitors and battery energy supplies was considered, and an ADP-based DPM strategy was presented. Simulation results show that the proposed ADP-based approach can find a DPM strategy that copes with uncertain workload demands well, and its performance is better compared to the conventional Q-learning-based DPM method. Our plans for future work include more detailed modeling. For example, in the current model of the paper, the losses in the storage systems and in the power converters were neglected, and we are planning to take into account such losses as well with a more detailed model. Also, we will perform more extensive simulation studies, which should reveal the strengths and weaknesses of ADP-based methods, and applications to other types of HPSSs.

Acknowledgments

This work was supported by a Korea University Grant.

Author Contributions

J.P. and G.C. conceived and designed the methodology of the paper; J.P., D.Y. and J.L. wrote the computer program and performed the numerical experiments; J.P. and G.C. wrote the paper.

Conflict of Interest

The authors declare no conflicts of interest. The funding sponsors had no role in the design of the study; in the collection, analyses, or interpretation of data; in the writing of the manuscript; or in the decision to publish the results.

References

1. Burke, A. Ultracapacitors: Why, How, and Where is the Technology. *J. Power Sources* **2000**, *91*, 37–50.
2. Yap, H.T.; Schofield, N.; Bingham, C.M. Hybrid energy/power sources for electric vehicle traction systems. In Proceedings of the IEEE Power Electronics, Machines and Drives (PEMD) Conference, Edinburgh, UK, 31 March–2 April 2004; Volume 1, pp. 61–66.
3. Miller, J.M.; Deshpande, U.; Dougherty, T.J.; Bohn, T. Power electronic enabled active hybrid energy storage system and its economic viability. In Proceedings of the IEEE Applied Power Electronics Conference and Exposition, Washington, DC, USA, 15–19 February 2009; pp. 190–198.
4. Wang, L.; Liu, X.; Li, H.; Im, W.S.; Kim, J.M. Power electronics enabled energy management for energy storage with extended cycle life and improved fuel economy in a PHEV. In Proceedings of the IEEE Energy Conversion Congress and Exposition (ECCE), Atlanta, GA, USA, 12–16 September 2010; pp. 3917–3922.
5. Wang, Y.; Boyd, S. Approximate dynamic programming via iterated bellman inequalities. *Int. J. Robust Nonlinear Control* **2015**, *25*, 1472–1496.
6. Bertsekas, D. *Dynamic Programming and Optimal Control: Volume 1*; Athena Scientific: Belmont, MA, USA, 2005.
7. Bertsekas, D. *Dynamic Programming and Optimal Control: Volume 2*; Athena Scientific: Belmont, MA, USA, 2007.
8. O'Donoghue, D.; Wang, Y.; Boyd, S. Min-Max approximate dynamic programming. In Proceedings of the 2011 IEEE International Symposium on Computer-Aided Control System Design (CACSD), Denver, CO, USA, 20–23 September 2011; pp. 424–431.
9. O'Donoghue, B.; Wang, Y.; Boyd, S. Iterated approximate value functions. In Proceedings of the European Control Conference (ECC), Zurich, Switzerland, 17–19 July 2013; pp. 3882–3888.
10. Romaus, C.; Bocker, J.; Witting, K.; Seifried, A. Optimal energy management for a hybrid energy storage system combining batteries and double layer capacitors. In Proceedings of the Energy Conversion Congress and Exposition (ECCE), San Jose, CA, USA, 20–24 September 2009.

11. Chen, G.; Bao, Z.J.; Yang, Q.; Yan, W.J. Scheduling strategy of hybrid energy storage system for smoothing the output power of wind farm. In Proceedings of the IEEE Control and Automation (ICCA), Hangzhou, China, 12–14 June 2013; pp. 1874–1878.
12. Choi, M.E.; Kim, S.W.; Seo, S.W. Energy Management Optimization in a Battery/Supercapacitor Hybrid Energy Storage System. *IEEE Trans. Smart Grid* **2012**, *3*, 463–472.
13. Gee, A.M.; Robinson, F.V.P.; Dunn, R.W. Analysis of battery lifetime extension in a small-scale wind-energy system using supercapacitors. *IEEE Trans. Energy Convers.* **2013**, *28*, 24–33.
14. Blanes, J.M.; Gutierrez, R.; Garrigos, A.; Lizan, J.L.; Cuadrado, J.M. Electric vehicle battery life extension using ultracapacitors and an FPGA controlled interleaved buck-boost converter. *IEEE Trans. Power Electron.* **2013**, *28*, 5940–5948.
15. Min, S.W.; Kim, S.J. Optimized installation and operations of battery energy storage system and electric double layer capacitor modules for renewable energy based intermittent generation. *J. Electr. Eng. Technol.* **2013**, *8*, 238–243.
16. Gao, L.; Dougal, R.A.; Shengyi, L. Power enhancement of an actively controlled battery-ultracapacitor hybrid. *IEEE Trans. Power Electron.* **2005**, *20*, 236–243.
17. Mirzaei, A.; Farzanehfard, H.; Adib, E.; Jusoh, A.; Salam, Z. A fully soft switched two quadrant bidirectional soft switching converter for ultra capacitor interface circuits. *J. Power Electron.* **2011**, *11*, 1–9.
18. Mirhoseini, A.; Koushanfar, F. Learning to manage combined energy supply systems. In Proceedings of the IEEE/ACM International Symposium on Low-Power Electronics and Design, Fukuoka, Japan, 1–3 August 2011; pp. 229–234.
19. Plett, G.L. Extended Kalman Filtering for Battery Management Systems of LiPB-based HEV Battery pack: Part 1. Background. *J. Power Sources* **2004**, *134*, 277–292.
20. Sutton, R.S.; Barto, A.G. *Reinforcement Learning: An Introduction*; MIT Press: Cambridge, MA, USA, 1998.
21. Boyd, S.; Ghaoui, L.E.; Feron, E.; Balakrishman, V. *Linear Matrix Inequalities in System and Control Theory*; Society for Industrial and Applied Mathematics: Philadelphia, PA, USA, 1994.
22. Mirhoseini, A.; Koushanfar, F. HypoEnergy: Hybrid supercapacitor-battery power-supply optimization for energy efficiency. In Proceedings of the Design, Automation & Test in Europe Conference & Exhibition, Grenoble, France, 14–18 March 2012; pp. 1–4.
23. Boyd, S.; Mueller, M.T.; O'Donoghue, D.; Wang, Y. Performance bounds and suboptimal policies for multi-period investment. *Found. Trends Optim.* **2014**, *1*, 1–69.
24. Watkins, C. Learning from Delayed Rewards. Ph.D. Thesis, Cambridge University, Cambridge, UK, 1989.

Permissions

All chapters in this book were first published in Energies, by MDPI; hereby published with permission under the Creative Commons Attribution License or equivalent. Every chapter published in this book has been scrutinized by our experts. Their significance has been extensively debated. The topics covered herein carry significant findings which will fuel the growth of the discipline. They may even be implemented as practical applications or may be referred to as a beginning point for another development.

The contributors of this book come from diverse backgrounds, making this book a truly international effort. This book will bring forth new frontiers with its revolutionizing research information and detailed analysis of the nascent developments around the world.

We would like to thank all the contributing authors for lending their expertise to make the book truly unique. They have played a crucial role in the development of this book. Without their invaluable contributions this book wouldn't have been possible. They have made vital efforts to compile up to date information on the varied aspects of this subject to make this book a valuable addition to the collection of many professionals and students.

This book was conceptualized with the vision of imparting up-to-date information and advanced data in this field. To ensure the same, a matchless editorial board was set up. Every individual on the board went through rigorous rounds of assessment to prove their worth. After which they invested a large part of their time researching and compiling the most relevant data for our readers.

The editorial board has been involved in producing this book since its inception. They have spent rigorous hours researching and exploring the diverse topics which have resulted in the successful publishing of this book. They have passed on their knowledge of decades through this book. To expedite this challenging task, the publisher supported the team at every step. A small team of assistant editors was also appointed to further simplify the editing procedure and attain best results for the readers.

Apart from the editorial board, the designing team has also invested a significant amount of their time in understanding the subject and creating the most relevant covers. They scrutinized every image to scout for the most suitable representation of the subject and create an appropriate cover for the book.

The publishing team has been an ardent support to the editorial, designing and production team. Their endless efforts to recruit the best for this project, has resulted in the accomplishment of this book. They are a veteran in the field of academics and their pool of knowledge is as vast as their experience in printing. Their expertise and guidance has proved useful at every step. Their uncompromising quality standards have made this book an exceptional effort. Their encouragement from time to time has been an inspiration for everyone.

The publisher and the editorial board hope that this book will prove to be a valuable piece of knowledge for researchers, students, practitioners and scholars across the globe.

List of Contributors

Jaw-Kuen Shiau
Department of Aerospace Engineering, Tamkang University, Tamsui, New Taipei City 25137, Taiwan

Yu-Chen Wei
Department of Aerospace Engineering, Tamkang University, Tamsui, New Taipei City 25137, Taiwan

Min-Yi Lee
Department of Aerospace Engineering, Tamkang University, Tamsui, New Taipei City 25137, Taiwan

Luigi Piegari
Department of Electronics, Information and Bioengineering, Politecnico di Milano, Piazza Leonardo da Vinci 32, 20133 Milan, Italy

Renato Rizzo
Department of Electrical Engineering and Information Technologies, University of Naples Federico II, Via Claudio 21, 80125 Naples, Italy

Ivan Spina
Department of Electrical Engineering and Information Technologies, University of Naples Federico II, Via Claudio 21, 80125 Naples, Italy

Pietro Tricoli
School of Electronic, Electrical and Computer Engineering, University of Birmingham, Gisbert Kapp Building, Birmingham B15 2TT, UK

Wei Li
Department of Economics and Management, North China Electric Power University, No. 619 Yonghua Street, Baoding 071003, Hebei, China

Hao Li
Department of Economics and Management, North China Electric Power University, No. 619 Yonghua Street, Baoding 071003, Hebei, China

Shuang Sun
Department of Economics and Management, North China Electric Power University, No. 619 Yonghua Street, Baoding 071003, Hebei, China

Yiduan Chen
Department of Electrical Engineering, the Hong Kong Polytechnic University, Kowloon 999077, Hong Kong, China

Weinong Fu
Department of Electrical Engineering, the Hong Kong Polytechnic University, Kowloon 999077, Hong Kong, China

Huangqiu Zhu
School of Electrical and Information Engineering, Jiangsu University, Zhenjiang 212013, China

Hui Li
School of Electrical and Information Engineering, Jiangsu University, Zhenjiang 212013, China

Markus Klimscheffskij
Grexel Systems Oy, Lautatarhankatu 6, FI-00580 Helsinki, Finland

Thierry Van Craenenbroeck
Vlaamse regulator van de elektriciteits-en gasmarkt, 1000 Brussels, Belgium

Marko Lehtovaara
Grexel Systems Oy, Lautatarhankatu 6, FI-00580 Helsinki, Finland

Diane Lescot
Observatoire des énergies renouvelables, 146 rue de l'Université, 75007 Paris, France

Angela Tschernutter
Energie-Control Austria, Rudolfsplatz 13a, 1010 Vienna, Austria

Claudia Raimundo
IT Power Consulting Ltd., St. Brandon's House 29 Great George Street, Bristol BS1 5QT, UK

Dominik Seebach
Oeko Institut e.V., Merzhauser Strasse 173, PO Box 17 71, 79017 Freiburg, Germany

Christof Timpe
Oeko-Institut e.V., Merzhauser Strasse 173, PO Box 17 71, 79017 Freiburg, Germany

Anna Rita Di Fazio
Department of Electrical and Information Engineering, University of Cassino and Southern Lazio, via Di Biasio 43, I-03043 Cassino, Italy

Mario Russo
Department of Electrical and Information Engineering, University of Cassino and Southern Lazio, via Di Biasio 43, I-03043 Cassino, Italy

Sara Valeri
Department of Electrical and Information Engineering, University of Cassino and Southern Lazio, via Di Biasio 43, I-03043 Cassino, Italy

Yiqi Liu
School of Electrical Engineering and Automation, Harbin Institute of Technology, Harbin 150001, China

Jianze Wang
School of Electrical Engineering and Automation, Harbin Institute of Technology, Harbin 150001, China

Ningning Li
School of Electrical Engineering and Automation, Harbin Institute of Technology, Harbin 150001, China

Yu Fu
School of Electrical Engineering and Automation, Harbin Institute of Technology, Harbin 150001, China

Yanchao Ji
School of Electrical Engineering and Automation, Harbin Institute of Technology, Harbin 150001, China

Doohyun Kyung
School of Civil and Environmental Engineering, Yonsei University, 134 Shinchon-dong, Seodaemun-gu, Seoul 120-749, Korea

Youngho Choi
KTP Consultants Pte Ltd., E-Centre@Redhill, 3791 Jalan Bukit Merah, Singapore 159471, Singapore

Sangseom Jeong
School of Civil and Environmental Engineering, Yonsei University, 134 Shinchon-dong, Seodaemun-gu, Seoul 120-749, Korea

Junhwan Lee
School of Civil and Environmental Engineering, Yonsei University, 134 Shinchon-dong, Seodaemun-gu, Seoul 120-749, Korea

Fengbing Li
State Key Laboratory of Power Transmission Equipment & System Security, Chongqing University, Chongqing 400044, China
State Grid Sichuan Electric Power Company Meishan Power Supply Company, Meishan 620010, China

Kaigui Xie
State Key Laboratory of Power Transmission Equipment & System Security, Chongqing University, Chongqing 400044, China

Jiangping Yang
State Grid Sichuan Electric Power Company Meishan Power Supply Company, Meishan 620010, China

Young-Sik Jang
Seoul National University, Gwanak-ro 599, Gwanak-gu, Seoul 151-744, Korea

JoonHyung Park
Seoul National University, Gwanak-ro 599, Gwanak-gu, Seoul 151-744, Korea

Yong Tae Yoon
Seoul National University, Gwanak-ro 599, Gwanak-gu, Seoul 151-744, Korea

Jooyoung Park
Department of Control and Instrumentation Engineering, Korea University, Sejong 339-700, Korea

Gyo-Bum Chung
Department of Electronic & Electrical Engineering, Hongik University, Sejong 339-701, Korea

Jungdong Lim
Department of Control and Instrumentation Engineering, Korea University, Sejong 339-700, Korea

Dongsu Yang
LG Electronics, Seoul 135-860, Korea